D0216516

Current Topics in Microbiology and Immunology

181

Macrophage Biology and Activation

Edited by S.W. Russell and S. Gordon

With 42 Figures

Springer-Verlag
Berlin Heidelberg NewYork
London Paris Tokyo
Hong Kong Barcelona
Budapest

STEPHEN W. RUSSELL
Wilkinson Distinguished Professor
and Director
University of Kansas Cancer Ctr.
Kansas City, KA 66160-7312
USA

SIAMON GORDON
Sir William Dunn School of Pathology
University of Oxford
South Parks Road
Oxford OX1 3RE
U.K.

Cover illustration: See page 8, Fig. 1i.

ISBN 3-540-55293-6 Springer-Verlag Berlin Heidelberg New York
ISBN 0-387-55293-6 Springer-Verlag New York Berlin Heidelberg

Library of Congress Catalog Card Number 15-12910
Printed in Germany

Typesetting: Thomson Press (India) Ltd, New Delhi; Offsetprinting: Saladruck, Berlin; Bookbinding: Lüderitz & Bauer, Berlin.
23/3020-5 4 3 2 1 0 – Printed on acid-free paper.

Preface

The amazing world of the mononuclear phagocyte keeps expanding at a pace that, in current vernacular, is truly awesome. As a result, maintaining currency with the latest developments and controversies that pertain to this cell type is becoming increasingly difficult. Hopefully, what is contained in this volume will be of help in this regard.

The topics have been selected to provide an overview of subject areas that either have recently become much better understood or are ones that, in our opinion, hold the promise of new levels of understanding as they are developed in the future. The scope of what is included ranges from how these cells develop, through the means that are used to regulate them, to the roles that they have in different tissues and in a variety of infectious diseases.

In closing these brief introductory remarks, we want to thank the contributors to this volume, especially those who helped make our job easier by meeting their deadlines, and our coordinator at Springer-Verlag, Ms. Marga Botsch.

Stephen W. Russell and Siamon Gordon

List of Contents

List of Contributors

(Their addresses can be found at the beginning of their respective chapters)

Antigen Markers of Macrophage Differentiation in Murine Tissues

S. Gordon, L. Lawson, S. Rabinowitz, P. R. Crocker, L. Morris, and V. H. Perry

1 Introduction

Cell-restricted membrane antigens have made it possible to map the distribution of mature macrophages in many murine tissues. Monoclonal antibodies (mAbs) have been used to define the appearance of macrophages dusing foetal and postnatal development, to establish the anatomic relationships between macrophages and other cells in the normal and diseased adult, and to investigate cellular modulation and heterogeneity within different microenvironments. Current studies have illustrated the complex differentiation pathway of mononuclear phagocytes in vivo and have raised questions concerning the mechanisms that determine monocyte entry, migration and fate within tissues. Macrophages constitute a major, widely dispersed system of cells that regulate homeostasis in the normal host and respond to tissue injury by contributing essential functions

Sir William Dunn School of Pathology, University of Oxford, South Parks Road, Oxford OX1 3RE, United Kingdom

Current Topics in Microbiology and Immunology, Vol. 181

Table 1. Membrane glycoprotein antigens used to define macrophage distribution and heterogeneity in murine tissues

Antigen (Ag)	Antibody (Ab)	MW (kDa)	Tissue distribution	Comments	References
F4/80	Rat mAb Rabbit polyclonal	~150	Mature macrophages	Down-regulated in T-cell dependent areas. Polyclonal Ab is cytotoxic and cross-reacts with rat Ag.	Austyn and Gordon 1981 Starkey et al. 1987 Dri and Crocker, unpublished
CR_3	Rat mAb M1/70 5C6	150/90	PMNs, macrophages and NK cells Heterogeneity on macrophages in tissues	Leucocyte integrin. Both mAbs block iC_3b binding site; only 5C6 blocks adhesion in vitro and in vivo.	Springer et al. 1979 Beller et al. 1982 Rosen and Gordon 1987
7/4	Rat mAb	~40	PMNs and activated macrophages	Polymorphic, useful for PMN depletion, e.g. in bone marrow.	Hirsch and Gordon 1983 Tree, Hirsch and Rabinowitz, unpublished
Sialoadhesin	Rat mAb SER-4 Rabbit polyclonal	185	Stromal macrophages	Macrophage lectin for sialylated structures. Expression modulated by homologous plasma/serum inducer.	Crocker and Gordon 1989 Crocker, unpublished
FA.11	Rat mAb FA.11	85–90	Macrophage and dendritic cell endosome membrane	Differential glycosylation (WGA binding) in exudate macrophages	Rabinowitz et al. 1991; Rabinowitz, da Silva, Milon, Steinman and Austyn, unpublished

during inflammation and repair. In this review we consider several membrane marker antigens which have proved useful in studying the life history and biologic properties of macrophages, and relate immunochemical studies on antigen expression to lineage analysis and macrophage differentiation in vivo. We restrict our discussion to the mouse, in which it is possible to manipulate the system in its entirety. Where known, properties of macrophages in other species are broadly similar.

2 Antigen Markers

In this section we summarise the properties of antigens which we have used to study macrophage populations in situ and in vitro. Table 1 lists features that are relevant to their use as markers. Other mAbs that have been used to characterise murine macrophages will be referred to in the text. Details of antigen expression will be described and illustrated below and our standard protocol for immunocytochemical analysis is given in an appendix.

2.1 F4/80

Knowledge of the presence of mature macrophages in murine tissues derives mainly from studies with mAbs and monospecific polyclonal antibodies directed against the macrophage-specific plasma membrane differentiation antigen F4/80. The epitope defined by a rat mAb isolated by AUSTYN and GORDON (1981) proved stable to perfusion–fixation, thus permitting HUME, PERRY and others to identify macrophages in a variety of tissues (for earlier reviews see HUME and GORDON 1985; PERRY and GORDON 1988). Subsequently DRI (unpublished observations) used affinity purified F4/80 antigen to raise a potent polyclonal antiserum that reacts with additional epitopes on the F4/80 molecule. This monospecific reagent enhances the detection of macrophage plasma membrane within tissues (LAWSON et al. 1990). The F4/80 antigen is a single chain glycoprotein which displays microheterogeneity on Western blot analysis of lysates prepared from various tissues and cell lines. The molecule has been purified (STARKEY et al. 1987) but not yet cloned and its function is unknown. Expression of F4/80 by peritoneal macrophages is down-regulated by inflammatory stimuli, which induce the recruitment of immature cells, by short-term adhesion in cell culture and by exposure to lymphokines, especially interferon-γ (GORDON et al. 1986a). However, the F4/80 antigen can be readily detected on macrophages by immunocytochemistry after all these treatments. The labelling pattern in isolated macrophages is mainly (>80%) at the plasma membrane and is uniform over the cell surface, although F4/80 antigen expression is concentrated in ruffles and at the edge of spreading cells.

2.2 Complement Receptor, Type 3 (CR_3)

The M1/70 mAb originally described by SPRINGER et al. (1979) reacts with the α chain of this member of the $\beta 2$ leucocyte integrin family (SPRINGER 1990) and was subsequently found to inhibit binding of iC_3b-coated sheep erythrocytes to leucocytes (BELLER et al. 1982). Whilst strongly expressed on polymorphonuclear leucocytes (PMNs), circulating monocytes and NK cells, CR_3 expression is variable on tissue macrophages (LEE et al. 1986; CROCKER and GORDON 1985). Another mAb raised by ROSEN and GORDON (1987) to inhibit adhesion of peritoneal macrophages to artificial substrata such as bacteriologic plastic also reacts with CR_3. This reagent blocks iC_3b binding activity and in addition is a potent inhibitor of myelomonocytic adhesion to inflamed endothelium in vivo (ROSEN and GORDON 1990a), unlike M1/70. In contrast with its modulation in vivo, expression of CR_3 is relatively constant on isolated macrophages in culture and is predominantly at the plasma membrane, although it is detectable in endosomes after endocytosis (ROSEN, unpublished).

2.3 7/4

The polymorphic myelomonocytic antigen 7/4 is expressed at high levels on PMNs (HIRSCH and GORDON 1983), but is absent on resident tissue macrophages. Monocytes and immune-activated macrophages express low levels of antigen (TREE, RABINOWITZ and HIRSCH, unpublished). The 40-kDa antigen is stable to perfusion fixation and Western blotting (RABINOWITZ, unpublished), but has not been characterised further. mAb 7/4 has been useful as a depleting agent for myeloid cells in bone marrow (BERTONCELLO et al. 1989) and to enrich for undifferentiated 7/4-negative haemopoietic precursor cells, in combination with other lineage-restricted reagents (IKUTA et al. 1990; ZSEBO et al. 1990).

2.4 Sialoadhesin(SER)

Stromal macrophages express a lectin-like receptor that binds sialylated glycoconjugates on sheep erythrocytes (SER) (CROCKER and GORDON 1986) and other erythroid model systems (COCKER et al. 1991). Ligands for this receptor are also present on developing murine myeloid cells (CROCKER, MORRIS and GORDON, unpublished). Activity of the receptor, now termed sialoadhesin, can be induced on non-stromal peritoneal macrophages by cultivation in homologous serum (CROCKER et al. 1988a). This made it possible for CROCKER and GORDON (1989) to raise a specific inhibitory mAb, SER-4, and, after purification of the receptor by affinity chromatography, to raise a further panel of mAbs, and a monospecific polyclonal antibody (CROCKER, MCWILLIAM and GORDON, unpublished). The sialoadhesin molecule is a single chain glycoprotein which is mainly expressed on the plasma membrane, where it mediates binding but not ingestion of attached cells. In situ, SER antigen is concentrated at points of contact between bone marrow stromal macrophages and myeloid, but not erythroid cells (CROCKER et al. 1990). A distinct divalent cation-dependent receptor for adherent erythroblasts (EbR) is expressed by stromal macrophages in foetal (MORRIS et al. 1988a) and adult tissues (MORRIS et al. 1991), but has not been defined immunochemically.

2.5 FA.11

The mAb FA.11 isolated by SMITH and KOCH (1987) reacts with an intracellular membrane glycoprotein that is more widely expressed by tissue macrophages than is F4/80 (RABINOWITZ, PERRY and LAWSON, unpublished). It is also present on lymphoid dendritic cells (RABINOWITZ, MILON, STEINMAN, AUSTYN and GORDON, unpublished), but it is more tissue restricted than a family of lysosomal glycoproteins, present in many cell types, to which it may be related (DA SILVA and ROSEN, unpublished). Exudate (elicited and activated) but not resident peritoneal macrophages express a glycoform that binds wheat germ agglutinin through terminal sialic acid residues (RABINOWITZ et al. 1991a). The FA.11 molecule consists of a core polypeptide, recognised by the FA.11 mAb, and is heterogeneous as a result of extensive N and O glycosylation. Although it can be detected in the plasma membrane, the bulk of labelling in macrophages is within the prelysosomal compartment and phagolysosomes and FA.11 reactivity is absent from terminal lysosomes (RABINOWITZ et al. 1991b). The FA.11 antigen is present in resident and exudate macrophages, but its level of expression, as well as of glycosylation, is enhanced by endocytic stimuli (RABINOWITZ et al. 1991a; DA SILVA, unpublished).

3 Distribution and Turnover of Macrophages

3.1 Ontogeny

The ontogeny of macrophages during development was largely unknown before the introduction of antigen markers. It is now clear from studies with F4/80 (MORRIS et al. 1991b) that macrophages are among the earliest haemopoietic cells to appear in the embryo (Fig. 1). Macrophages are abundant during mid and late gestation in most organs and are likely to play an important role in organogenesis and tissue remodelling. HUME, PERRY and others first used mAb F4/80 to follow recruitment of monocytes from blood to the developing nervous system at the time of natural death of neurons and their axons, before and after birth (Fig. 2) (HUME et al. 1983a; PERRY et al. 1985). Recruited cells phagocytose neuronal debris and differentiate into progressively more arborised microglia to

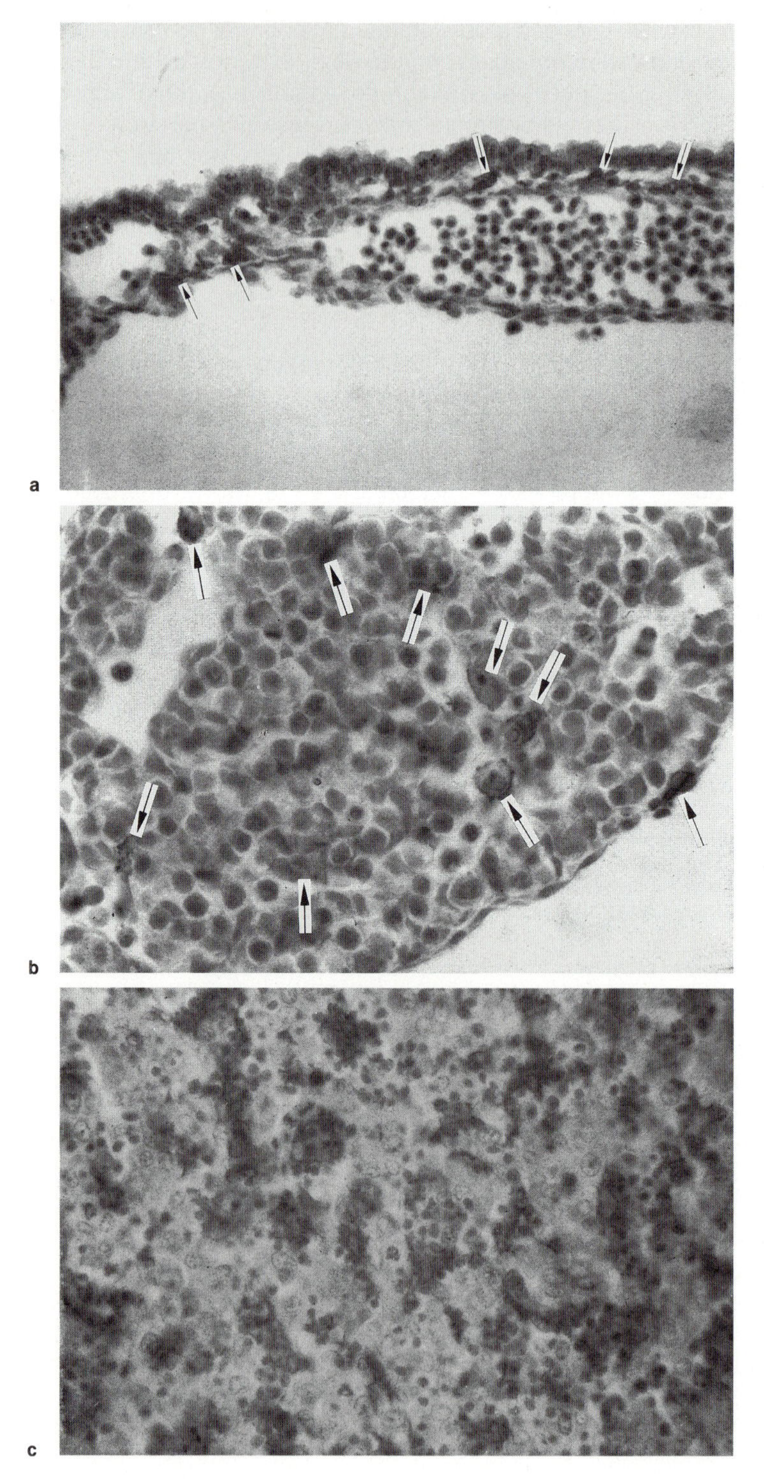

Fig. 1 (*cont.*)

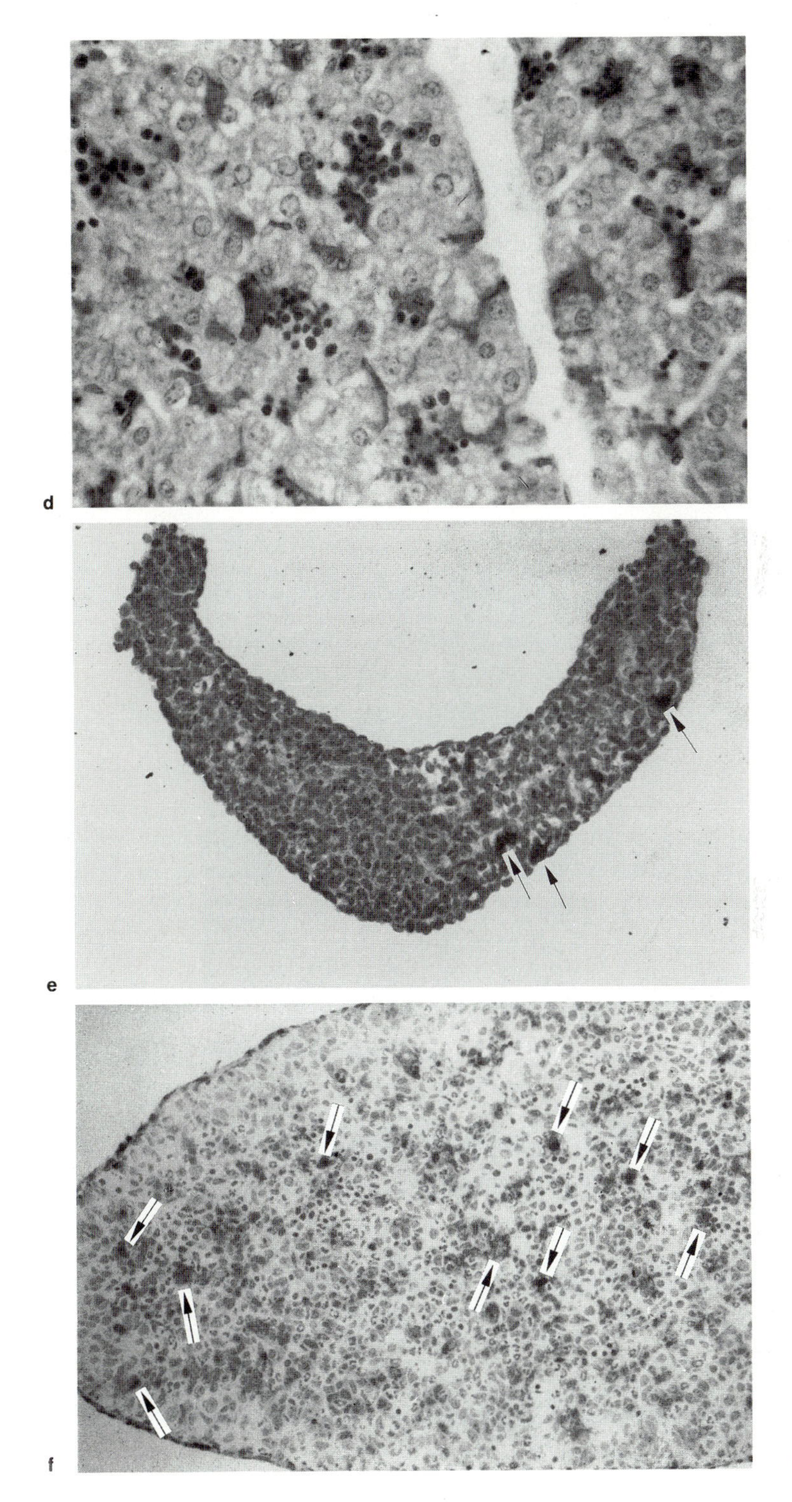

Fig. 1 (*cont.*)

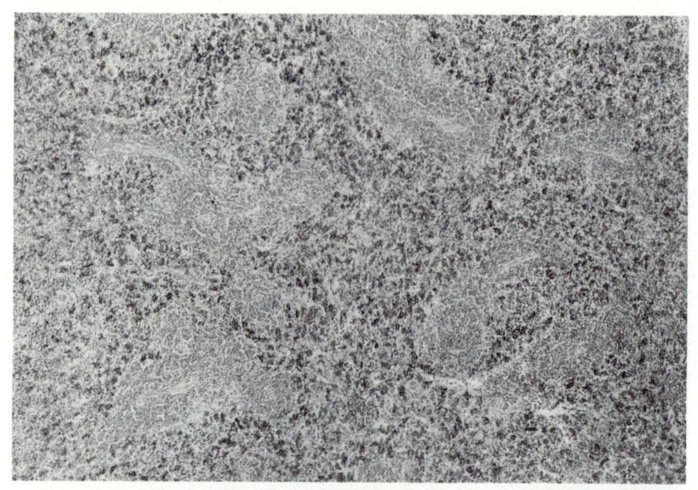
g

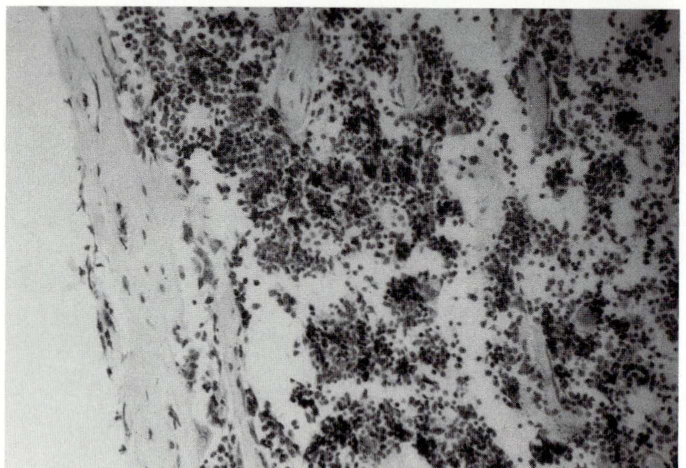
h

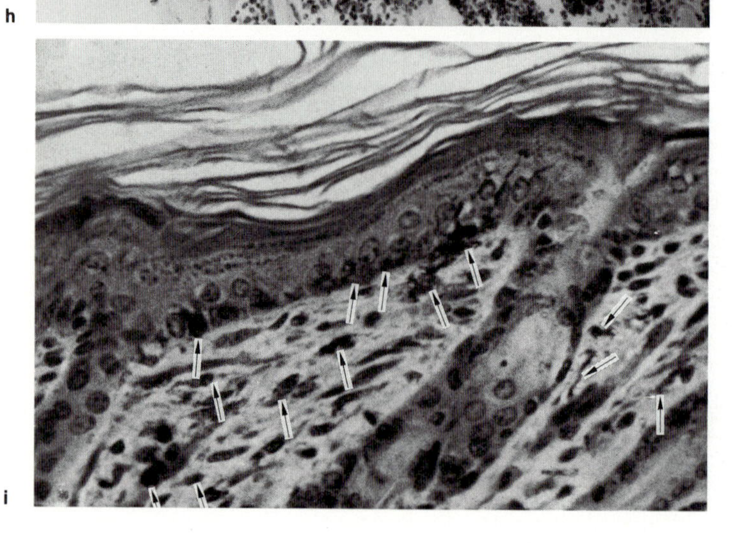
i

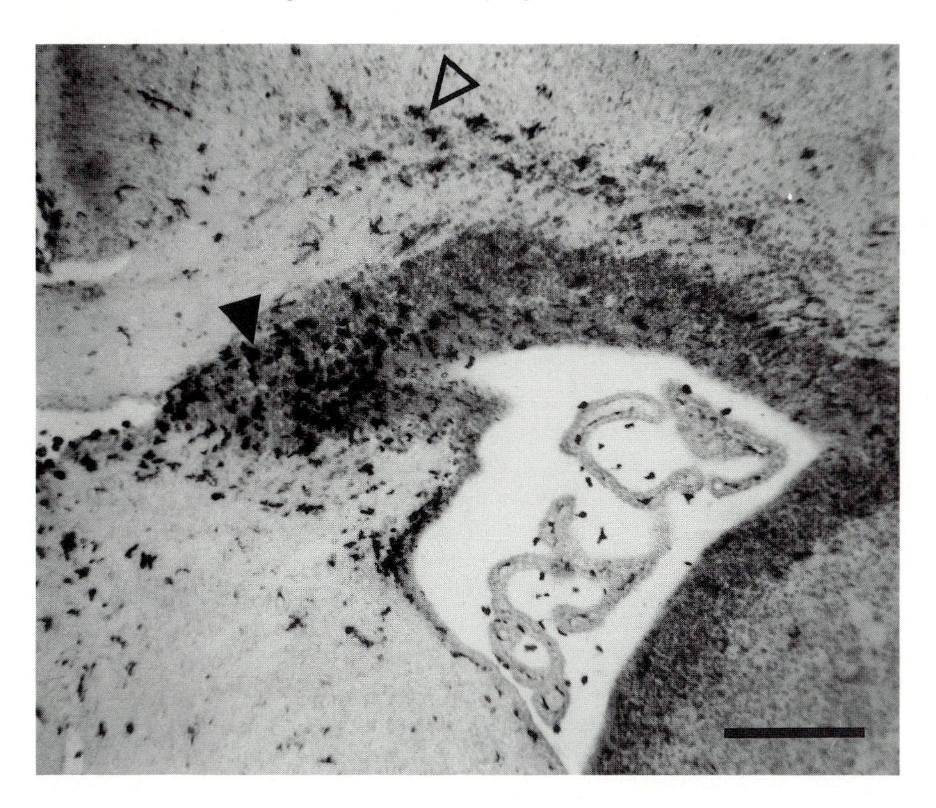

Fig. 2. Macrophages in the developing brain. F4/80 demonstrates invading "amoeboid microglia" (*solid arrowhead*) alongside ramified cells (*open arrowhead*) similar to resident microglia, in the septum of a 1-day-old mouse. *Scale bar* 200 μm; cresyl violet counterstain. (Prepared by L. LAWSON)

◀

Fig. 1 a–i. Macrophage ontogeny in lymphohaemopoietic organs. F4/80 labelling of murine tissues as described. See MORRIS et al. (1991b) for further details. During development, F4/80$^+$ monocytes and macrophages appear sequentially in yolk sac, liver, spleen and bone marrow and in non-lymphoid organs such as skin. **a** *Yolk sac*, day 10. F4/80$^+$ cells are found in small vessels and in interstitium, but are not associated with abundant erythroblasts in haemopoietic islands. **b–d** *Liver*: **b** Rounded F4/80$^+$ monocytes and early stellate macrophages appear in foetal liver (day 10) before local erythropoiesis is established; **c** stellate macrophages are found at peak levels in mainly erythropoietic islands by day 15; **d** by 3 days postnatally, haemopoiesis wanes and some sinus-lining F4/80 labelled macrophages resemble Kupffer cells. **e–g** *Spleen*: **e** day 12, earliest F4/80$^+$ macrophages; **f** birth, dispersed F4/80 labelled macrophages in haemopoietically active region containing erythroid and myeloid cells, but few lymphocytes; **g** 1 week, newborn. F4/80$^+$ macrophages are abundant in red pulp but are excluded from developing white pulp, which is rich in lymphoid cells. **h** *3-day newborn femur*. Stromal F4/80$^+$ macrophages have appeared at centre of haemopoietic cell clusters. **i** *3-day newborn skin*. Stellate F4/80$^+$ cells in developing epidermis resemble Langerhans cells. Labelled macrophages are also abundant in dermis

form a mosaic network of cells within the retina. Similar cell populations are recruited throughout the developing central and peripheral nervous system.

Subsequently, MORRIS et al. (1991b) studied the presence of F4/80$^+$ cells in the early embryo with special reference to haemopoietic tissues, mesenchyme and other developing organs. F4/80$^+$ cells appear in the yolk sac together with erythroid cells at day 9–10, although it is known from unpublished studies by SHIA that F4/80-negative precursors of mature macrophages, (Granulocyte-macrophage colony-forming units, GM-CFUc) can be detected in suspensions of embryos by day 5. In yolk sac, primitive erythroid cells and F4/80$^+$ macrophages are not physically associated in haemopoietic cell clusters, as in more mature tissues (CROCKER et al. 1988b). From day 10 haemopoietic activity shifts to the foetal liver, where a rapidly expanding population of F4/80$^+$ stromal macrophages is found in erythroblastic islands. Monocytes and more differentiated, definitive erythroid cells produced in foetal liver sinusoids seed developing organs and mesenchymal tissues widely from day 10, presumably via newly formed blood vessels which develop at the same time.

After peak levels of haemopoietic activity are reached in the liver at day 14, the spleen (high levels at day 17) and bone marrow (from day 19) become active in turn. Erythroblastic islands disappear from liver later in gestation and during further postnatal development, and sinus-lining hepatic macrophages assume the appearance of Kupffer cells. A striking feature of haemopoiesis in the foetus is the relative absence of granulocyte production in foetal liver at day 14, and myelopoiesis only becomes prominent in spleen and bone marrow during late gestation. Lymphocytes in spleen and thymus also develop late in foetal life and expansion of secondary lymphoid organs occurs mostly during the early weeks of postnatal development.

The F4/80 antigen serves as a sensitive, specific marker for mature macrophages in foetal tissues throughout gestation. By contrast, the sialoadhesin antigen (SER) appears later in development and is restricted to subpopulations of F4/80$^+$ macrophages in lymphohaemopoietic tissues (MORRIS et al. 1992). The sialoadhesin receptor was first identified by CROCKER and GORDON (1986) on mature macrophages present in adult bone marrow clusters in situ and in vitro. Immunochemical and rosetting studies with the inhibitory mAb, SER-4, have shown that sialoadhesin is not expressed by day 14 foetal liver macrophages although these F4/80$^+$ cells express the distinct divalent cation-dependent receptor for erythroblasts (EbR) which appears to be the major adhesion receptor involved in erythroblast island formation (MORRIS et al. 1991a). Sialoadhesin appears on stromal macrophages in developing lymphohaemopoietic organs from ~ day 17, at the time of myelopoiesis, and the striking pattern of strongly SER-labelled cells observed within the marginal metallophil zone of adult spleen (CROCKER and GORDON 1989) appears 2–4 weeks postnatally, in parallel with development of lymphoid cells and white pulp.

These studies have established that different macrophage- and subset-specific antigens are regulated independently during development. Stromal macrophages in foetal liver, spleen and bone marrow are likely to regulate adhesion, growth and differentiation of various haemopoietic cells, although the role of different haemagglutinins and the nature of the ligands and haemopoietic cellular interactions remain to be defined. The non-haemopoietic functions of macrophages in the foetus are obscure, apart from a possible role in remodelling of the nervous system. Foetal macrophages express characteristic endocytic receptors such as FcR (CLINE and MOORE 1972) and proliferate vigorously in response to autocrine and paracrine stimuli. It is likely that they are an important source of growth factors for a range of other cell types during development (see chapter by RAPPOLEE and WERB elsewhere in this volume).

3.2 Normal Adult

After the distribution of blood monocytes to many tissues during development, recruitment of haematogenous cells continues throughout adult life. Their entry, lifespan and rate of turnover vary in different tissues (for review see GORDON 1986). Enhanced recruitment of monocytes in response to inflammatory and infectious stimuli depends largely on production within the bone marrow. However, some resident tissue macrophage populations turn over independently, e.g. in lung, and are renewed by local proliferation. Macrophages are able to persist as relatively long-lived cells in tissues such as the adult nervous system, or migrate from peripheral sites such as skin and gut to lymph nodes, where they become trapped. It is often difficult to distinguish newly recruited from resident tissue cells since adaptation of monocytes and macrophages to each specialised microenvironment makes it impossible to use morphologic or antigenic markers by themselves to draw such a distinction. Antigen markers can be combined with labels for DNA synthesis to trace the kinetics of tissue entry. Haemopoietic reconstitution of x-irradiated animals has been used to trace the life history of macrophages, and the Y chromosome may provide a useful marker to distinguish macrophages of donor male origin from recipient female cells (PERRY and LAWSON, unpublished). Recently, fluorescent hydrophobic dyes such as dil (1.1′ dioctadecyl 3,3,3′,3′ tetra methyl indocarbocyanine perchlorate) have proved useful for ex vivo labelling of peritoneal macrophages before adoptive transfer to recipient animals (ROSEN and GORDON 1990b). Dil-labelled peritoneal macrophages can be used as surrogate monocytes since they migrate from blood to specialised regions of spleen as well as to lungs and liver in unstimulated animals, and are recruited into peritoneal exudates. After transfer to the peritoneal cavity, labelled resident peritoneal macrophages (RPM) migrate rapidly to regional draining lymph nodes (ROSEN and HUGHES, unpublished).

Antigen markers have helped to identify macrophages in tissues, but differences in regional expression of antigens reveal considerable heterogeneity of cell phenotype. We first describe the distribution of F4/80-labelled cells in tissues and then note variations in marker expression revealed by the use of SER-4, FA.11 and CR_3. Figure 3 summarises the antigen phenotype of resident macrophages in various tissues, as described below.

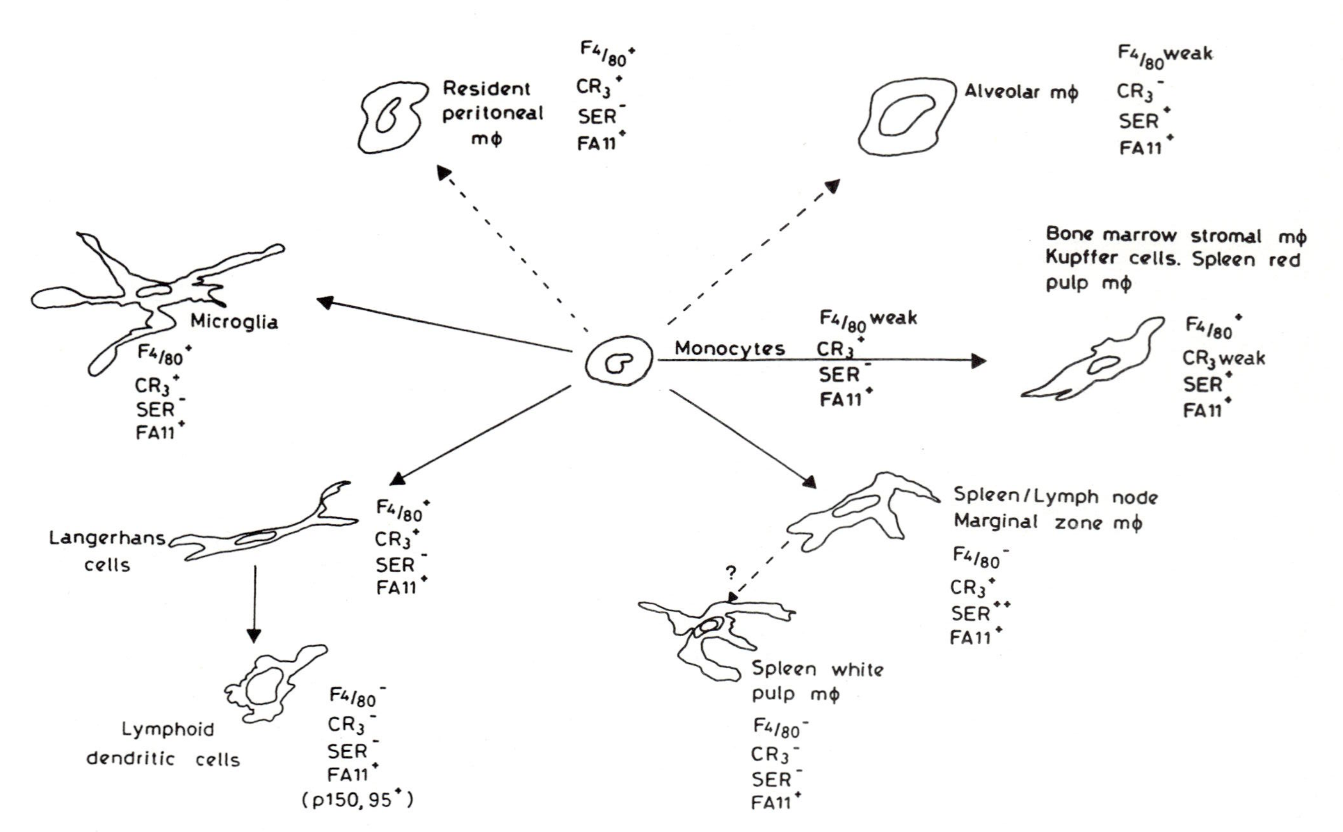

Fig. 3. Schematic representation of differentiation of tissue macrophages and lymphoid dendritic cells in adult mouse. Diagram shows antigen phenotype and pathways of proven (*solid line arrows*) or presumed (*broken line arrows*) precursor cells

3.2.1 Expression of Antigen Markers by Tissue Macrophages

3.2.1.1 F4/80

A detailed description of F4/80$^+$ cells in normal murine tissues has been given in a series of immunocytochemical studies (summarised by HUME and GORDON 1985). Examples of the varied morphology and cellular associations of resident macrophages in adult lymphohaemopoietic organs and brain are illustrated in Figs. 4–6. Quantitative analysis of antigen levels in different tissues is broadly in agreement with the immunocytochemical findings (LEE et al. 1985). Substantial populations of sinusoidal and interstitial F4/80$^+$ cells are found in liver (Kupffer cells), red pulp of spleen, bone marrow stroma (within haemopoietic cell clusters) and subcapsular regions of lymph nodes (HUME et al. 1983b). Thymus and T lymphocyte-dependent regions (white pulp of spleen and lymph nodes, Peyer's patches) are conspicuously free of F4/80 label whereas other mAbs reveal macrophage subpopulations at these sites (RABINOWITZ et al. 1991a). Populations of F4/80$^+$cells are present throughout the lamina propria of the gastrointestinal tract, in epidermis (Langerhans cells) as well as in subcutaneous tissues (HUME et al. 1984a), and in the parenchyma of the nervous system (LAWSON et al. 1990). Delicately arborised F4/80$^+$ cells are found regularly dispersed within epithelium (Langerhans cells) and in the brain (microglia), whereas other stellate cells with shorter plasma membrane processes are present beneath the basement membrane of epithelial cells in small intestine and in renal medulla (HUME and GORDON 1983).

The airway, a major portal of entry to the body, contains a population of rounded alveolar macrophages which express F4/80 only weakly (HUME and GORDON 1985; GORDON, unpublished). F4/80$^+$ macrophages are ubiquitous in connective tissues but are absent within bone and cartilage matrix (HUME et al. 1984b) and infrequent in normal muscle and heart (GORDON, unpublished). Interstitial or sinusoidal F4/80$^+$ cells can be readily detected in several endocrine organs (testis, ovary, adrenal, pituitary) whereas they are sparse in normal pancreatic islets and thyroid (HUME et al. 1984c; HUTCHINGS et al. 1990; POW et al. 1989). In the ovary, F4/80 labelling has revealed striking changes in macrophage number and morphology during the reproductive cycle (HUME et al. 1984c), and in posterior pituitary F4/80 labelled microglia selectively endocytose terminals of neuroendocrine cells containing oxytocin/vasopressin (POW et al. 1989), suggestive of functional responses to hormonal stimulation.

F4/80 labelling has revealed considerable heterogeneity among macrophages in and outside the brain and made it possible for LAWSON et al (1990) to prepare a map of the distribution and morphology of microglia in the adult murine CNS (compare Figs. 5 and 6). Microglia form a network of F4/80$^+$ plasma membrane processes throughout white and grey matter; the morphology of F4/80$^+$ cells varies in different regions of the parenchyma, and distinguishes microglia from other macrophage populations in the choroid plexus and leptomeninges. F4/80 antigen is expressed at high levels on

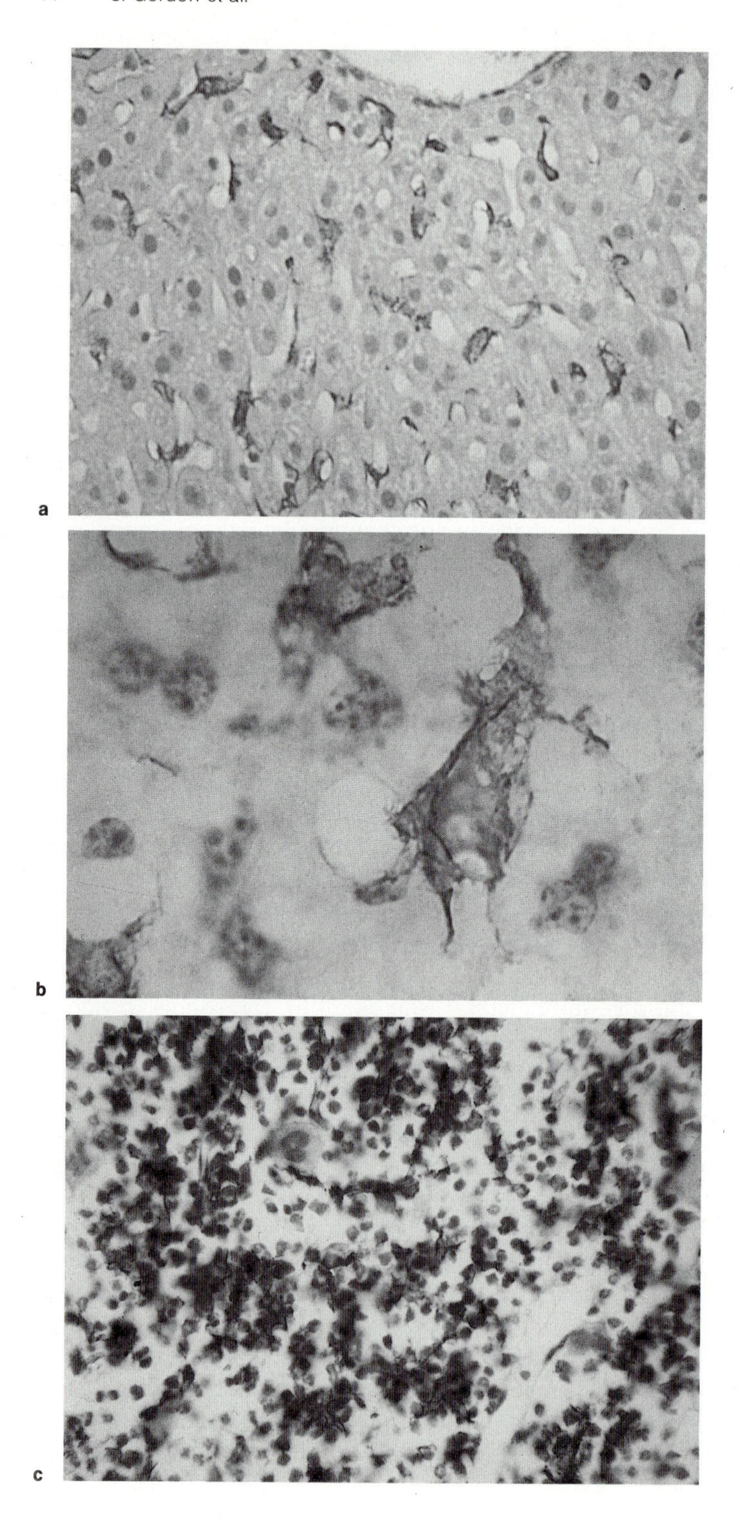

Fig. 4 (*cont.*)

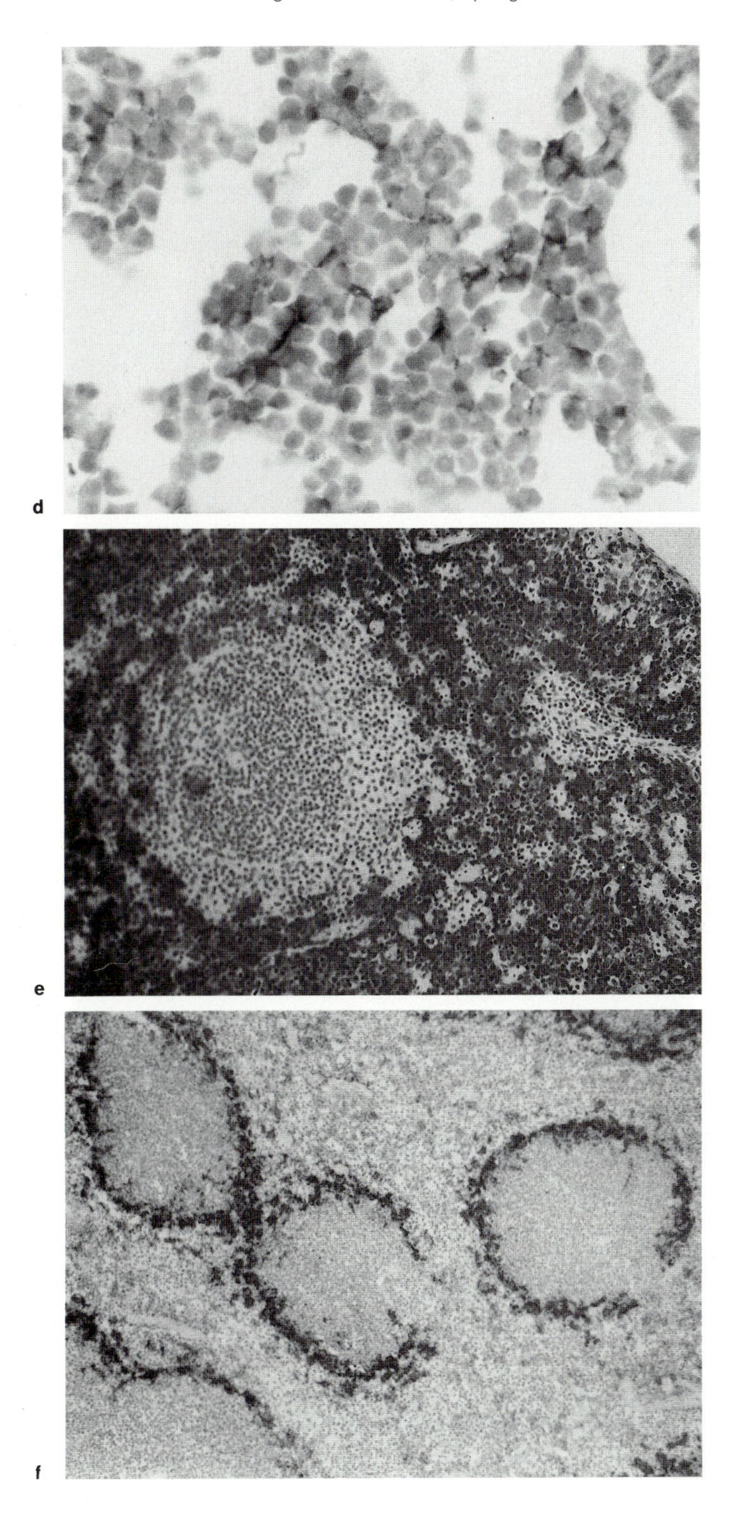

Fig. 4 (*cont.*)

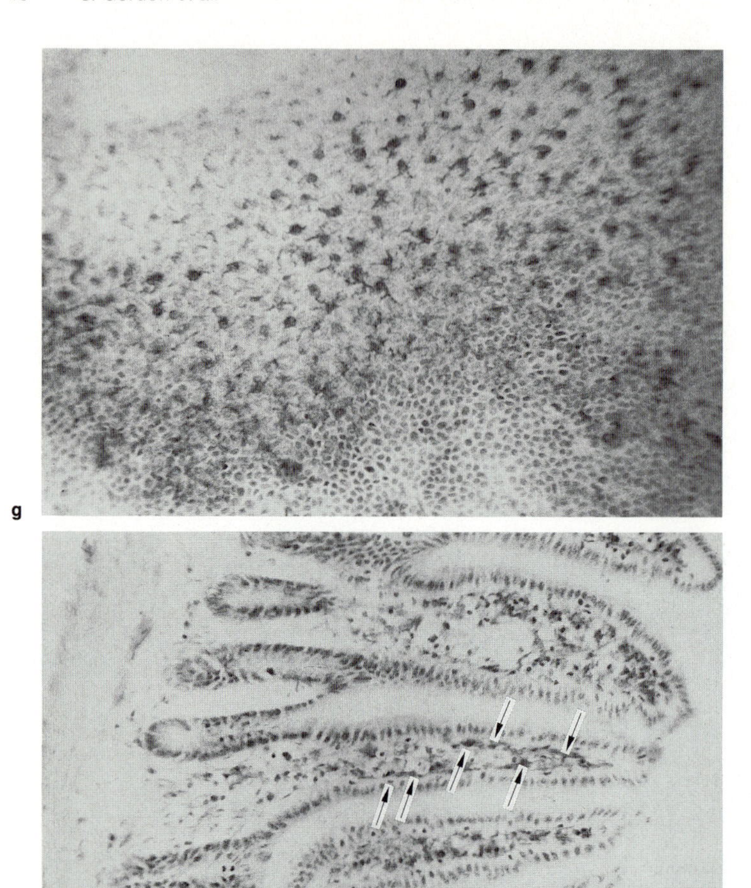

Fig. 4 a–h. Macrophages in adult murine lymphohaemopoietic tissues exhibit varied phenotype and cellular associations. **a, b** *Liver*. Kupffer cells are relatively large and have simple stellate morphology. Hepatocytes and endothelial cells are F4/80 negative, **c, d** *Bone marrow*. Stromal macrophages in haemopoietic clusters express both F4/80 (**c**) and SER-4 antigen (**d**), whereas monocytes are only labelled by F4/80. Stromal macrophages are finely branched and contact developing erythroid and myeloid cells. **e** *Spleen*. F4/80 labels red pulp macrophages intensely, but not white pulp. SER-4 expression is intense in marginal matallophil zone, weak in red pulp and absent in white pulp (**f**). **g** *Skin*. The cytoplasm is drawn out into many fine processes with some secondary branching. The F4/80 labelled cells occur in a regular array and associate with keratinocytes, shown by nuclear counterstain. **h** *Small intestine*. Macrophages are found throughout the lamina propria in gut, forming an almost continuous core within villi. Macrophages lie beneath epithelium, often enveloping capillaries and lymphatics.

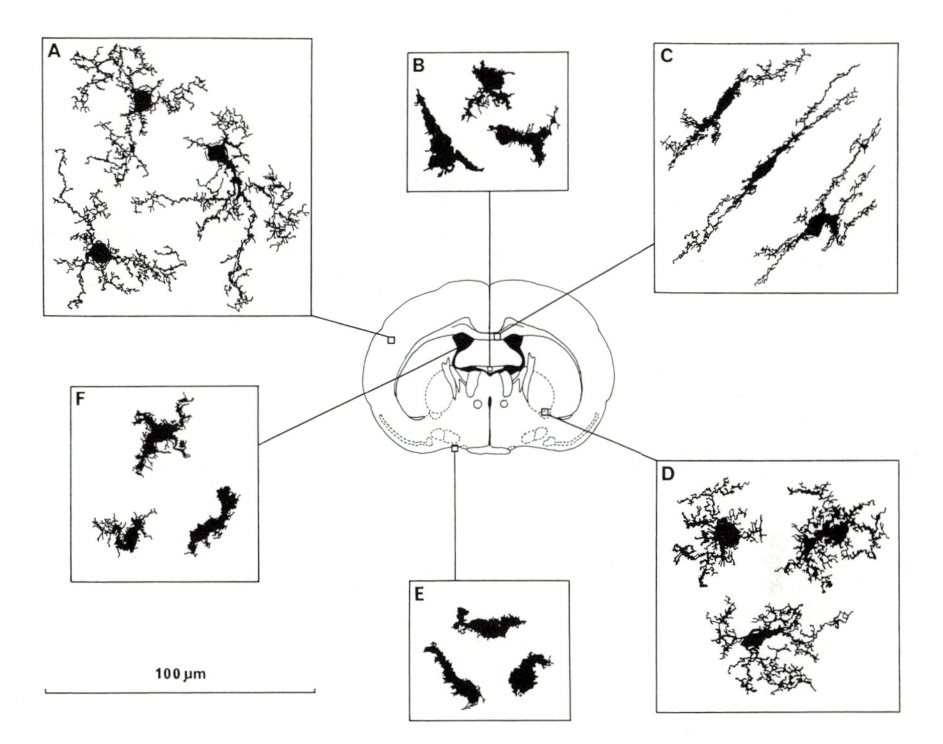

Fig. 5 A–F. Camera lucida drawings from F4/80 stained coronal sections of mouse brain: **A** cortex; **B** subfornical organ; **C** corpus callosum; **D** basal ganglia; **E** leptomeninges; **F** choroid plexus. Macrophages of the adult central nervous system vary in morphology and antigenic phenotype according to their location. Those lying outside the blood brain barrier (**B**, **E**, **F**) tend to be simpler stellate cells. Cells within the parenchyma are extensively arborised, with a radial pattern if they occur in grey matter (**A**, **D**) or a longitudinal one if in white matter (**C**). SER and CD4 antigens are down-regulated on these cells

microglia, whereas several other antigens (e.g. leucocyte common antigen, CD_4) are down-regulated once cells pass through the blood-brain barrier (PERRY and GORDON 1987, 1991), as discussed further below.

In summary, F4/80 has proved to be a reliable and specific marker for the presence of macrophages in many adult tissues apart from specialised T cell-dependent regions and the alveolar space. F4/80 labelling has defined the association of tissue macrophages with endothelial, epithelial and other cell types, and has made it possible to begin to reconstruct the migration and differentiation of monocytes and macrophages in different compartments of the body.

3.2.1.2 Sialoadhesin (SER)

The SER antigen is restricted to selected tissue macrophages in lymphohaemopoietic organs and scattered macrophages elsewhere (CROCKER and GORDON

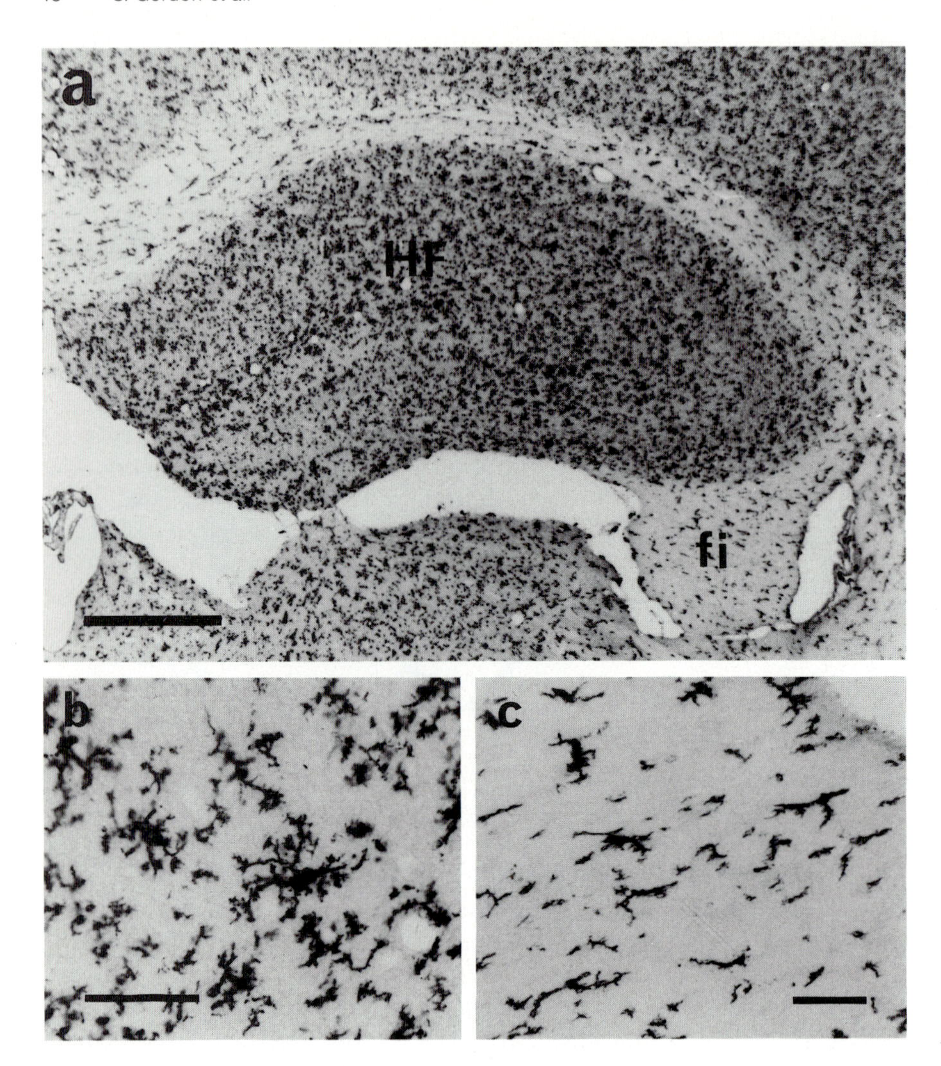

Fig. 6 a–c. Microglia (resident macrophages) are numerous, ubiquitous but heterogeneous in their distribution and morphology in the normal adult mouse brain. **a** Mouse hippocampus stained with F4/80 shows the large amount of microglial membrane present and its heterogeneous distribution. *Scale bar* 500 μm; *HF*, hippocampal formation; *fi*, fimbria. **b**, **c** Detail of microglial morphology in hippocampus (**b**) and fimbria (**c**). *Scale bars* 60 μm, no counterstain

1989). Strongly SER^+ stellate macrophages are found in bone marrow stroma (where F4/80 but not SER mAb also labels developing monocytes), in a characteristic region of the spleen (marginal metallophil zone), where F4/80 is weak or absent, and in subcapsular stroma in lymph nodes. Weaker SER labelling is observed on Kupffer cells and spleen red pulp macrophages which are strongly $F4/80^+$. Conversely, F4/80-poor alveolar macrophages can express sialoadhesin (Crocker, McWilliam and Gordon, unpublished). One reason for differential expression of sialoadhesin is its dependence on exogenous inducer protein(s) which are present in normal mouse plasma and lymph (CROCKER et al. 1988a; MCWILLIAM and TREE, unpublished). Mouse serum contains similar levels of inducer activity to that found in plasma, and there is species restriction with regard to inducer source and target. Although the inducer proteins have not been characterised, exposure to plasma is correlated with sialoadhesin expression in vivo. In the central nervous system, for example, the SER antigen is absent where microglia are shielded from contact with plasma by the blood-brain barrier. Microglia in circumventricular organs outside the blood-brain barrier are exposed to plasma proteins and express sialoadhesin; leakage of plasma proteins after injury to the nervous system also enhances SER antigen expression on microglia (PERRY et al. 1992). Monocytes and other SER-poor macrophages can be induced to express high levels of SER in vitro, although cells vary in the rate and extent of induction. Preliminary studies (MCWILLIAM and TREE, unpublished) indicate that cytokines, including interferons and interleukin-4, modulate sialoadhesin expression. These findings indicate that macrophage sialoadhesin is regulated by a complex network of signals that link immune responses and haemopoiesis. Although the role of sialoadhesin in haemopoietic cell interactions is not understood, its distinctive regulation in adult and foetal tissues makes it a valuable marker of specialised macrophages.

3.2.1.3 FA.11

The FA.11 mAb reacts more broadly in tissues than F4/80 and is a candidate pan-macrophage reagent. However, since its expression is mostly intracellular, it is less satisfactory than mAbs that react with plasma membrane markers. FA.11 labels most of the cells which express F4/80 antigen in tissues, although some resident macrophage populations such as microglia are labelled lightly (LAWSON, unpublished), possibly reflecting down-regulation of endocytic activity in these cells. In contrast with F4/80 and sialoadhesin, FA.11 also reacts with scattered macrophage-like cells in T lymphocyte-dependent regions in spleen white pulp (RABINOWITZ et al. 1991a). Its labelling of dendritic and related cells in tissues will be discussed further below. Alveolar macrophages, another population which expresses low levels of F4/80, are strongly FA.11^{+} (RABINOWITZ and CROCKER, unpublished). The distribution of FA.11 antigen therefore reflects endocytic stimulation, and its presence could provide a possible marker of macrophage activity in situ.

3.2.1.4 CR_3

Whilst the role of CR_3 in induced myelomonocytic recruitment has been studied in a variety of murine models of inflammation, as discussed below, expression on normal murine tissue macrophages is heterogeneous (FLOTTE et al. 1983). CR_3 and the other leucocyte integrins that share a common β_2 chain, LFA-1 and p150/95, are regulated independently. Non-leucocyte integrins have been poorly characterised in murine tissues because of lack of suitable reagents, but monocytes and macrophages in other species are known to interact with fibronectin, fibrinogen, laminin and vitronectin (BEVILACQUA et al. 1981; BROWN and GOODWIN 1988; GRESHAM et al. 1989; SHAW et al. 1990; KRISANSEN et al. 1990).

Several anti-murine CR_3 mAbs label peritoneal macrophages and microglia. However, other major tissue macrophage populations, including Kupffer cells (LEE et al. 1985, 1986) and bone marrow stromal macrophages (CROCKER and GORDON 1985), express low levels of CR_3. Alveolar macrophages lack CR_3 (Blusse van OUD ALBLAS and VAN FURTH 1979; GORDON, unpublished) but express high levels of LFA-1 (CROCKER, unpublished), a marker enhanced by macrophage activation (STRAUSMAN et al. 1986). In spleen, CR_3 is mainly present on PMNs in red pulp, but marginal zone macrophages are also labelled. The expression of p150/95 in murine tissues may provide a marker for lymphoid dendritic and related cells in spleen, as described below. Expression of the α chains of leucocyte β_2 integrins is therefore differentially regulated by their microenvironment in situ; CR_3 expression is rapidly up-regulated once CR_3-negative macrophages are isolated from tissues and maintained in culture.

3.3 Elicited and Activated Macrophages

Entry, accumulation and turnover of macrophages are markedly enhanced by many forms of tissue injury, inflammation and repair, including metabolic and neoplastic diseases. Plasma membrane molecules contribute to, and serve as useful markers of, many of the functional changes of macrophages involved in these processes. It is convenient to distinguish between exudate macrophages elicited by immunologically non-specific inflammatory stimuli and cells that are activated to display enhanced microbicidal and cytocidal properties by the actions of antigen-stimulated T lymphocytes (EZEKOWITZ and GORDON 1984). Both types of recruited cell also express common properties that distinguish them from resident macrophage populations in the peritoneal cavity and elsewhere (GORDON et al. 1988a). However, responses of macrophages in different tissues to local and systemic stimuli vary considerably and result in phenotypic heterogeneity that is difficult to interpret in regard to mechanism or functional significance (GORDON et al. 1988b).

Several broad generalisations can be made in this regard. Induced recruitment of monocytes following microbial infection and other forms of tissue injury is often accompanied by that of other myeloid cells, especially PMNs (myelomonocytic responses), whereas many forms of CNS injury, viral infection or malignancy recruit monocytes with T lymphocytes (mononuclear responses). Macrophages play a central role in initiating, perpetuating and resolving inflammation through their extensive repertoire of specific plasma membrane receptors (GORDON et al. 1988b) and secretory products (RAPPOLEE and WERB 1991). However, it is not always clear to what extent these activities are mediated by resident macrophages already present in tissues, or by newly recruited monocytes. Resident macrophages in liver, spleen and bone marrow, for example, express plasma membrane receptors for sugar-specific recognition of foreign agents, but there is little information on the ability of these cells, possibly involved in first-line interactions with invading organisms, to produce monokines and other chemotactic and vasoactive mediators of inflammation. Resident tissue macrophages may be refractory to stimuli such as lipopolysaccharide (LPS), and host defence depends on rapid mobilization of blood monocytes that can be induced to release cytotoxic molecules (LEPAY et al. 1985a, b).

Surface receptors play a key role in macrophage production, migration and interactions with cellular and humoral ligands. The leucocyte integrin CR_3 is known to be essential for induced myelomonocytic recruitment in response to a wide range of inflammatory stimuli (ROSEN and GORDON 1990a). The anti-CR_3 mAb 5C6 is a potent inhibitor of myelomonocytic cell adhesion to inflamed endothelium after non-specific stimulation (thioglycollate broth, LPS) and acute listerial infection (ROSEN et al. 1989) and partially inhibits monocyte recruitment in T cell-dependent delayed-type hypersensitivity (ROSEN et al. 1988) and autoimmune islet cell damage in non-obese diabetic mice (HUTCHINGS et al. 1990) However, other murine models of infection (e.g. BCG, PLASMODIUM yoelii) are resistant to 5C6 mAb (ROSEN, unpublished) and presumably involve CR_3-independent pathways of monocyte recruitment.

Once recruited cells arrive at a site of inflammation they undergo discrete stages of further differentiation before becoming fully activated, reflected by complex alterations of phenotype. For example, in a relatively simple model of inflammation, monocytes can be recruited to the peritoneal cavity of mice by local injection of biogel polyacrylamide beads (FAUVE et al. 1983; RABINOWITZ, unpublished; STEIN and GORDON 1991). These are too large to be ingested, but evoke the influx of macrophages which can be stimulated further by lymphokines or a phagocytic trigger to release high levels of tumour necrosis factor α (TNF-α), a characteristic property of immunologically activated as well as of thioglycollate-broth elicited macrophages (STEIN and GORDON 1991). Plasma membrane receptors for mannosylated ligands (MFR) or for Fc regions of IgG are efficient triggers for secretion of a range of mediators by primed macrophages, whereas CR_3 and other membrane antigens are inert in this regard (reviewed by GORDON et al. 1988b).

How useful are antigen markers in characterising the functional state of excudate macrophages in different local microenvironments? Several antigens are retained on recruited monocytes and/or induced in exudate macrophages, irrespective of the site at which the cells localise (LEE et al. 1986). These include F4/80, which is also present on Kupffer cells, and CR_3, 7/4 and Ia (MHC class II), which are expressed mainly by recruited macrophages, for example in BCG-induced liver granulomata (Fig. 7) (RABINOWITZ, unpublished) Scattered cells in liver sinusoids which express these antigens in infected animals represent newly

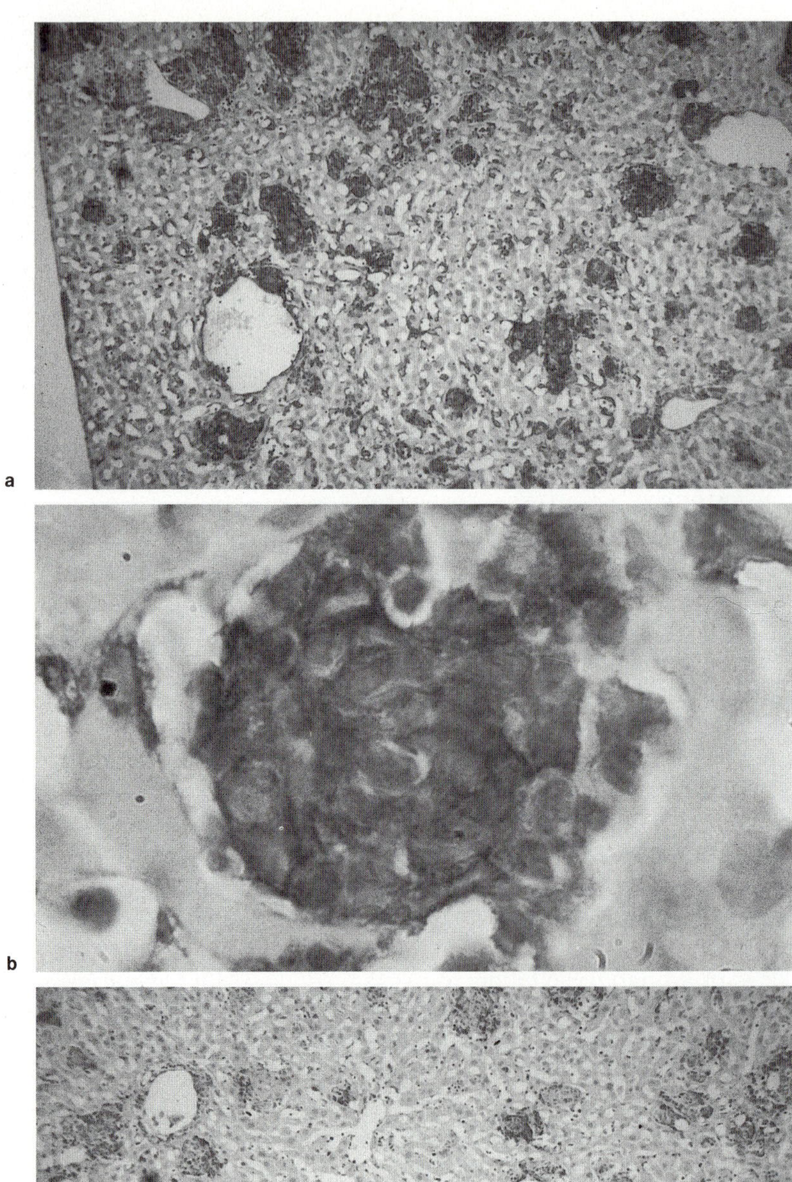
a

b

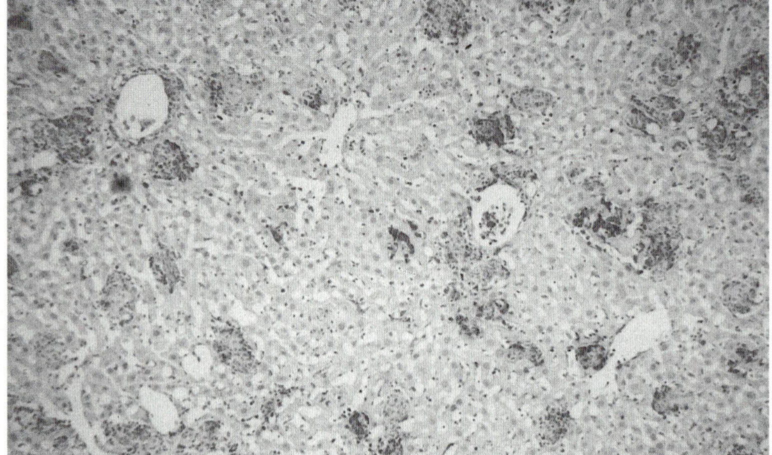
c

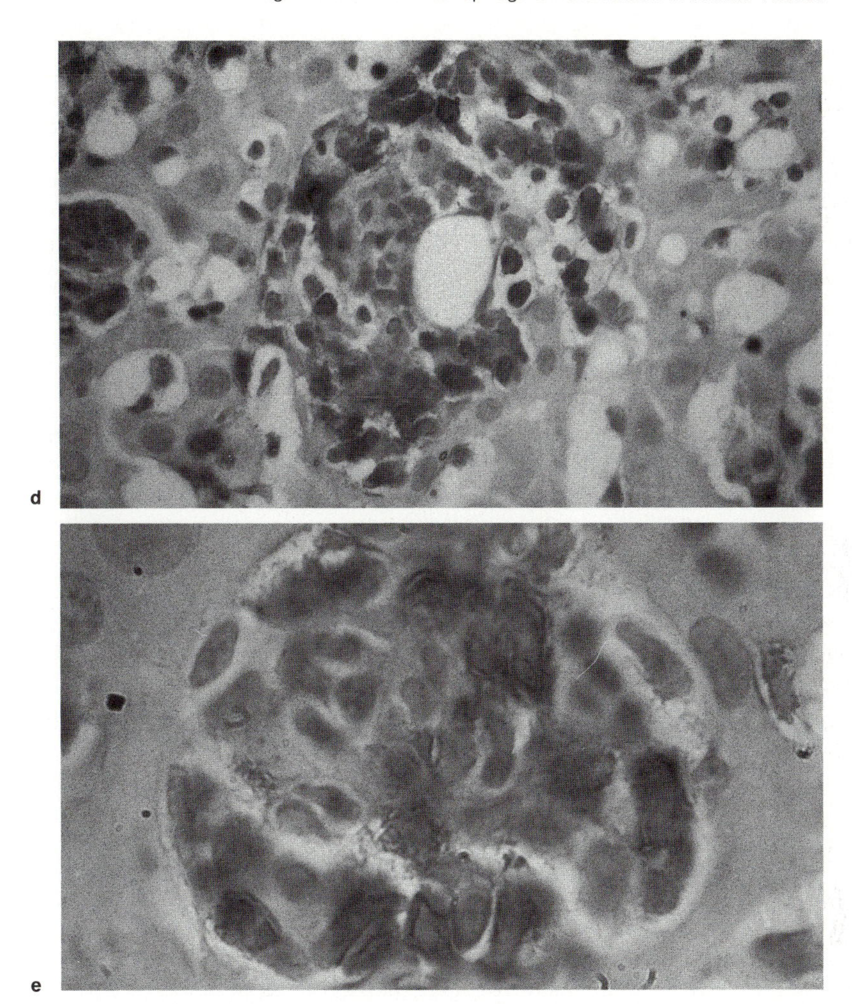

Fig. 7 a–e. Heterogeneous expression of antigens by macrophages in murine liver during granuloma formation. Sections were labelled at the peak of the inflammatory response (day 10) after intravenous injection of bacille Calmette-Guérin (BCG) (RABINOWITZ, unpublished). **a**, **b** Low and high power views. F4/80 antibody labels all macrophages in granulomata, as well as Kupffer cells and trafficking monocytes in sinusoids. **c–e** Low, intermediate and high power views. Antibody 7/4 labels a subpopulation of activated macrophages in granulomata, as well as monocytes and neutrophils in blood vessels and sinusoids. Kupffer cells lining sinusoids are unlabelled

recruited migrating cells. In tissues such as liver (LEE et al. 1986) and the CNS (ANDERSSON et al. 1991) it becomes increasingly difficult to distinguish newly recruited monocytes that adopt the characteristic phenotype of macrophages in each microenvironment from resident cells that have been reactivated by local inflammation to express previously down-regulated markers.

It has also proved difficult to distinguish immunologically activated macrophages (i.e. those recruited cells which display an enhanced capacity to kill intracellular pathogens) from non-cytocidal exudate macrophages by antigen marker analysis in situ. Cytokines produced by macrophages themselves, by activated T lymphocytes or by other cells are potent modulators of the macrophage phenotype. There is little information on the effects of locally produced cytokines on antigen expression at sites of immune and inflammatory reactions and on the range of cellular targets induced. Interferon-γ, for example, is a major, although not unique, inducer of Ia in lymphokine-activated macrophages. The 7/4 antigen is expressed by BCG-activated macrophages in situ; however, it is not inducible in vitro by interferon-γ alone, but only in combination with other T cell dependent and other growth factors (TREE, unpublished; MAUDSLEY et al. 1991). Ia (MUNRO et al. 1989) and an intracellular antigen (IP10) induced in macrophages by interferon-γ are also widely expressed in neighbouring endothelial and epithelial cells (KAPLAN et al. 1987). Immunomodulatory agents which deactivate macrophages in situ include glucocorticoids, transforming growth factor β (DING et al. 1990) and interferon-α/β (EZEKOWITZ et al. 1986). The profile of antigen markers expressed by macrophages therefore varies considerably at different stages of an inflammatory or immune process.

At present there are surprisingly few antigen markers available to assess macrophage functional status in situ, in spite of many attempts to produce such reagents. Further studies are needed to correlate antigen expression with changes in secretory repertoire (respiratory burst, TNF etc.) and to define the link between cell differentiation and activation more clearly. Cellular antigens could prove useful in analysing the effects of chemotactic and phagocytic stimuli, cytokines and hormones on macrophages if more discriminating mAbs were isolated and antigen expression correlated more precisely with function.

4 Relationship of Macrophages to Cells of Related Lineages

Several cell types derived from bone marrow progenitors pass through the blood at some stage of their life history and express macrophage-like characteristics in tissues. Antigen markers have been of great value in distinguishing cells of clearly different haemopoietic lineages, but the relationship of macrophages (the cells of the mononuclear phagocyte system) to other cells of bone marrow origin such as lymphoid dendritic cells (LDCs) and osteoclasts remains unclear. LDCs belong to an ill-defined group of specialised accessory cells that include veiled, interdigitating and Langerhans cells (for reviews see AUSTYN 1987; MACPHERSON 1989). Isolated LDCs are uniquely potent in presenting antigen to T lymphocytes to initiate primary immune responses (STEINMAN et al. 1986). Macrophages and B lymphocytes also serve as accessory cells in secondary immune responses, but lack the ability to stimulate naive T lymphocytes efficiently. Dendritic cells are found mainly in lymphoid organs, although not necessarily in the same sites as macrophages, and both cell types can be present in epithelia. LDCs migrate through blood as well as lymph (AUSTYN 1989; LARSEN et al. 1990), and may recirculate from peripheral lymphoid organs, unlike macrophages. Both cell types express common markers during part of their life history and it is not clear whether, or at what stage, these lineages have separated from each other. Similarly, the mature osteoclasts of bone derive from circulating mononuclear cells (for review see CHAMBERS 1989), but it is not known whether they are distinct from monocytes. Recent studies to characterise these various cells in tissues and in culture have provided insights into their specialisation without resolving the question of their interrelationship. Antigen marker expression will be considered in the context of their other properties and of general methods of lineage analysis.

In principle, precursor–product relationships can be established for particular cell types by examining the progeny of suitably marked progenitor cells in vivo and in vitro. One difficulty has been the lack of defined culture systems to grow and maintain cells with the appropriate phenotype, another the shortage of specific markers. Although macrophages adapt readily to cell culture, they do not necessarily retain or remain able to acquire a specialised tissue phenotype, especially when derived from proliferating precursors. Studies on the stromal markers sialoadhesin and EbR have demonstrated the deficiencies of cell lines and standard culture systems in maintaining macrophage differentiation in vivo (MCWILLIAM and FRASER, unpublished). Lineage analysis of these macrophage-related cells has proved difficult in vitro. It has not been possible to generate LDCs from haemopoietic progenitor cells in culture although cells of the LDC phenotype have been derived from isolated Langerhans cells, as discussed further below. Osteoclasts can be generated in small numbers as part of multilineage colonies derived from bone marrow or spleen progenitors (HATTERSLEY, KERBY and CHAMBERS, 1991; UDAGAWA et al. 1989; KURIHARA et al. 1991; KERBY et al. 1991). Stromal fibroblasts are required, as well as multispecific growth factors, and osteoclasts are often found in these cultures with macrophages, which proliferate more vigorously (KODAMA et al. 1991b). Tumour-derived or transformed haemopoietic cell lines may provide alternative precursors for clonal analysis to primary sources in foetal liver, bone marrow or spleen, but have not yet contributed to the elucidation of this problem.

It might be thought that expression of plasma membrane receptors for lineage-restricted growth factors could provide ideal markers to distinguish macrophages from their close relatives. For example, receptors for CSF-1 (c-*fms*) mediate monocytic differentiation when primary precursors (see chapter by STANLEY in this volume) or undifferentiated myeloid cell lines (PIERCE et al. 1990; WU et al. 1990) are treated with CSF-1 in vitro. However, analysis of CSF-1 receptor expression on tissue macrophages, osteoclasts and LDCs is sketchy. Expression of CSF-1 receptors in situ is not restricted to macrophages, but is found on trophoblast epithelium, and receptor expression on macrophages is subject to down-regulation.

Recent studies with osteopetrotic mice, which display deficient osteoclast function, provide evidence that CSF-1 is essential for development of osteoclasts

(WIKTOR-JEDRZEJCZAK et al. 1990; YOSHIDA et al. 1990). Homozygote *op/op* mice carry a mutation in CSF-1 and in the adult contain reduced numbers of blood monocytes and selected tissue macrophages (FELIX et al. 1990a). Some of these deficiencies can be corrected by exogenous CSF-1 (KODAMA et al. 1991a; FELIX et al. 1990b). The role of CSF-1 in LDC differentiation has not been defined. Receptors for GM-CSF are also not sufficiently selective for fine lineage resolution. Apart from various myeloid cells, there is evidence that dendritic cells respond to this factor in cell culture (KOCH et al. 1990). Other receptors for known cytokines are often broadly distributed on haemopoietic and non-haemopoietic cells.

At present analysis of lineage depends on antigen markers and phenotypic properties of cells that have been studied in situ, or after isolation from various tissues. Interpretation of these studies should take into account the differentiation, migration and modulation of cells within different microenvironments. The problem of variability of cellular phenotype can be illustrated by considering the properties of LDCs and Langerhans cells (LENZ et al. 1989). Isolated LDCs lack F4/80, but constitutively express high levels of Ia and are potent stimulators of a mixed leucocyte reaction (MLR) when cocultivated with resting, allogeneic T lymphocytes. Earlier attempts to identify these cells in situ were handicapped by lack of dendritic cell-specific markers suitable for immunocytochemistry. Recent studies have reported that dendritic interdigitating cells in mouse spleen express a p150/95 integrin (METLAY et al. 1990) which is less widely expressed on other tissue macrophages than in man (HOGG et al. 1986), and which may therefore be a selective marker for murine LDCs. A possible relationship between altered integrin expression and induced migration of Langerhans cells and LDCs has not been reported.

Langerhans cells are widely distributed in the epidermis and in other complex epithelia and express properties which link them to tissue macrophages and LDCs. They contain morphologically distinct Birbeck granules and have been implicated in antigen responses in skin and in draining lymph nodes, to which they migrate. Langerhans cells express F4/80 and other plasma membrane antigens in a characteristic pattern of regularly spaced stellate cells surrounded by keratinocytes. When isolated from epidermal sheets, Langerhans cells lose F4/80 antigen, but acquire enhanced MLR activity in culture (SCHULER and STEINMAN 1985). GM-CSF improves survival of Langerhans cells in vitro (WITMER-PACK et al. 1987), but the cells do not retain F4/80, unlike macrophages derived from bone marrow precursors in the same culture medium. These findings are compatible with the hypothesis that Langerhans cells mature into functional LDCs.

Expression of the FA.11 antigen provides further evidence that LDCs and macrophages are related cells. FA.11 labelling is restricted to macrophages and dendritic cells (RABINOWITZ, MILON, STEINMAN and AUSTYN, unpublished), unlike other endosomal/lysosomal membrane glycoproteins which are more widely distributed in haemopoietic and non-haemopoietic cells (DA SILVA, LAWSON and ROSEN, unpublished). The FA.11 antigen is found in solitary granules in dendritic cells, perhaps reflecting a rudimentary or specialised vacuolar apparatus, compared with its distribution in macrophages, which contain numerous FA.11$^+$ vesicles in their cytoplasm. Recent studies have confirmed that endocytic organelles become reduced when Langerhans cells are isolated from epidermis and differentiate into LDC-like cells in culture (STÖSSEL et al. 1990).

Osteoclasts lack F4/80 antigen, unlike macrophages on adjacent periosteal or endosteal surfaces in bone (HUME et al. 1984b), but express other leucocyte antigens (ATHANASOU et al. 1987). Although macrophages are potent catabolic cells able to degrade a range of connective tissue elements, including fragments of ingested bone, only true osteoclasts excavate intact bone by local secretion. Bone resorption is inhibited by calcitonin and osteoclasts express abundant receptors for calcitonin (NICHOLSON et al. 1986; TAYLOR et al. 1989). Giant cell formation is uninformative in distinguishing between osteoclasts, which can be mononuclear, and macrophages, which are often multinucleated, the extent of polykaryocytosis depending on the species. Osteoclasts and macrophages share a H^+ electrogenic vacuolar proton pump, but it is not known whether endosomal antigens such as FA.11 are present in both cell types.

These studies indicate that there is considerable overlap among these different lineages. Improved cell-restricted marker antigen would be helpful in defining in vivo precursors, branch points and possible interconversions among these cells, and their migration pathway within tissues.

5 Antigens and Macrophage Heterogeneity

We have demonstrated that macrophages and related cells are constitutively present in different compartments of the body, that they migrate via blood and lymph and that they undergo complex changes in phenotype in different tissues even in the absence of inflammation. The induced recruitment of blood monocytes as a result of inflammation and injury brings cells from a common pool to tissue environments in which extrinsic and local factors further modulate the macrophage phenotype.

Mature macrophages interact with most other cell types in the body, including endothelium, epithelium, fibroblasts, lymphohaemopoietic and neuroendocine cells. These cells and extracellular matrix influence expression of antigens and secretory products by macrophages within each microenvironment. Regional specialisation in macrophage biosynthetic activity can be detected by in situ hybridisation for mRNA products such as lysozyme (CHUNG et al. 1988), TNF-α (KESHAV et al. 1991) and other monokines. Does the extensive heterogeneity of cell phenotype reflect a single major lineage of mononuclear phagocytes or are there subsets of macrophages, perhaps analogous to the diversity of T lymphocytes? To what extent can cell differentiation and modulation within tissues account for current knowledge of macrophage heterogeneity and how successfully can we replicate in culture the unique phenotype expressed by macrophage populations in situ?

Antigen markers can be used to follow modulation of mature macrophages directly after adoptive transfer to different sites in the body. The effects on macrophage antigen expression of specialised microenvironments in spleen, liver and lung, for example, can be studied by introducing suitably marked cells via different routes in the intact animal. A more general method is to use mAbs directed against various plasma membrane antigens to separate progenitor cells from mature haemopoietic cells in bone marrow, foetal liver or spleen, and to study their progeny. This approach has been little used to analyse macrophage differentiation in vivo, but was used by HIRSCH et al. (1981) to examine expression of F4/80 during macrophage differentiation in cell culture. Bone marrow progenitors were sorted on the basis of F4/80 labelling and incubated with L cell conditioned medium as a source of CSF-1. FACS analysis showed that GM-CFUc lack F4/80, which first appeared on immature promonocytes at the time cells became adherent. Studies with anti CR_3 and 7/4 mAbs (HIRSCH and GORDON, unpublished) indicate that, as expected for myelomonocytic markers, these antigens are expressed earlier than F4/80 on GM presursors that form clusters of both macrophages and PMNs in the presence of GM-CSF. Since SER and FA.11 are not expressed by PMNs, they are probably also acquired after these lineages separate. Monocytes and less mature stages of macrophage development express FA.11 (MILON and GORDON, unpublished), but not SER. F4/80 expression is weak and variable on monocytes so that it is difficult to use this marker to study blood mononuclear cell heterogeneity by fluorescence sorting experiments. Moreover, the yield of blood monocytes from the mouse is small and these cells have a more limited proliferative capacity than progenitor cells.

In addition to relating antigen expression to cell maturity, it is possible to study clonal heterogeneity in CSF-supplemented cultures. Bone marrow CFUc cultivated in L cell-conditioned medium yielded colonies in which all macrophages in all colonies expressed F4/80 uniformly (HIRSCH et al. 1981). In other studies, the activation marker Ia was induced on all independent colonies, although variable numbers of macrophages in each colony expressed this antigen. Early studies in which only a limited range of precursors, CSFs and antigen markers were examined, failed to reveal clonal heterogeneity among mononuclear phagocytes, and the conclusion was drawn that we are dealing with a single cell lineage. Similar further clonal studies are needed with specific markers and functional assays for LDCs and osteoclasts.

Several variables need to be taken into account to perform this type of analysis. It is necessary to define the role of cell growth in influencing expression of a stable specialised phenotype. Clonal analysis might also bring to light a requirement for another cell type (macrophage or fibroblast) which could play an accessory role in haemopoietic cell differentiation. Improved culture systems are needed to study relationships among macrophages, LDCs and osteoclasts, and among mononuclear phagocytes, such as stromal macrophages, alveolar macrophages and Langerhans cells which display a tissue-specific phenotype. In particular, are different cell phenotypes stable or interconvertible, can they be modulated during a proliferative phase or only once cells become terminally differentiated? A related requirement is to define the environmental factors which regulate the phenotype of these cells in situ. Expression of markers such as CR_3 and sialoadhesin on macrophages is tightly controlled in different tissues, but regulation is lost or only partly reproduced in present culture systems. Although some of the variables that regulate marker expression in vivo have been identified, such as the plasma inducer of sialoadhesin, other modulating factors in the microenvironment such as the role of extracellular matrix require further study. The extensive down-regulation of antigens and other markers in highly differentiated macrophages such as resident microglia cannot be reproduced in vitro. This may result from inhibitory interactions with specialised cells in brain parenchyma (neurons, other glia) acting through macrophage surface receptors, or from the absence of elements found outside the CNS. The role of foreign antigens, LPS and other exogenous agents in modulating macrophage differentiation in tissues should also be borne in mind.

The lymphohaemopoietic system provides a variety of model systems to study the interplay between pluripotent stem cells and specific cytokines that regulate their growth and differentiation. Bipotential precursors for macrophages and granulocytes have been of particular interest in defining events that accompany commitment to specialised lineages. The heterogeneity of tissue macrophages and closely related cells, as discussed here, offers a unique opportunity to study modulation and differentiation of mature cells, and the interplay of environmental signals and intracellular events that regulate cell-specific functions. Development and characterisation of well-defined antigens should provide markers to analyse the mechanisms by which macrophage diversity is generated.

6 Conclusion

The combined use of antigen markers and in situ hybridisation has delineated the life history of macrophages and their functions in tissues. The distribution of cells has been followed during development, and through adult life, in the steady state and in response to physiologic changes and pathologic perturbations. The F4/80 mAb provides a tool to map cell distribution, to identify microheterogeneity among macrophages within a single organ, and to reveal accumulation of newly recruited monocytes at sites of injury and their adaptation within a specific tissue environment. F4/80 antigen expression is absent on macrophages, defined by other marker antigens, in specialised regions of lymphoid organs. The FA.11 antigen is more widely expressed on macrophages and on LDCs and is responsive to endocytic stimuli. The sialoadhesin (SER) receptor marks a subpopulation of stromal macrophages in lymphohaemopoietic tissues which are involved in non-phagocytic, possibly trophic, cellular interactions. Antigen markers have revealed adaptations and novel functions of macrophages, and have provided tools to manipulate their migration and behaviour in tissues. Membrane antigens combined with other marker molecules make it possible to study the role of these versatile cells in a wide range of disease processes, and provide model systems to study fundamental questions of cellular differentiation.

Acknowledgements. This work was supported by grants from the Medical Research Council, UK, The Wellcome Trust and the Multiple Sclerosis Society. We thank Margaret Sherrington for help in preparing the manuscript, Professor T. Chambers for helpful discussions and colleagues for allowing us to quote unpublished results.

Appendix

Immunohistochemistry with F4/80 mAb (adapted from HUME et al. 1983b; PERRY et al. 1985). Method is based on the biotin-avidin-peroxidase method of HSU et al. (1981).

Reagents

a) F4/80 hybridoma supernatant can be purchased from:

UK:	Serotec 22 Bankside Station Approach Kidlington Oxford OX5 1JE	Tel: UK 08675 79941 Fax: UK 08675 3899
USA:	Bioproducts for Science Inc. PO Box 29176 Indiana 46229	Tel: (317) 894 7536 Fax: (317) 894 4473
Japan:	Dai Nipon Pharmaceuticals Co. Ltd. 6–8 Doshomachi 2-Chome Chuo Ku Osaka 541	

Use 1:20 in phosphate-buffered saline (PBS)

b) Biotinylated rabbit anti-rat IgG, preferably adsorbed to remove cross-reactivity with mouse IgG, 1:100 (Vector Laboratories Ltd. 16 Wulfric Square, Bretton, Peterborough, Cambridgeshire, UK).
c) 1% normal rabbit serum (Vector Laboratories Ltd).
d) Detection complex (Elite ABC; Vector Laboratories Ltd), 2 drops of each reagent in 5 ml PBS. This needs to be made up at least 30 min before use, but should not be kept longer than 72 h.

e) Chromogen (DAB)
   0.125 g diaminobenzidine tetrahydrochloride
   0.2 g imidazole (optional)
   250 ml 0.1 M phosphate buffer, pH 7.2
   125 μl 30% hydrogen peroxide, added just before use.

Methods

1. Dewax wax sections and take to water. Thaw frozen sections and ensure they are thoroughly dry.
2. Wash sections in PBS for ca. 10 min to remove embedding medium.
3. Incubate with 1% normal rabbit serum for 30 min at room temperature, in a humidity chamber. Do not allow any reagents to dry on sections.
4. Remove excess serum and incubate with F4/80 mAb for 60 min at room temperature.
5. Wash sections in PBS; at least 2 × 10-min washes are desirable.
6. Incubate in biotinylated secondary antiserum for 45 min.
7. Wash sections (at least 2 × 10 min).
8. To quench endogenous peroxidase activity, place sections in 0.3% hydrogen peroxide in methanol (alternatively 96% alcohol) for 20 min.
9. Wash sections (2 × 10 min).
10. Incubate in Elite ABC for 45 min at room temperature.
11. Wash well (at least 2 × 10 min, preferably 3 washes).
12. React in DAB, observing the progress of the reaction. If often takes only around 20–30 s.
13. Wash well in PBS or 0.1 M phosphate buffer.
14. The DAB reaction product may be intensified by incubation in 0.01% osmium tetroxide in 0.1 M phosphate buffer for about 30 s, followed by careful washing.
15. Counterstain as required, dehydrate and mount sections.

Notes

1. The commonest cause of failure of F4/80 staining is poor fixation of the tissue. We routinely use 2% paraformaldehyde, lysine, periodate (PLP) perfusion-fixed material (McLean and Nakane 1974) with or without 0.05%–0.1% glutaraldehyde for use with mAb F4/80, ensuring that the final pH is 7–7.4. The antigen is also stable to 0.5% buffered glutaraldehyde for use with mAb F4/80, Carnoy's fixative or acid alcohol. Buffered paraformaldehyde, Karnovsky's fixative and picric acid are not satisfactory fixatives.
2. F4/80 staining may be carried out on fixed frozen sections or on material embedded in paraffin or polyester wax. Morphologic preservation is better with wax-embedded material but there may be some loss of antigen during prolonged infiltration steps with paraffin wax at 60 °C.

3. It may be necessary to extend the incubation time for the primary antibody or to increase its concentration.

   The perfusion fixation protocol has been successfully used with the other mAbs described, although the appropriate antibody concentration and incubation time have to be determined for each. For intracellular antigens such as FA.11 it may be necessary to extend the incubation time with mAb.
4. 0.1% Triton X-100 may be added to wash buffers, to improve washing and render sections more wettable. It is also useful for enhancing penet[illegible] of mAbs directed at intracellular antigen.
5. An alternative method for carrying out colour reaction is to use glucose oxidase and glucose to generate H_2O_2:

   To 250 ml 0.1 M phosphate buffer add:—0.125 g DAB
   —0.4 g β-D-glucose
   —0.08 g NH_4Cl
   —A few mg glucose oxidase

   Incubate sections with this mixture and observe progress of reaction. It takes about 8–10 min with 10 μM sections depending on the amount of enzyme added.

   A similar principle may be applied to the quenching of endogenous peroxidase activity. In this case a few mg glucose oxidase are added to 0.1 M phosphate buffer containing 1 mM sodium azide and 10 mM glucose. Sections are incubated with this mixture for 15 min at 37 °C with shaking.
6. Negative controls, where the primary antibody is replaced with diluted normal serum, PBS or irrelevant antibody should be included with each batch of experimental slides. Causes of false positive reactions include the presence of melanin, endogenous peroxidase activity, endogenous biotin, non-specific binding of the primary antibody (including binding by its Fc moiety) and cross-reactivity of the secondary antibody between rat and mouse IgG which may be present in the tissue.

References

Andersson P-B, Perry VH, Gordon S (1991) The kinetics and morphological characteristics of the macrophage-microglial response to kainic acid-induced neuronal degeneration. Neuroscience 42:201–214

Athanasou NA, Quinn J, McGee JO'D (1987) Leucocyte common antigen is present on osteoclasts. J Pathol 153: 121–126

Austyn JM (1987) Lymphoid dendritic cells. Immunology 62: 161–170

Austyn JM (1989) Migration patterns of dendritic leukocytes. Res Immunol 140: 898–902

Austyn JM, Gordon S (1981) F4/80, a monoclonal antibody directed specifically against the mouse macrophage. Eur J Immunol 11: 805–815

Beller DI, Springer TA, Schreiber RD (1982) Anti Mac-1 selectively inhibits the human type three complement receptor. J Exp Med 156: 1000–1009

Bertoncello I, Bradley TR, Hodgson GS (1989) The concentration and resolution of primitive hemopoietic cells from normal mouse bone marrow by negative selection using monoclonal antibodies and dynabead monodisperse magnetic microspheres. Exp Hematol 17: 171–179

Bevilacqua MP, Amcani D, Mosesson MW, Bianco C (1981) Receptors for cold insoluble globin (plasma fibronectin) on human monocytes. J Exp Med 153: 42–60
Blusse van Oud Alblas A, van Furth R (1979) The origin, kinetics and characteristics of pulmonary macrophages in the normal steady state. J Exp Med 149: 1504–1518
Brown EJ, Goodwin JL (1988) Fibronectin receptors of phagocytes. Characterisation of the Arg-Gly-Asp binding protein of human monocytes and of polymorphonuclear leucocytes. J Exp Med 167: 772–793
Chambers TJ (1989) The origin of the osteoclast. In: Peck W (ed) Bone and mineral research annual. Elsevier, Amsterdam 6: 1–25
Chung L-P, Keshav S, Gordon S (1988) Cloning of the human lysozyme cDNA: inverted Alu repeat in the mRNA and in situ hybridization for macrophages and Paneth cells. Proc Natl Acad Sci USA 85: 6227–6231
Cline MJ, Moore MAS (1972) Embryonic origin of the mouse macrophage. Blood 39: 842–849
Crocker PR, Gordon S (1985) Isolation and characterization of resident stromal macrophages and haematopoietic cell clusters from mouse bone marrow. J Exp Med 162: 993–1014
Crocker PR, Gordon S (1986) Properties and distribution of a lectin-like haemagglutinin differentially expressed by stromal tissue macrophages. J Exp Med 164: 1862–1875
Crocker PR, Gordon S (1989) Mouse macrophage haemagglutinin (sheep erythrocyte receptor) with specificity for sialylated glycoconjugates characterised by a monoclonal antibody. J Exp Med 169: 1333–1346
Crocker PR, Hill M, Gordon S (1988a) Regulation of a murine macrophage haemagglutinin (sheep erythrocyte receptor) by a species-restricted serum factor. Immunology 65: 516–522
Crocker PR, Morris L, Gordon S (1988b) Novel cell surface adhesion receptors involved in interactions between stromal macrophages and haematopoietic cells. J Cell Sci [Suppl] 9: 185–206
Crocker Pr, Werb Z, Gordon S, Bainton DF (1990) Ultrastructural localisation of a macrophage-restricted sialic acid binding hemagglutinin, SER, in macrophage-hemopoietic cell clusters. Blood 76: 1131–1138
Crocker PR, Kelm S, Dubois C et al. (1991) Purification and properties of sialoadhesin, a sialic acid-binding receptor of tissue macrophages. EMBO J 10: 1661–1669
Ding A, Nathan CF, Graycar J, Derynck R, Stuehr DJ, Srimal S (1990) Macrophage deactivating factor and transforming growth factors-beta 1,-beta 2 and -beta 3 inhibit induction of macrophage nitrogen oxide synthesis by IFN-gamma. J Immunol 145: 940–944
Ezekowitz RAB, Gordon S (1984) Alterations of surface properties by macrophage activation. Expression of receptors for Fc and mannose-terminal glycoproteins and differentiation antigens. Contemp Top Immunobiol 18: 33–56
Ezekowitz RAB, Hill M, Gordon S (1986) Selective antagonism by interferon α/β of interferon-τ-induced marker of macrophage activation. Biochem Biophys Res Commun 136: 737–744
Fauve RM, Jusforgues H, Hevin B (1983) Maintenance of granuloma macrophages in serum-free medium. J Immunol Methods 64: 345–351
Felix R, Cecchini MG, Hofstetter W, Elford PR, Stutzer A, Fleisch H (1990a) Impairment of macrophage colony-stimulating factor production and lack of resident bone marrow macrophages in the osteopetrotic *op/op* mouse. J Bone Miner Res 5: 781
Felix R, Cecchini MG, Fleisch H (1990b) Macrophage colony stimulating factor restores in vitro bone resorption in the *op/op* osteopetrotic mouse. Endocrinology 127: 2592
Flotte TJ, Springer TA, Thorbecke GJ (1983) Dendritic cell and macrophage staining by monoclonal antibodies in tissue sections and epidermal sheets. Am J Pathol 111: 112–124
Gordon S (1986) Biology of the macrophage. J Cell Sci [Suppl] 4: 267–286
Gordon S, Crocker PR, Lee S-H, Morris L, Rabinowitz S (1986a) Trophic and defense functions of murine macrophages. In: Steinman RM, North RJ (eds) Host-resistance mechanisms to infectious agents, tumors and allografts. Rockefeller University Press, New York, pp 121–137
Gordon S, Starkey P, Hume D, Ezekowitz RAB, Hirsch S, Austyn J (1986b) Plasma membrane markers to study differentiation, activation and localisation of murine macrophages. AgF4/80 and the mannosyl fucosyl receptor. In: Weir DM, Herzenberg LA, Handbook of experimental immunology, 4th edn. Blackwell Scientific, Oxford, 43.1–43.14
Gordon S, Keshav S, Chung L-P (1988a) Mononuclear phagocytes: tissue distribution and functional heterogeneity. Current Opinion in Immunology 1: 26–35
Gordon S, Perry VH, Rabinowitz S, Chung L-P, Rosen H (1988b) Plasma membrane receptors of the Mononuclear Phagocyte System. J Cell Sci [Suppl] 9: 1–26

Gresham HD, Goodwin JC, Allen PM, Anderson DC, Brown J (1989) A novel member of the integrin receptor family mediates Arg-Gly-Asp stimulated neutrophil phagocytosis. J Cell Biol 108: 1935–1945

Hattersley G, Kerby JA, Chambers TJ (1991) Identification of osteoclast precursors in multilineage hemopoietic colonies. Endocrinology 128: 259–262

Hirsch S, Gordon S (1983) Polymorphic expression of a neutrophil differentiation antigen revealed by monoclonal antibody 7/4. Immunogenetics 18: 229–239

Hirsch S, Austyn JM, Gordon S (1981) Expression of the macrophage-specific antigen F4/80 during differentiation of mouse bone marrow cells in culture. J Exp Med 154: 713–725

Hogg N, Takacs L, Palmer DG, Selvendran Y, Allen C (1986) The p150, 95 molecule is a marker for mononuclear phagocytes: comparison and expression of class II molecules. Eur J Immunol 16: 240–248

Hsu SM, Raine L, Fanger H (1981) The use of avidin biotin peroxidase (ABC) complex in immunoperoxidase techniques. A comparison between ABC and unlabelled antibody (PAP) procedures. J Histochem Cytochem 29: 577–580

Hume DA, Gordon S (1983) The mononuclear phagocyte system of the mouse defined by immunohistochemical localisation of antigen F4/80. Identification of resident macrophages in renal medullary and cortical interstitium and the juxtaglomerular complex. J Exp Med 157: 1704–1709

Hume DA, Gordon S (1985) The mononuclear phagocyte system of the mouse defined by immunohistochemical localisation of antigen F4/80. In: Van Furth R (ed) Mononuclear phagocytes. Characteristics, physiology and function. Martinus Nijhoff, Boston, pp 9–17

Hume DA, Perry VH, Gordon S (1983a) The histochemical localisation of a macrophage-specific antigen in developing mouse retina. Phagocytosis of dying neurons and differentiation of microglial cells to form a regular array in the plexiform layers. J Cell Biol 97: 253–257

Hume DA, Robinson AP, MacPherson GG, Gordon S (1983b) The immunohistochemical localisation of antigen F4/80. The relationship between macrophages, Langerhans cells, reticular cells and dendritic cells in lymphoid and hematopoietic organs. J Exp Med 158: 1522–1536

Hume DA, Perry VH, Gordon S (1984a) The mononuclear phagocyte system of the mouse defined by immunohistochemical localisation of antigen F4/80. Macrophages associated with epithelia. Anat Rec 210: 503–572

Hume DA, Loutit JF, Gordon S (1984b) The mononuclear phagocyte system of the mouse defined by immunohistochemical localisation of antigen F4/80. Macrophages of bone and associated connective tissue. J Cell Sci 66: 189–194

Hume DA, Halpin D, Charlton H, Gordon S (1984c) The mononuclear phagocyte system of the mouse defined by immunohistochemical localisation of antigen F4/80. Macrophages of endocrine organs. Proc Natl Acad Sci USA 81: 4174–4177

Hutchings P, Rosen H, O'Reilly L, Simpson E, Gordon S, Cooke A (1990) Blockade of adhesion promoting receptor on macrophages prevents transfer of IDDM in NOD mice. Nature 348: 639–642

Ikuta K, Kina T, MacNeil I, Uchida N, Peault B, Chien Y-h, Weissman IL (1990) A developmental switch in thymic lymphocyte maturation potential occurs at the level of hematopoietic stem cells. Cell 60: 863–874

Kaplan G, Luster AD, Hancock G, Cohn ZA (1987) The expression of a gamma interferon-induced protein (IP-10) in delayed immune responses in human skin. J Exp Med 166: 1098–1108

Kerby JA, Hattersley G, Collins DA, Chambers TJ (1992) Derivation of osteoclasts from hemopoietic colony-forming cells in culture. J Bone Miner Res: 7:353–362

Keshav S, Chung L-P, Milon G, Gordon S (1991) Lysozyme is an inducible marker of macrophage activation in murine tissues as demonstrated by in situ hybridization. J Exp Med 174: 1049–1058

Koch F, Heufler C, Kampgen E, Schneeweiss D, Bock G, Schuler G (1990) Tumor necrosis factor alpha maintains the viability of murine epidermal Langerhans cells in culture, but in contrast to granulocyte/macrophage colony-stimulating factor, without inducing their functional maturation. J Exp Med 171: 159–171

Kodama H, Yamasaki A, Nose M et al. (1991a) Congenital osteoclast deficiency in osteopetrotic (*op/op*) mice is cured by injections of macrophage colony-stimulating factor. J Exp Med 173: 269–272

Kodama H, Nose M, Niida S, Yamasaki A (1991b) Essential role of macrophage colony-stimulating factor in the osteoclast differentiation supported by stromal cells. J Exp Med 173:1291–1297

Krisansen GW, Elliot MJ, Lucas CM et al. (1990) Identification of a novel integrin beta subunit expressed on cultured monocytes (macrophages). Evidence that one alpha subunit can associate with multiple beta subunits. J Biochem 265: 823–830

Kurihara N, Suda T, Miura Y et al. (1989) Generation of osteoclasts from isolated hematopoietic progenitor cells. Blood 74: 1295–1302

Larsen CP, Morris PJ, Austyn JM (1990) Migration of dendritic leukocytes from cardiac allografts into host spleens. A novel pathway for initiation of rejection. J Exp Med 171: 307–314

Lawson LJ, Perry VH, Dri P, Gordon S (1990) Heterogeneity in the distribution and morphology of microglia in the normal adult mouse brain. Neuroscience 39: 151–170

Lee S-H, Starkey P, Gordon S (1985) Quantitative analysis of total macrophage content in adult mouse tissues. Immunochemical studies with monoclonal antibody F4/80. J Exp Med 161: 475–489

Lee S-H, Crocker P, Gordon S (1986) Macrophage plasma membrane and secretory properties in murine malaria. Effects of *Plasmodium yoelii* infection on macrophages in the liver, spleen and blood. J Exp Med 163: 54–74

Lenz A, Heufler C, Rammensee HG, Gassl H, Koch F, Romani N, Schuler G (1989) Murine epidermal Langerhans cells express significant amounts of class I major histocompatibility complex antigens. Proc Natl Acad Sci USA 86: 7527–7531

Lepay DA, Nathan CF, Steinman RM, Murray HW, Cohn ZA (1985a) Murine Kupffer cells. Mononuclear phagocytes deficient in the generation of reactive oxygen intermediates. J Exp Med 161: 1079–1096

Lepay DA, Steinman RM, Nathan CF, Murray HW, Cohn ZA (1985b) Liver macrophages in murine listeriosis. Cell-mediated immunity is correlated with an influx of macrophages capable of generating reactive oxygen intermediates. J Exp Med 161: 1503–1512

MacPherson GG (1989) Lymphoid dendritic cells. Their life history and roles in immune responses. Res Immunol 140: 877–926

Maudsley DJ, Cook N, Whetton AD (1992) Expression of macrophage (F4/80) and neutrophil (7/4) differentiation antigens on GM-CFC cultured in Il-3, GM-CSF, M-CSF or G-CSF: effect of interferon gamma. (Manuscript in preparation)

McLean JW, Nakane PK (1974) Periodate-lysine-paraformaldehyde fixative. A new fixative for immuno-electronmicroscopy. J Histochem Cytochem 22: 1077–1083

Metlay JP, Wittmer-Pack MP, Agger R, Growley MT, Lawless D, Steinman RM (1990) The distinct leucocyte integrins of mouse spleen dendritic cells identified with new hamster monoclonal antibodies. J Exp Med 171: 1753–1771

Morris L, Crocker PR, Gordon S (1988) Murine foetal liver macrophages bind developing erythroblasts by a divalent cation-dependent haemagglutinin. J Cell Biol 106: 649–656

Morris L, Crocker PR, Fraser I, Hill M, Gordon S (1991a) Expression of a divalent cation-dependent erythroblast adhesion receptor by stromal macrophages from murine bone marrow. J Cell Sci 99: 141–147

Morris L, Graham CF, Gordon S (1991b) Macrophages in haemopoietic and other tissues of the developing mouse detected by monoclonal antibody F4/80. Development 112: 517–526

Morris L, Crocker PC, Hill M, Gordon S (1992) Developmental regulation of sialoadhesin (sheep erythrocyte receptor), a macrophage cell interaction molecule expressed in lymphohemopoietic tissues. Dev Immunol 2: 7–17

Munro JM, Pober JS, Cotran RS (1989) Tumour necrosis factor and interferon gamma induce distinct patterns of endothelial activation and leucocyte accumulation in skin of *Papio anobis.* Am J Pathol 35: 121–133

Nicholson GC, Moseley JM, Sexton PM, Mendelsohn FAD, Martin TJ (1986) Abundant calcitonin receptors in isolated rat osteoclasts. J Clin Invest 78: 355–360

Perry VH, Hume DA, Gordon S (1985) Immunohistochemical localization of macrophages and microglia in the adult and developing mouse brain. Neuroscience 15: 313–326

Perry VH, Gordon S (1987) Modulation of CD4 antigens on macrophages and microglia in rat brain. J Exp Med 166: 1138–1143

Perry VH, Gordon S (1988) Macrophages and microglia in the nervous system. Trends Neurosci 11: 273–277

Perry VH, Gordon S (1991) Macrophages and the nervous system. Int Rev Cytol 125: 203–244

Perry VH, Crocker PR, Gordon S (1992) The blood-brain barrier regulates expression of a macrophage sialic acid binding receptor modulated by plasma. J Cell Sci 101: 201–207

Pierce JA, di Marco E, Cox GW et al. (1990) Macrophage-colony- stimulating-factor (CSF-1) induces proliferation, chemotaxis, and reversible monocytic differentiation in myeloid progenitor cells transfected with the human c-fms CSF-1 receptor cDNA. Proc Natl Acad Sci USA 87: 5613–5617
Pow DV, Perry VH, Morris JF, Gordon S (1989) Microglia in the neurohypophysis associate with and endocytose terminal portions of neurosecretory neurons. Neuroscience 33: 567–578
Rabinowitz S, Gordon S (1991) FA.11, a macrophage-restricted membrane sialoprotein differentially glycosylated in response to inflammatory stimuli. J Exp Med 174: 821–836
Rabinowitz S, Horstmann H, Gordon S, Griffiths G (1992) Meeting of the endocytic and phagocytic pathways in murine macrophages. J Cell Biol 116:95–112
Rappolee DA, Werb Z (1992) Macrophage-derived growth factors. In: Russell SW, Gordon S (eds) Macrophage biology and activation. Springer, Berlin Heidelberg New York (Current Topics in Microbiology and Immunology) Vol. 181, 87–140
Rosen H, Gordon S (1987) Monoclonal antibody to the murine type 3 complement receptor inhibits adhesion of myelomonocytic cells in vitro and inflammatory cell recruitment in vivo. J Exp Med 166: 1685–1701
Rosen H, Gordon S (1990a) The role of the type 3 complement receptor in the induced recruitment of myelomonocytic cells to inflammatory sites in the mouse. Am J Resp Cell Mol Biol 3: 3–10
Rosen H, Gordon S (1990b) Adoptive transfer of fluorescent labelled cells shows that resident peritoneal macrophages are able to migrate into specialized lymphoid organs and inflammatory sites in the mouse. Eur J Immunol 20: 1251–1258
Rosen H, Milon G, Gordon S (1988) Antibody to the murine type 3 complement receptor inhibits T-lymphocyte-dependent recruitment of myelomonocytic cells in vivo. J Exp Med 169:535–549
Rosen H, Gordon S, North RJ (1989) Exacerbation of murine listeriosis by a monoclonal antibody specific for the type 3 complement receptor of myelomonocytic cells. Absence of monocytes at infective foci allows *Listeria* to multiply in nonphagocytic cells. J Exp Med 170: 27–37
Schuler G, Steinman RM (1985) Murine epidermal Langerhans cells mature into potent immunostimulating dendritic cells in vitro. J Exp Med 161: 526–546
Shaw LM, Messier, JM, Mercurio AM (1990) The activation dependent adhesion of macrophages to laminin involves cytoskeletal anchoring and phosphorylation of the $\alpha 6 \beta$integrin. J Cell Biol 110: 2167–2174
Smith MJ, Koch GCE (1987) Differential expression of murine macrophage surface glycoprotein antigens in intracellular membranes. J Cell Sci 87: 113–119
Springer T, Galfre G, Secher DS, Milstein C (1979) Mac-1 macrophage differentiation antigen identified by a monoclonal antibody. Eur J Immunol 9: 301–306
Springer TA (1990) Adhesion receptors of the immune system. Nature 346: 425–434
Starkey PM, Turley L, Gordon S (1987) The mouse macrophage-specific glycoprotein defined by monoclonal antibody F4/80: characterization, biosynthesis and demonstration of a rat analogue. Immunology 60: 117–122
Stein M, Gordon S (1991) Regulation of TNF release by murine peritoneal macrophages: role of cell stimulation and specific phagocytic plasma membrane receptors. Eur J Immunol 21: 431–437
Steinman RM, Inaba K, Schuler G, Witmer M (1986) Stimulation of the immune response: contributions of dendritic cells. In: Steinman RM, North RJ (eds) Mechanisms of host resistance to infectious agents, tumors and allografts. Rockefeller University Press, New York, pp 71–97
Stössel H, Koch F, Kämpgen E, Stöger P, Lenz A, Heufler C, Romani N, Schuler G (1990) Disappearance of certain acidic organelles (endosome and Langerhans cell granules) accompanies loss of antigen processing capacity upon culture of epidermal Langerhans cells. J Exp Med 172: 1471–1482
Straussman G, Somers SD, Springer TA, Adams DO, Hamilton TA (1986) Biochemical models of interferon gamma mediated macrophage activation: independent regulation of lymphocyte function associated antigen (LFA-1) and Ia antigen on murine peritoneal macrophages. Cell Immunol 97: 110–120
Taylor LM, Tertinegg I, Okuda A, Heersche JNM (1989) Expression of calcitonin receptors during osteoclast differentiation in mouse metatarsals. J Bone Miner Res 4 No 5: 751–758
Udagawa N, Takahashi N, Akatsu T et al. (1989) The bone marrow-derived stromal cell lines MC3T3-G2/PA6 and ST2 support osteoclast-like cell differentiation in cocultures with mouse spleen cells. Endocrinology 125: 1805
Wiktor-Jedrzejczak W, Bartocci WA, Ferrante AW Jr, Ahmed-Ansari A, Sell KW, Pollard JW, Stanley ER (1990) Total absence of colony-stimulating factor 1 in the macrophage deficient osteopetrotic (*op*/*op*) mouse. Proc Natl Acad Sci USA 87: 4828–4832

Witmer-Pack MD, Olivier W, Valinsky J, Schuler G, Steinman RM (1987) Granulocyte macrophage colony stimulating factor is essential for the viability and function of cultured murine epidermal Langerhans cells. J Exp Med 166: 1484–1498

Wu J, Zhu JQ, Han KK, Zhu DX (1990) The role of the c-*fms* oncogene in the regulation of HL-60 cell differentiation. Oncogene 5: 873–877

Yoshida H, Hayashi S, Kunisada T, Ogawa M, Nishinkawa S, Okumura H, Sudo T, Shultz LD, Nishikawa S (1990) The murine mutation osteopetrosis is in the coding region of the macrophage colony stimulating factor gene. Nature 345: 442–443

Zsebo KM, Wypych J, McNiece IK et al. (1990) Identification, purification, and biological characterization of hematopoietic stem cell factor from buffalo rat liver-conditioned medium. Cell 63: 195–201

Macrophages in the Uterus and Placenta

J. S. HUNT[1] and J. W. POLLARD[2]

1 Introduction

Macrophages are abundant in the mesenchymal and connective tissue stroma of the cycling and pregnant uterus, and constitute a significant proportion of the villous or labyrinthine mesenchymal cells in the human and murid placenta. In other contexts, the activities of these multifunctional cells are strongly influenced by regulatory molecules such as steroid hormones, polypeptide growth factors, and bioactive lipids. All of these are present at particularly high concentrations in the pregnant uterus and placenta. Thus, uterine and placental macrophages stimulated by endogenous factors could contribute to the complex cellular and molecular interactions that result in successful pregnancy.

[1] Department of Pathology and Oncology, University of Kansas Medical Center, 39th Street and Rainbow Boulevard, Kansas City, Kansas 66103, USA
[2] Department of Developmental Biology and Cancer and Department of Obstetrics and Gynecology, Albert Einstein College of Medicine, 1300 Morris Park Avenue, Bronx, New York 10461, USA

Current Topics in Microbiology and Immunology, Vol. 181

This article will focus on uterine and placental macrophages in the human, mouse, and rat species, in which placentation displays some similarities and considerable experimentation has been done. Studies are presented that document the preferred locations of these cells in uterine and placental compartments, explore chemoattractants, describe differentiation and activation in resident cells, and suggest some specific uterine and placental macrophage functions that might contribute to the success of pregnancy. The discussion will conclude with a commentary on our findings in the colony stimulating factor-1 (CSF-1)-less, macrophage-deficient osteopetrotic (*op/op*) mouse, which illustrate for the first time that a specific hormone-dependent uterine growth factor, CSF-1, has a major influence on the properties of macrophages in the uterus.

2 Anatomic Arrangements of Uterine and Placental Tissues and Hormonal Influences

In mammals, the uterus comprises two distinct layers of tissue, the endometrium and the myometrium, as shown for rats in Fig. 1a. The endometrium is composed of a single layer of epithelial cells forming the uterine lumen and leading to glands that ramify through the supporting mesenchymal stroma. The myometrium is composed of circular and longitudinal muscle layers interspersed with connective tissue stroma. In the mesometrial region of rat and mouse uterus, the stromal area between the longitudinal and circular muscles is termed the mesometrial triangle, the region in which the metrial gland arises during pregnancy.

In humans, rats, and mice, implantation of the blastocyst is accompanied by differentiation of adjacent endometrial stromal cells into decidual cells, as shown for rats in Fig. 1b. In humans, decidualization also takes place in the cycling uterus. During pregnancy in all three species, decidual cells form the maternal cellular component of the maternal–fetal interface. The fetal component of this interface comprises tropoblast cells, which arise from the trophectoderm layer of the blastocyst to form the placenta, the position of which is shown for rats in Fig. 1c. The inner cell mass of the blastocyst contributes cells to the underlying placental mesenchyme that is contiguous with the cord, to the membranes that surround the embryo, and to the embryo. Although the anatomic arrangements of the placental cell layers are not identical, hemochordial placentation, where maternal blood circulates through the placenta in direct contact with trophoblast cells, is common to humans, rats, and mice. The murids have therefore been used extensively for experimental purposes.

In the cycling or pseudopregnant uterus, fluctuating levels of estrogens and progesterone cause dramatic changes in the uterine endometrium. These include proliferation of the uterine epithelial cells in response to estradiol-17-β(E_2), differentiation stimulated by progesterone, and altered production of

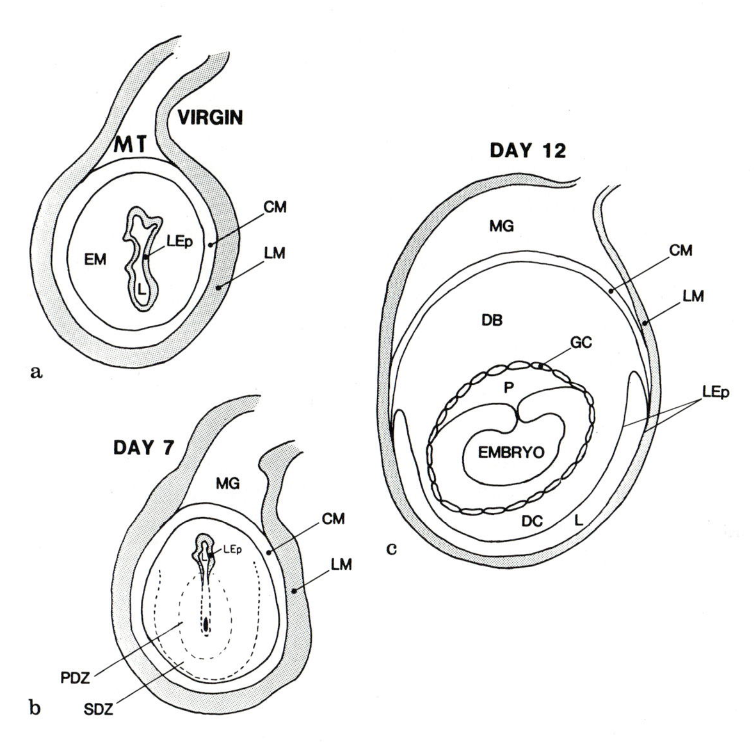

Fig. 1 a–c. Anatomic compartments of the rat virgin uterus (**a**), and the pregnant uterus at days 7 (**b**) and 12 (**c**) of gestation. The figures are oriented such that the mesometrial region is at the top and the antimesometrial region is at the bottom. The positions of the blastocyst within the lumen (day 7) and the embryo (day 12) are shown. *CM*, circular muscle of the myometrium; *DB*, decidua basalis; *DC*, decidua capsularis; *EM*, stroma of the endometrium; *GC*, giant trophoblast cell layer of the placenta; *L*, lumen; *LEp*, luminal epithelium; *LM*, longitudinal muscle of the myometrium; *MG*, metrial gland; *MT*, mesometrial triangle; *P*, placenta; *PDZ*, primary decidual zone; *SZD*, secondary decidual zone. (CHEN et al. 1991)

polypeptide growth factors (POLLARD 1990). In the endometrial stroma, progesterone sensitizes the cells to respond to E_2 by proliferation (MARTIN et al. 1973; TABIBZADEH 1990), and cyclic changes in proportions of stromal cells expressing specific hematopoietic cell markers have been noted (KING et al. 1989; LAGUENS et al. 1990). In humans, estrogen and progesterone concentrations remain high during pregnancy. In rat and mouse uterus, whilst progesterone levels are high, the estrogen levels only reach high concentrations late in pregnancy. New hormones such as chorionic gonadotropin (in humans), placental prolactins, luteotropins, and prolactin-like molecules are also synthesized in the decidua and placenta.

Alterations in cellular behavior observed in the cycling and pregnant uterus in response to steroid hormones may be an indirect effect of these hormones on the synthesis or release of locally acting growth factors and bioactive lipids (HILL 1989; BRIGSTOCK et al. 1989; POLLARD 1990; SIMMEN and SIMMEN 1991). For example, in mice, stimulation of uterine epithelial cells by ovarian hormones induces synthesis of CSF-1 (BARTOCCI et al. 1986; POLLARD et al. 1987; ROTH and STANLEY, this volume), a growth factor for macrophages (TUSHINSKI et al. 1982; BOOCOCK et al. 1989) that is targeted, in addition to macrophages, to both decidual cells and trophoblast in the pregnant uterus (MULLER et al. 1983; REGENSTREIF and ROSSANT 1989; ARCECI et al. 1989; POLLARD 1990). Other uterine growth factors that give indications of ovarian hormone regulation, but are less well documented than CSF-1, are transforming growth factor-β1 (TGF-β1) (TAMADA et al. 1990) and tumor necrosis factor-α (TNF-α) (YELAVARTHI et al. 1991), molecules with pleiotropic effects on the growth and differentiation of cells, including macrophages (RIZZINO 1988; SPORN and ROBERTS 1989; BEUTLER and CERAMI 1989).

Hormonally stimulated epithelial–mesenchymal cell interactions have been identified in the uterus and other tissues (CUNHA et al. 1983). Macrophages are abundant in the connective tissue and mesenchymal stroma of the cycling and pregnant uterus (NICOL 1935; TACHI et al. 1981), and populate the mesenchyme of the placental villi, particularly in humans (MOSKALEWSKI et al. 1974, 1975; FOX 1978). Uterine and placental macrophages are therefore likely to be participants in these epithelial–mesenchymal cell interactions. Their products could influence developmental events in the uterus, placenta, and, possibly, the embryo (TACHI et al. 1981; HUNT 1989a, 1990; SOKOL et al. 1990).

3 Uterine Macrophages

Migration of blood monocytes into the uterus is increased during pregnancy and the cells home to specific anatomic compartments, where residency stimulates differentiation into phenotypically distinct subpopulations, and local conditions induce the expression of activation-associated markers. Uterine macrophages seem to have multiple functions, many of which are related to their secretory products. (Table 1)

3.1 Distribution and Chemotaxis

The anatomic locations of macrophages are highly predictable in both the cycling and the pregnant uterus. In the virgin mouse uterus, macrophages are distributed throughout the endometrium. A recent study suggests the proximity to the epithelial cell layer may be related to stage of the cycle (DE and WOOD 1990). In the myometrium, the cells are present in the connective tissue stroma and are closely associated with the serous membrane (HUNT et al. 1985). Quantitative studies on rat macrophages in situ show that macrophages account for approximately 10% of the total cells in virgin rat endometrium, and 5% of the cells in the myometrium (YELAVARTHI et al. 1991).

Table 1. Potential functions of uterine and placental macrophages

Uterine macrophages	
Phagocytosis	Tachi et al. 1981 (rat) Redline et al. 1990 (mouse)
Immunoregulation	Hunt et al. 1984b (mouse) Tawfik et al. 1986a, b (mouse) Matthews and Searle 1987 (mouse) Lala et al. 1988 (human) Oksenberg et al. 1988 (human)
Growth factors	
CSF-1	Not documented
TGF-β1	Tamada et al. 1990 (mouse)[a]
TNF-α	Yelavarthi et al. 1991 (rat)[a]
IL-1	Hu et al. 1992 (monocytes, human)
IL-6	Tabibzadeh et al. 1988 (human)[a]
Tissue remodeling	Not documented
Placental macrophages	
Phagocytosis	Fox 1978 (human) Loke et al. 1982 (human)
Immunostimulation	Hunt et al. 1984a (human)
Growth factors	
CSF-1	Daiter et al. 1992 (human)[a]
TGF-β1	Not documented
TNF-α	Chen et al. 1991 (human)[a]
IL-1	Flynn et al. 1982 (human)[a] Hu et al. 1992 (human)
IL-6	Not documented

[a] The results of the these studies provide circumstantial evidence that mononuclear phagocytes are sources of the particular uterine and placental polypeptide growth factors

Studies on dispersed cells show that the proportion of uterine cells bearing macrophage markers doubles during pregnancy (Hunt et al. 1985), and immunocytochemical analysis of rat tissues indicates the same (Yelavarthi et al. 1991). In early gestation tissues, the distribution of macrophages in the myometrium and nondecidualized endometrial stroma is unchanged. However, where decidualization takes place in the stroma immediately surrounding the blastocyst (Fig. 1b), the macrophages are redistributed. There are virtually no cells bearing the usual macrophage-specific markers in the rat (Tachi et al. 1981; Yelavarthi et al. 1991) or mouse (Pollard et al. 1991a) primary decidual zone. Instead, macrophages are relegated to the nondecidualized endometrium immediately beneath the circular muscle of the myometrium. Exclusion of macrophages from the decidua basalis continues into late stages of pregnancy (Redline and Lu 1988, 1989; Yelavarthi et al. 1991). Mouse decidual substratum is apparently not conducive to macrophage migration, which could account for the lack of these cells in late gestation decidua (Redline et al. 1990).

At mid to late stages of pregnancy, macrophages remain abundant in the myometrium [15%–20% of the total cells in the rat (Yelavarthi et al. 1991)].

Macrophages account for 10–20% of the cells in the rat metrial gland, a specialized structure in the mesometrial triangle (Fig. 1b, c) that is not present in human tissues, whereas they are scarcer in this compartment in mice (PARR et al. 1990).

The population densities of uterine macrophages as a function of the menstrual cycle have not been determined with certainty in human tissues because of the obvious difficulties in obtaining normal tissues, and the size of the uterus, which could cause sampling errors. Although marker studies indicate that the relative proportions of bone marrow-derived cells in the endometrial stroma fluctuate during the cycle (KING et al. 1989; LAGUENS et al. 1990), the proportions of macrophages seem to remain stable (KING et al. 1989). Macrophages are present in the cycling and pregnant human myometrium (KHONG 1987) and, during pregnancy, in the decidua. Unlike the situation in mice and rats, macrophages are among the bone marrow-derived cells that are most common in the human decidua, being found near the implantation site (KABAWAT et al. 1985) and in close proximity to trophoblast cells at both early and late stages of pregnancy (LESSIN et al. 1988; BULMER et al. 1988, 1989; HUNT 1989b).

Concentration gradients of chemotactic factors, discussed below, might dictate the final anatomic locations of monocytes migrating into the pregnant uterus. Synergistic interactions among some of these factors, CSF-1 and TNF-α for example (BRANCH et al. 1989), could also stimulate in situ proliferation. In mice, there is clear evidence that migration of blood leukocytes into the uterus is hormonally regulated (FINN and POPE 1986). An influx of leukocytes into the immature uterus in response to exogenously administered estrogens has been documented in rats, and monocytes constitute a significant proportion of the migrating cells (ZHENG et al. 1988). Pregnancy hormones such as progesterone may be directly chemotactic for monocytes (YANG et al. 1989) or might stimulate the production of factors that influence their migration (FINN and POPE 1986). This latter pathway seems more likely given the fact that at least one chemotactic molecule, CSF-1 (WANG et al. 1988), is hormonally regulated (BARTOCCI et al. 1986; POLLARD et al. 1987), and uterine levels of this factor increase markedly from day 1 of mouse pregnancy (POLLARD et al. 1987; POLLARD, unpublished data). Other potential chemoattractants are TGF-β1, TNF-α, and colony stimulating factor for granulocytes and macrophages (GM-CSF) (WAHL et al. 1987; MING et al. 1987; WANG et al. 1987), all of which are present in higher concentrations in the pregnant than in the cycling uterus (TAMADA et al. 1990; CHEN et al. 1991; YELAVARTHI et al. 1991; ROBERTSON and SEAMARK 1990) and might be stimulated by ovarian or other hormones. However, no studies have as yet documented directly that higher levels of any of these growth factors stimulate monocyte migration into the uterus.

3.2 Differentiation

Monoclonal antibodies to rat macrophages, ED1 and ED2 (DIJKSTRA et al. 1985; SMINIA and JEURISSEN 1986), have been instrumental in subdividing rat uterine

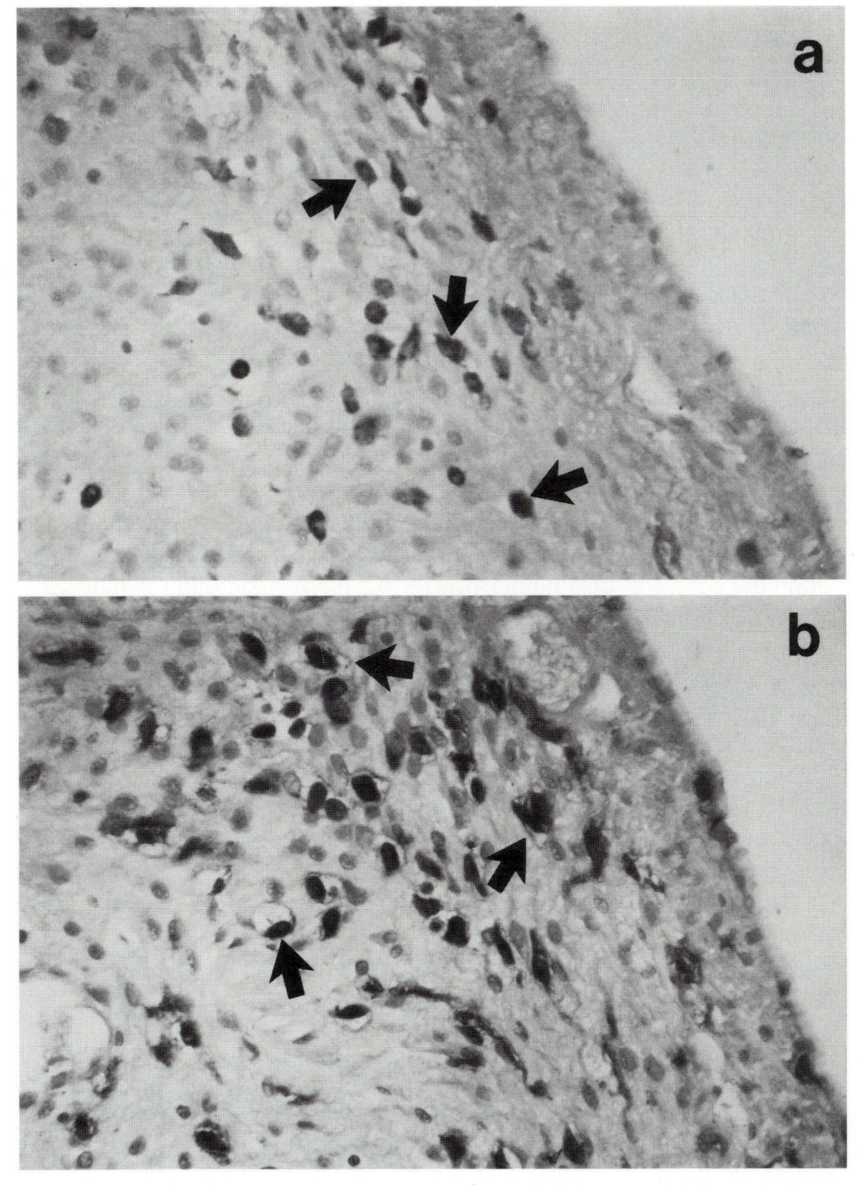

Fig. 2 a,b. Subpopulations of macrophages in the rat mesometrial myometrium at day 12 of gestation identified with the mouse monoclonal antibodies ED1 and ED2 (Bioproducts for Science). The tissue was fixed in an alcohol-based fixative (OmniFix, Xenetics Biomedical). **a** $ED1^+$ cells; **b** $ED2^+$ cells. *Arrows* mark some of the positive cells. Note that $ED1^+$ cells are small and round whereas $ED2^+$ cells are larger and highly vacuolated. x 313

macrophages into two categories, small round cells that closely resemble monocytes ($ED1^+$ cells) and fully differentiated, highly vacuolated tissue macrophages ($ED2^+$ cells) (Fig. 2). Studies on $ED1^+$ and $ED2^+$ subpopulations in the pregnant rat uterus show that $ED1^+$ cells are more common in the undecidualized endometrial stroma than $ED2^+$ cells, and that the few macrophages that are found in the decidua express only this marker (YELAVARTHI et al. 1991). $ED1^+$ and $ED2^+$ cells are present in approximately equal numbers in the metrial gland whereas the myometrial stroma contains predominantly $ED2^+$ cells. TACHI and TACHI (1989) have shown that $ED1^+$ and $ED2^+$ cells secrete different patterns of bioactive lipids. Thus, the $ED1^+$ and $ED2^+$ cells in the uterus might function differently. Differentiated subpopulations have not been identified in the mouse uterus—only a polyclonal reagent and the antimacrophage monoclonal antibody F4/80 (HUME et al. 1983), which marks a 150–kDa glycoprotein present on a proportion of bone marrow cells, monocytes, and macrophages (STARKEY et al. 1987), have been used (HUNT et al. 1985; REDLINE and LU 1988, PARR et al. 1990; POLLARD et al. 1991a).

There have not as yet been any studies on specific differentiation-associated macrophage markers expressed by the cells in the human uterus through the course of gestation. However, we have noted that the monoclonal antibody 63D3 binds to different macrophage subpopulations in the decidua adjacent to the chorion membrane than does the monoclonal antibody OKM1, which identifies C3 receptors (HUNT, unpublished data), and others have reported similar findings with different antimacrophage reagents (BULMER and JOHNSON1985).

Colony stimulating factor-1, GM-CSF, and TNF-α are major candidates for the factors that cause differentiation of new arrivals into morphologically and functionally distinct uterine tissue macrophages. These factors, found in the uterus, decidua, and placenta (BARTOCCI et al. 1986; ROBERTSON and SEAMARK 1990; CRAINIE et al. 1990; YELAVARTHI et al. 1991; CHEN et al. 1991), can influence macrophage maturation (TUSHINSKI et al. 1982; BOOCOCK et al. 1989; BRANCH et al. 1989).

3.3 Activation

Approximately half of the rat and mouse uterine macrophages express class II major histocompatibility (Ia) antigens (HUNT et al. 1985; HEAD and GAEDE 1986; REDLINE and LU 1988), a marker that is strongly associated with activation, and nearly all human decidual macrophages are class II HLA-D positive (LESSIN et al. 1988; BULMER et al. 1988). This is in contrast to the $\sim 10\%$ of blood monocytes that are Ia positive in all of these species.

Although subclasses of Ia antigens have not been identified on rat or mouse uterine macrophages, in first trimester human tissues the decidual macrophages express only HLA-DR whereas during the second trimester and throughout the balance of pregnancy, the cells exhibit both HLA-DR and HLA-DQ (LESSIN et al.

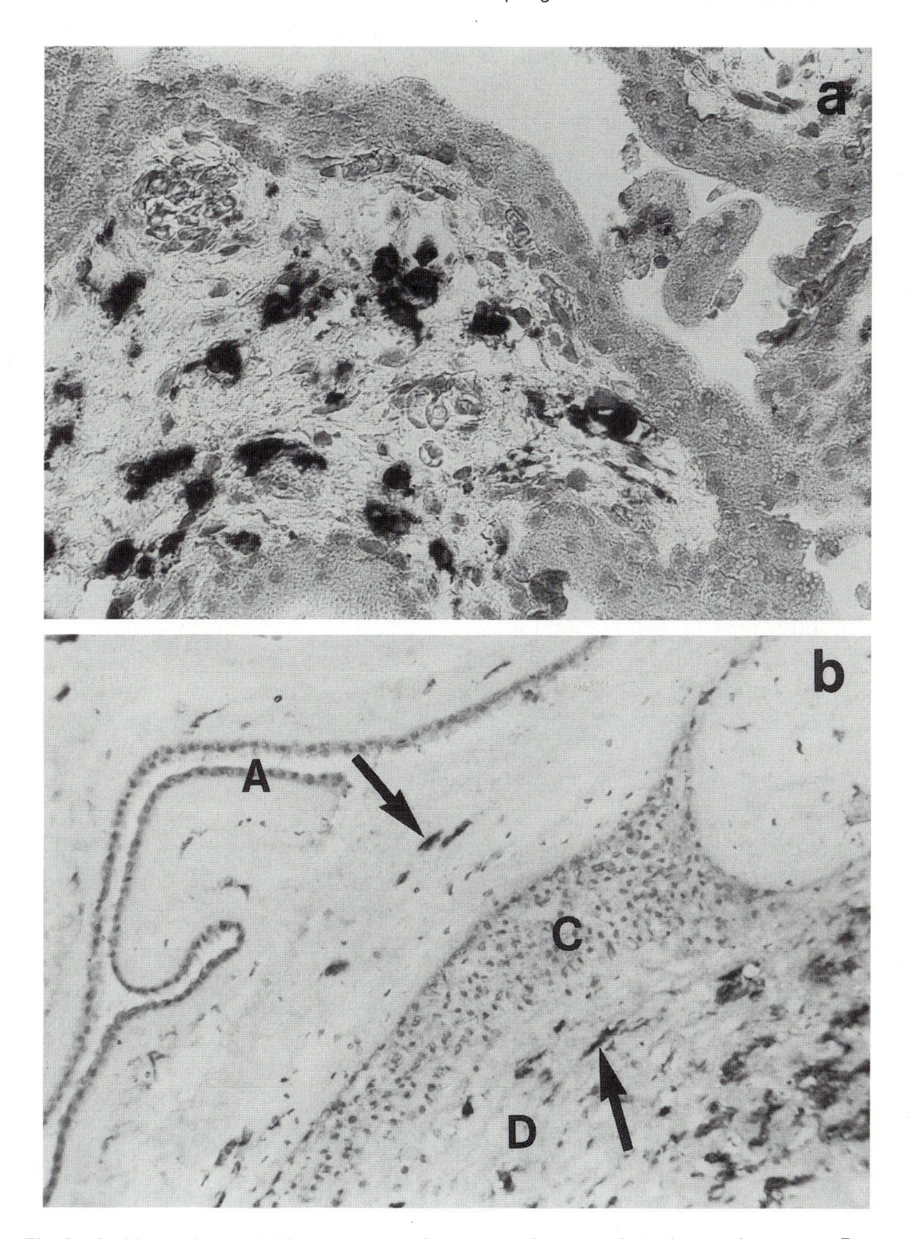

Fig. 3 a, b. Macrophages in human term placenta and extraembryonic membranes. **a** Paraformaldehyde-fixed first trimester placenta stained with a mouse monoclonal antimacrophage reagent from ENZO Diagnostics, clone HAM56. Positive cells are present in the villous stroma. x 313. **b** A section of frozen term extraembryonic membranes stained with a monoclonal antibody to HLA-DR (clone 243, Becton–Dickinson) shows activated macrophages in the decidua and the mesenchymal stroma between the amnion and chorion (*arrows*). *A*, amnion membrane: *C*, chorion membrane; *D*, decidua. x 125

1988). Figure 3 shows HLA-DR-positive macrophages in the decidua adjacent to the chorion membrane component of the term extraplacental membranes.

Interferons (IFN), well-described modulators of macrophage activation (RUSSELL and PACE 1987), are likely to be among the endogenous molecules that stimulate uterine macrophages. IFNs are present in both mouse (FOWLER et al. 1980) and human (CHARD et al. 1986; CHARD 1989) tissues; trophoblast cells produce type I IFN in response to double-stranded RNA (TOTH et al. 1990), and both type I and type II IFN have been identified in human trophoblast cells (BULMER et al. 1990).

Macrophages are also activated by endotoxins, which are present when tissues are infected with gram-negative bacteria. This causes increased rates of transcription and subsequent elaboration of various proteins and biologically active lipids such as interleukin-1 (IL-1) (DINARALLO 1988), TNF-α (BEUTLER et al. 1985), TGF-β (ASSOIAN et al. 1987), and prostaglandin E_2 (PGE_2) (ADEREM et al. 1986). Concentrations of some of these factors, particularly PGE_2, IL-1, interleukin-6 (IL-6), and TNF-α (ROMERO et al. 1987, 1989a, b, 1990), are higher in cases of preterm delivery associated with infection. In vitro, endotoxin stimulates TNF-α synthesis by decidua (CASEY et al. 1989). Such high concentrations of polypeptide growth factors might affect placental cell functions by altering DNA synthesis (HUNT et al. 1989) or the expression of membrane proteins (HUNT et al. 1990) by trophoblast cells, and prostaglandins might induce premature contractions in the uterus (OKAZAKI et al. 1981; MCGREGOR et al. 1988). Thus endotoxin-stimulated uterine macrophages may bear some of the responsibility for early pregnancy termination in cases of infection (HUNT 1989a).

Although many uterine macrophages are in a state that is associated with enhanced phagocytosis, increased antigen-presenting ability, increased cytotoxic potential, and enhanced synthesis of polypeptide growth factors and bioactive lipids (NATHAN 1987), the ability of the activated macrophages to perform specific tasks may be influenced by endogenous cytokines. For example, TGF-β1, which is present in the uterus (TAMADA et al. 1990), has been shown to diminish intracellular killing of parasites while having little effect on tumor cell killing (NELSON et al. 1991).

3.4 Potential Functions

Functional studies on uterine macrophages suggest that these cells contribute in a variety of ways to survival of the embryo. Of particular note, the cells appear to serve as immunoregulators, defenders against microbial invasion, and sources of growth factors (Table 1).

3.4.1 Immunosuppression, Antigen Presentation, and Phagocytosis

It has long been recognized that grafts to the pregnant uterus enjoy prolonged survival in comparison to grafts in other locations (BEER and BILLINGHAM 1974).

This immunosuppressive environment is believed to be mediated by soluble substances, and to aid in protection of the semiallogeneic fetus by preventing colonization of the uterus by potentially harmful maternal antifetal cytotoxic lymphocytes (HUNT et al. 1984b, 1991). Mouse uterine macrophages synthesize high levels of PGE_2 (TAWFIK et al. 1986a, b; MATTHEWS and SEARLE 1987), which inhibits lymphocyte proliferation by modulating interleukin-2 synthesis and receptor expression (CHOUAIB et al. 1985). In human tissues, both macrophages and decidual cells produce PGE_2 (PARHAR et al. LALA et al. 1988). PGE_2-mediated immunosuppression might be augmented by the TGF-β2-like substance that has been identified as a product of small mouse uterine cells (CLARK et al. 1988), by TGF-β1 (WAHL et al. 1988; TAMADA et al. 1990), or by TNF-α (UMEDA et al. 1983; YELAVARTHI et al. 1991; CHEN et al. 1991). Complexities in factor interaction cannot be underestimated. For example, PGE_2 modulates macrophage synthesis of TNF-α (RENZ et al. 1988).

Efficient presentation of antigens to T lymphocytes, an initial step in the sequence of events that leads to clonal expansion and development of an immune response, seems not to be accomplished by uterine macrophages in situ, despite their display of class II MHC antigens. For example, chemical denaturation is required for antibody stimulation by a hapten-adjuvant preparation administered into the uterus but not other tissues (LANDE 1986). Although the male-specific H–Y antigen stimulates cytotoxic T lymphocytes in the spleen and lymph nodes when systemically administered, multiple inseminations of male cells at natural mating do not prime cytotoxic cells (HANCOCK and FARUKI 1986). Evidence has been presented which suggests that human decidual accessory cells exposed to fetal cells may, in fact, stimulate suppressor T cells (OKSENBERG et al. 1988). On balance, therefore, the evidence favors an immunosuppressive rather than immunostimulatory role for uterine macrophages.

A third protective function of uterine macrophages, phagocytosis of debris and microbial invaders, is well documented in situ in rats (TACHI and TACHI 1981). Of particular interest is a study showing that in mice infected with *Listeria monocytogenes*, the uterine macrophages contain the organisms (REDLINE and LU 1988). Interestingly, these investigators have postulated that vulnerability of the placenta to infection by *Listeria* is due to the lack of macrophages in the mouse decidua, which, by virtue of their phagocytic capacity, would have otherwise constituted an effective barrier to transmission.

3.4.2 Growth Factors

The particular contributions of macrophages to uterine growth factor networks have been difficult to dissect. The cells are not morphologically distinct from other types of stromal cells, and selective harvesting of tissue macrophages could easily alter their patterns of gene expression (TANIGUCHI 1988). Thus, much of the evidence for growth factor production cited below is indirect and circumstantial. Further complexities have been introduced by the finding that, during pregnancy, other types of uterine and placental cells are sites of synthesis of many factors that have been traditionally associated with macrophages.

Members of the TGF-β family are well-documented regulatory molecules (SPORN and ROBERTS 1989; WAHL et al. 1989) that have pleiotropic effects on cell differentiation, influence formation of extracellular matrix (RIZZINO 1988), and are known to play a major role in mouse embryonic development (HEINE et al. 1987). Although activated macrophages synthesize TGF-β1 (ASSOIAN et al. 1987), in the postimplantation mouse uterus TGF-β1 originates primarily with epithelial and decidual cells (TAMADA et al. 1990). Results in this study also indicated that endometrial stromal cells, which might be macrophages, contained TGF-β1 mRNA. Correlation of transcription and translation of the TGF-β1 gene by specific cells in the human uterus has not yet been accomplished, although the factor has been purified from human placenta (FROLIK et al. 1983).

Interleukin-6 is another example of a growth factor that has been identified as a product of macrophages (VAN DAMME et al. 1988; KATO et al. 1990; VAN SNICK 1990). In the human cycling uterus, this factor is synthesized by unidentified, undifferentiated endometrial stromal cells, some of which might be macrophage precursors, and induces IL-1 synthesis by macrophages (TABIBZADEH et al. 1988). In mouse decidua, IL-6 mRNA has been localized by in situ hybridization to the cords of endothelial cells that line the maternal blood spaces, and has been postulated to influence angiogenesis (MOTRO et al. 1990).

A third growth factor that arises from activated macrophages is TNF-α (BEUTLER et al. 1985). While originally this factor was described as an inhibitor of tumor cell proliferation produced by activated macrophages (SUGARMAN et al. 1985), recent studies suggest that low levels of TNF-α in normal tissues may contribute to cellular renewal (ULICH et al. 1989), which might be accomplished in part by stimulation of other growth factors (KAUSHANSKY et al. 1988), and that TNF-α is produced by a number of cell types (ROBBINS et al. 1987; KESHAV et al. 1990; BARATH et al. 1990). In both rats and humans, TNF-α is present in the uterus throughout gestation (YELAVARTHI et al. 1991; CHEN et al. 1991). Data collected in a rat model show that although epithelial and decidual cells are the most abundant TNF-α mRNA-containing cells in the uterus, cells in the myometrial connective tissue that resemble macrophages by morphology also contain specific messages (YELAVARTHI et al. 1991). Interestingly, only the nonmacrophages contain high levels of the protein. Thus, macrophages in the rat myometrium may be similar to tumor-infiltrating macrophages, which transcribe this gene but do not translate the messages into protein (BEISSERT et al. 1989). In early gestation human tissues, TNF-α mRNA and protein are found in epithelial and decidual cells, and, late in gestation, in cells that reside near the chorion membrane that resemble macrophages by morphology and anatomic location (CHEN et al. 1991). TNF-α is contained in human amniotic fluid and is produced by human uterine and placental cells in vitro (JAATTELA et al. 1988).

The growth factor IL-1, which has overlapping functions and synergistic interactions with TNF-α (LE and VILCEK 1987; ELIAS et al. 1987; DINARELLO 1988; AKIRA et al. 1990), is present in human amniotic fluid (TAMATANI et al. 1988), and

IL-1 mRNA-positive cells have been reported in the mouse endometrium (TAKACS et al. 1988). Recent immunocytochemical studies in our laboratory (HU et al. 1992) that used two sets of polyclonal antibodies to the two species of IL-1, IL-1α and IL-1β, indicated that although maternal leukocytes in human first trimester and term tissues contained the proteins, staining intensities were higher with antibodies to the latter than to the former. Double-labeling experiments showed that the IL-1-positive blood leukocytes were monocytes, cells that are well-documented sources of IL-1 with a preference for synthesizing IL-1β (BEESLEY et al. 1990). Although IL-1α-positive cells were present in decidua, these may have been infiltrating extravillous cytotrophoblastic cells rather than maternal cells.

In mice, CSF-1, which is synthesized exclusively in the uterine epithelium in response to E_2 and progesterone, peaks on day 14 of pregnancy (BARTOCCI et al. 1986; POLLARD et al. 1987; REGENSTREIF and ROSSANT 1989; ARCECI et al. 1989; POLLARD 1990; POLLARD et al. 1991B; ROTH and STANLEY, this volume). Recent studies in our laboratory (DAITER et al. 1992) show that human tissues are similar to mice in many respects; uterine epithelium is the major source of CSF-1 in both the cycling and the pregnant human uterus, CSF-1 mRNA appears to be hormonally regulated in the cycling uterus, and CSF-1 concentrations in the endometrium increase with the onset of pregnancy. However, in contrast to the findings in mice, human endometrial CSF-1 concentrations are highest in the first trimester and decline as gestation progresses to term. Although there is no evidence at present that murine macrophages are capable of synthesizing CSF-1, human macrophages produce this substance (RAMBALDI et al. 1987), and the possibility that uterine macrophages in human endometrium contribute a portion of the CSF-1 has not been eliminated.

Macrophage-derived growth factors may be most influential in specific uterine microenvironments because the effects of polypeptide growth factors are highly concentration dependent. The observations accumulated to date on the spatial relationships between macrophages and other types of uterine cells suggest that in humans, rats, and mice myometrial macrophages have ample opportunity for tissue-specific effects, whereas there may be species differences in the ability of macrophages to influence decidual and placental cells.

3.4.3 Other Functions

The paragraphs above describe some studies that have led to a better understanding of the contributions of uterine macrophages to protection from cytotoxic lymphocytes, to defense against microbial invasion, and to growth factor production. Macrophages might have other critical functions; NATHAN (1987) lists 13 categories of macrophage products. Among these are complement components and coagulation factors, various enzymes including plasminogen activator and collagenase, and proteins such as fibronectin that compose extracellular matrix.

Uterine macrophages have not been examined specifically for production of any of these molecules, yet some might be highly important. To take just one example, in the myometrium, where the cells are consistently present in the connective tissue stroma and are closely associated with the serosa, proteolytic enzymes such as collagenase and plasminogen activator could be useful for tissue remodeling during uterine expansion and post-partum involution. As with other uterine macrophage products, synthesis of these enzymes would probably be influenced by endogenous growth factors, collagenase by uterine TNF-α (DAYER et al. 1985), and plasminogen activator by CSF-1 (HAMILTON et al. 1980) and GM-CSF (EVANS et al. 1989). Plasminogen activator enzyme activity is also stimulated by colony stimulating factor for granulocytes (KOJIMA et al. 1989), and receptors for this factor have been identified in the human placenta (UZUMAKI et al. 1989).

4 Placental Macrophages

Patterns of distribution and activation markers have now been identified for fetal placental macrophages, and a few potential functions for the cells have been described (Table 1). These studies indicate that macrophages gradually mature in the placenta until, at parturition, they are sensitive to activation signals and are capable of performing some of the functions that have been attributed to macrophages in adult tissues.

4.1 Distribution and Marker Expression

Eearly gestation human placental villous stroma contains large, highly vacuolated Hofbauer cells (FOX 1978), which are one morphologic form of fetal placental macrophages. These were, until the development of monoclonal antibodies that could identify cells that are less distinct by morphology, considered to be the only macrophage-like cells in the human placenta. However, in 1975, MOSKALEWSKI, PTAK, and CZARNIK showed that human placentae contain many Fc receptor-positive cells, and further studies verified the mononuclear phagocyte lineage of many of these cells (WOOD et al. 1978). Human fetal placental macrophages have now been tested extensively by immunohistology (BULMER and JOHNSON 1984; GOLDSTEIN et al. 1988; MUES et al. 1989). The morphologic characteristics of these cells in first trimester placental villous stroma are shown in Fig. 3a. In term placentas, stromal macrophages appear slightly smaller and are often marginated to the capillary endothelium (HUNT, unpublished data)

Cells bearing macrophage markers increase in density in the placental villous stroma as gestation progresses, and the cells gradually develop HLA-D activation antigens (LESSIN et al. 1988; BULMER et al. 1988). The same is true of fetal macrophages in the mesenchymal stroma between the amnion and chorion membrane, as shown in Fig. 3b.

Although macrophages can be identified by immunohistology in term mouse placentae (WOOD 1980), and are also present in dispersed cell suspensions from this tissue (MOSKALEWSKI et al. 1974; MATTHEWS et al. 1985), early gestation murid placentae seem to contain few macrophages. Immunohistologic studies in rats indicate that macrophage antigen-positive cells are rare in the fetally derived components of early- to mid-gestation placentae, whereas near parturition, positive cells can be identified in the chorioallantoic plate at the base of the placenta (HUNT, unpublished data). The rat experiments show that embryo-derived macrophages in the placenta bear only the marker for monocyte/macrophages, EDI, and that the cells only occasionally express detectable levels of class II MHC antigens.

It is entirely possible that many more macrophages are present in the human placental villous mesenchyme and the rat and mouse labyrinthine placenta than can be identified, particularly at early stages of gestation, because of their gradual development of lineage-specific markers. The studies have been reported at this date would also indicate that fetally derived murid macrophages in extraembryonic tissues are less numerous and develop their markers more slowly than do their counterparts, despite the presence of high concentrations of CSF-1 (BARTOCCI et al. 1986).

4.2 Potential Functions

Placental macrophages in situ often contain phagocytosed material (FOX 1978), and in vitro they participate in both immune and nonimmune phagocytosis (LOKE et al. 1982). These cells are therefore likely to supplement the activities of uterine macrophages, providing additional protection from microbial invasion. It is worthy of note that many of the organisms that are transmitted from the mother to the fetus are intracellular parasites such as *Listeria, Toxoplasma*, and viruses (KLEIN, and REMINGTON 1990). The role that macrophages might play in the transmission of HIV-1 from the mother to the embryo is a current cause for concern (LEWIS et al. 1990).

Although it is not known whether fetal placental macrophages contribute to the immunosuppressive environment during pregnancy, other functions have been postulated for the cells. Macrophages taken from term human placentae bear both class I and class II major histocompatibility antigens, and are capable of stimulating maternal lymphocyte proliferation (HUNT et al. 1984a). Because of the close proximity of the fetal placental macrophages to the maternal circulation, these cells might provide the immunogenic stimulus for the antibodies to paternal class I and class II antigens that are common in pregnant women (VAN ROOD et al. 1958). Human fetal placental macrophages have been postulated to protect the fetus from these potentially harmful antibodies by phagocytosing immune complexes (WOOD and KING 1982).

As with uterine macrophages, fetally derived placental macrophages could be important sources of polypeptide growth factors. Recent studies in our laboratory have shown that although placental stromal cells in human first trimester tissues do not contain TNF-α mRNA, similar cells in term placentae are strongly positive when tested by in situ hybridization (CHEN et al. 1991). The positive cells are probably macrophages, which constitute approximately 25% of the stromal cells (HUNT et al. 1984a). Biologically active IL-1 has been reported as a product of macrophages harvested from human term placentae (FLYNN et al. 1982). Recent immunocytochemical experiments in our laboratory (HU et al. 1992) have shown that in both first trimester and term human placentas, mesenchymal cells contain immunoreactive IL-1α and that fetal leukocytes in the blood vessels of term placentas contain IL-1β. Double labeling studies showed that many of these IL-1-positive cells were of mononuclear phagocyte lineage. CSF-1 is a product of villous mesenchymal cells in second trimester human placentas, suggesting that this growth factor might also be synthesized by fetal placental macrophages (DAITER et al. 1992). Thus, the data collected to date indicate that human fetal placental macrophages are sources of several of the polypeptide growth factors that have been identified as products of the same types of cells in adult tissues.

As in the uterus, other types of cells in the placenta synthesize growth factors that are usually associated with macrophages. These include IL-1, IL-6, CSF-1, and TNF-α, all of which have been localized to trophoblast cells in human, rat, or mouse placentas (MAIN et al. 1987; KAMEDA et al. 1990; MEAGHER et al. 1990; YELAVARTHI et al. 1991; CHEN et al. 1991; DAITER et al. 1992; HU et al. 1992). New tools consisting of monoclonal antibodies that specifically recognize human fetally derived placental macrophages and can be used to harvest the cells (NASH et al. 1989) should allow the performance of experiments that might shed further light on the functions of these cells.

5 Experiments in the Macrophage-Deficient Osteopetrotic Mouse Model System

The experiments described above have led to speculation that hormones influence growth factor production by uterine cells which, in turn, influences uterine macrophage population density, distribution, differentiation, and performance of differentiated cell functions. The osteopetrotic (*op/op*) mouse provides an opportunity to study regulation of uterine macrophages by a single growth factor, CSF-1. In the *op/op* mouse, an inactivating mulation in the CSF-1 gene results in the total absence of CSF-1 (YOSHIDA et al. 1990; WIKTOR-JEDRZEJCZAK et al. 1990). As a consequence these mice have less than one-tenth the normal number of bone marrow cells, greatly reduced numbers of blood monocytes, and severe deficiencies in the proportions of peripheral macrophages in locations such as the pleural and peritoneal cavities (WITKOR-JEDRZEJCZAK et al. 1982, 1990).

Experiments in this model system have shown that hormonally regulated uterine epithelial cell CSF-1 regulates the density and properties of mouse uterine macrophages, a relationship that was established by comparing F4/80-positive macrophages in the cycling and pregnant uteri of homozygous (*op/op*) and heterozygous (*op/+*) mice (POLLARD et al. 1991a). The observations were as follows: (a) the cycling uterus in *op/op* mice demonstrated a virtual absence of F4/80-positive cells whereas positive cells were, as expected, abundant in the endometrium, myometrium, and mesometrial triangle of *op/+* mice (Fig. 4a, b); (b) early postimplantation *op/op* mouse tissues (gestation days 7 and 8) contained macrophages, but fewer than the same tissues from *op/+* mice; (c)

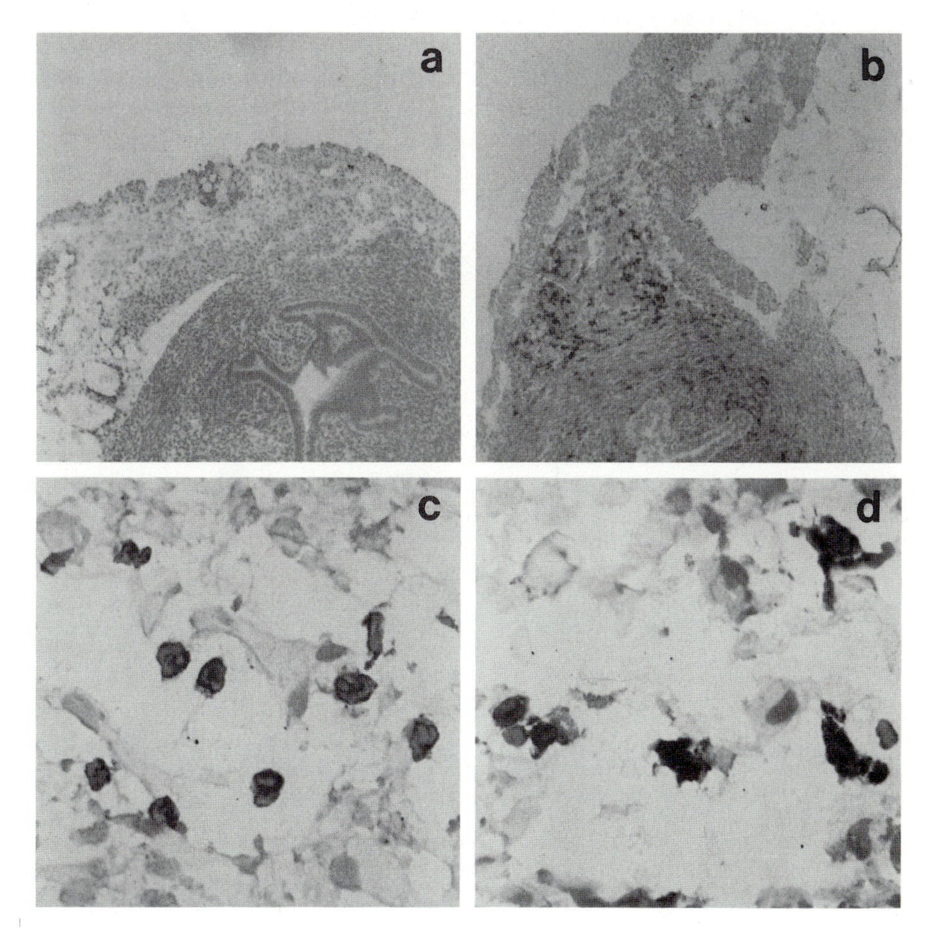

Fig. 4a–d. Macrophages are abundant in the uterus of an *op/+* mouse (**a**) but are virtually absent from the uterus of an *op/op* mouse (**b**) × 40. **c,d** Macrophages in the myometrial stroma of an *op/+* mouse uterus (**c**) and an *op/op* uterus. (**d**) At day 7 of gestation. Note that positive cells in the *op/+* uterus are large and have cytoplasmic extensions whereas the cells in *op/op* myometrium are smaller and rounded. x 626. Cryostat sections of paraformaldehyde-fixed (**a** and **b**) or acetone-fixed (**c** and **d**) frozen tissues were stained using the anti-mouse macrophage reagent F4/80

the macrophages in the early gestation *op/op* tissues remained small and round, whereas, those in the *op/+* tissues gave clear morphologic evidence of spreading (Fig. 4c, d); (d) the macrophages gradually disappeared in *op/op* uteri until, at day 14 of gestation, few macrophages could be identified in F4/80 stains of the *op/op* uterus; in contrast, numerous positive cells were present in the myometrium of *op/+* mouse tissues at the same stage.

Although these data demonstrate clearly that CSF-1 is required for maintenance and differentiation of uterine macrophages during mouse pregnancy, it seems reasonable to conclude from our observations that some molecule other than CSF-1 that is present in the post-implantation tissues of the *op/op* mice influences monocyte chemotaxis. However, the chemoattractant has less effect than CSF-1 on induction of differentiation. If macrophage spreading, as has been shown with macrophage adherence (HASKILL et al. 1988), influences transcription of their growth factor genes and proto-oncogenes, this failure could have important ramifications in pregnancy. GM-CSF, which is chemotactic for monocytes (WANG et al. 1987), may be a suitable candidate for (one of) the replacement factor(s). ROBERTSON and SEAMARK (1990) have shown that synthesis of GM-CSF by mouse uterine epithelial cells is initiated on day 1 of pregnancy, seemingly stimulated by the presence of seminal vesical fluid, and that GM-CSF production remains at high levels through day 10 pregnancy.

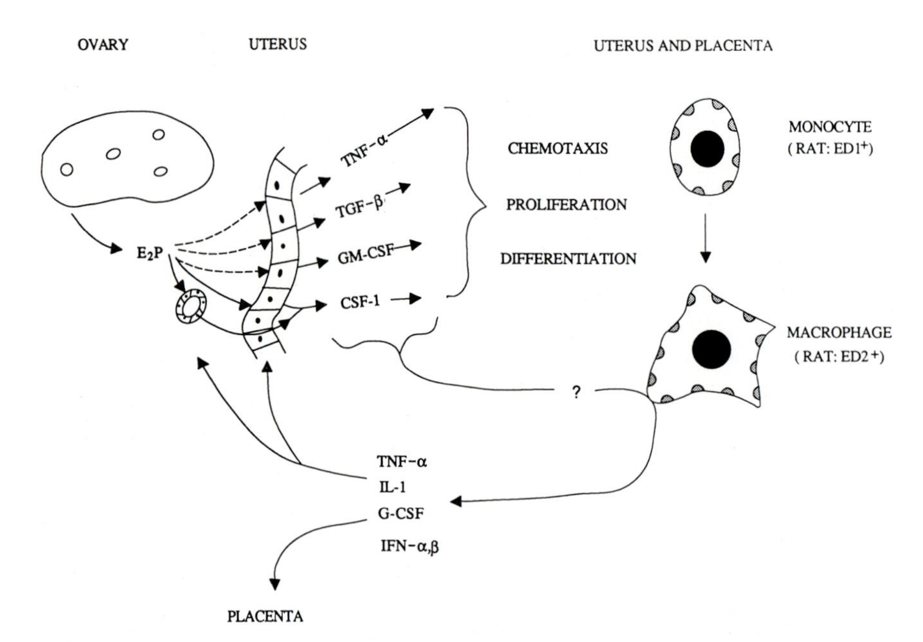

Fig. 5. A schematic representation of the potential influences of ovarian hormone-stimulated, epithelial cell-derived CSF-1 and other cytokines on uterine macrophage chemotaxis, differentiation, and growth factor synthesis. Pathways for reciprocal influences on uterine epithelium and modulation of placental cells by products of differentiated macrophages are indicated

GM-CSF or other natural replacement factors are not entirely successful in overcoming the CSF-1 deficiency. In the *op/op* mice, pregnancy is severely compromised and even in those embryos that proceed to term the development of the extraembryonic tissue appears retarded, thus confirming a causal role for CSF-1 in gestation. Systemic reconstitution of the mice with human recombinant CSF-1 during pregnancy fails to restore fertility (WIKTOR-JEDRZEJCZAK et al. 1991), suggesting that local production of CSF-1 is required for normal fertility.

All of the functions of uterine CSF-1 are not known, but it is clear that along with trophoblasts and decidual cells, uterine macrophages are targets for this molecule. Figure 5 shows a potential pathway for CSF-1 and other cytokine influences on uterine macrophage density, differentiation, and factor production. Appropriate manipulation of the *op/op* mice should provide a unique opportunity to study the regulation and functions of macrophages in pregnancy.

6 Conclusions

Macrophages are recruited to the uterus during pregnancy, are present in the placenta, and are likely to perform important pregnancy-associated functions. These appear to include immunoregulation, phagocytosis, and growth factor production. When evaluating the potential functions of uterine and placental macrophages, however, it is appropriate to recognize that: (a) other types of uterine and placental cells are, particularly during pregnancy, capable of synthesizing some factors that are traditionally associated with macrophages, and (b) some macrophage functions are probably assumed by other cell types when conditions disallow normal proportions of macrophages in the uterus and placenta.

The studies cited in this article indicate that highly complex interactions take place among pregnancy hormones, polypeptide growth factors, and macrophages in the uterus and placenta. Causal relationships are best illustrated for CSF-1, where experiments in the *op/op* mouse model show that this hormonally stimulated polypeptide growth factor is responsible in large part for uterine macrophage chemotaxis and differentiation, and, in all likelihood, their synthesis of other factors that promote pregnancy.

Note added in press. Three new reports have been added to the literature since the writing of this article. One describes rat placental macrophages [van Oostveen DC, van den Berg TK, Damoiseaux JGMC, van Rees EP (1992) Macrophage subpopulations and reticulum cells in rat placenta. Cell Tissue Res, in press], a second documents macrophages in preimplantation rat uterus [Kachkache M, Acker GM, Chaouat G, Noun A, Garabedian M (1991) Hormonal and local factors control the immunohistochemical distribution of immunocytes in the rat uterus before conceptus implantation: effects of ovariectomy, Fallopian tube section, and injection. Biol Reprod 45:860–868], and a third describes macrophages in mouse uterus [De M, Wood GW (1991) Analysis of the number and distribution of macrophages, lymphocytes and granulocytes in the mouse uterus from implantation through parturition. J Leukocyte Biol 50:381–392].

Acknowledgements. This work was supported by grants from the National Institutes of Health (HD24212, HD26429) and the Wesley Foundation, Wichita, Kansas, to Dr. Hunt, and by a grant from the National Institutes of Health (HD25074) and the Albert Einstein Core Center Grant, P30-CA13330, to Dr. Pollard.

References

Aderem AA, Cohen DS, Wright SD, Cohn ZA (1986) Bacterial lipopolysaccharides prime macrophages for enhanced release of arachadonic acid metabolites. J Exp Med 164: 165–179

Akira S, Hirano T, Taga T, Kishimoto T (1990) Biology of multifunctional cytokines: IL 6 and related molecules (IL-1 and TNF). FASEB J 4: 2860–2867

Arceci RJ, Shanahan F, Stanley ER, Pollard JW (1989) The temporal expression and location of colony stimulating factor-1 (CSF-1) and its receptor in the female reproductive tract are consistent with CSF-1 regulated placental development. Proc Natl Acad Sci USA 86: 8818–8822

Assoian RK, Fleurdelys BE, Stevenson HC et al. (1987) Expression and secretion of type β transforming growth factor by activated human macrophages. Proc Natl Acad Sci USA 84: 6020–6024

Barath P, Fishbein MC, Cao J, Berenson J, Helfant RH, Forrester JS (1990) Tumor necrosis factor gene expression in human vascular intimal smooth muscle cells detected by in situ hybridization. Am J Pathol 137: 503–509

Bartocci A, Pollard JW, Stanley ER (1986) Regulation of colony-stimulating factor 1 during pregnancy. J Exp Med 164: 956–961

Beer AE, Billingham RE (1974) Host responses to intra-uterine, tissue, cellular and fetal allografts. J Reprod Fertil [Suppl] 21: 59–88

Beesley JE, Bomford R, Schmidt JA (1990) Ultrastructural localization of interleukin 1 in human peripheral blood monocytes; evidence for IL-1β in mitochondria. Histochem J 22: 234–244

Beissert S, Bergholz M, Waase I, Lepsien G, Schauer A, Pfizenmaier K, Kronke M (1989) Regulation of tumor necrosis factor gene expression in colorectal adenocarcinoma: in vivo analysis by in situ hybridization. Proc Natl Acad Sci USA 86: 5064–5068

Beutler B, Cerami A (1989) The biology of cachectin/TNF—primary mediator of the host response. Annu Rev Immunol 7: 625–655

Beutler B, Mahoney J, Trang NL, Pekala P, Cerami A (1985) Purification of cachectin, a lipoprotein lipase-suppressing hormone secreted by endotoxin-induced RAW 264.7 cells. J Exp Med 161: 984–995

Boocock CA, Jones GE, Stanley ER, Pollard JW (1989) Colony-stimulating factor-1 induces rapid behavioural responses in the mouse macrophage cell line, BAC1.2F5. J Cell Sci 93: 447–456

Branch DR, Turner AR, Guilbert LJ (1989) Synergistic stimulation of macrophage proliferation by the monokines tumor necrosis factor-alpha and colony-stimulating factor 1. Blood 73: 307–311

Brigstock DR, Heap RP, Brown KD (1989) Polypeptide growth factors in uterine tissues and secretions. J Reprod Fertil 85: 747–758

Bulmer JN, Johnson PM (1984) Macrophage populations in the human placenta and amniochorion. Clin Exp Immunol 57: 393–403

Bulmer JN, Johnson PM (1985) Identification of leucocytes within the human chorion laeve. J Reprod Immunol 7: 89–92

Bulmer JN, Morrison L, Smith JC (1988) Expression of class II MHC gene products by macrophages in human uteroplacental tissue. Immunology 63: 707–714

Bulmer JN, Smith J, Morrison L, Wells M (1989) Maternal and fetal cellular relationships in the human placental basal plate. Placenta 9: 237–246

Bulmer JN, Morrison L, Johnson PM, Meager A (1990) Immunohistochemical localization of interferons in human placental tissues in normal, ectopic, and molar pregnancy. Am J Reprod Immunol 22: 109–116

Casey ML, Cox SM, Beutler B, Milewich L, MacDonald PC (1989) Cachectin/tumor necrosis factor-α formation in human decidua. J Clin Invest 83: 430–436

Chard T (1989) Interferon in pregnancy. J Dev Physiol 11: 271–276

Chard T, Craig PH, Menabawey M, Lee C (1986) Alpha interferon in human pregnancy. Br J Obstet Gynaecol 98: 1145–1149

Chen H-L, Yang Y, Hu X-L, Yelavarthi KK, Fishback JL, Hunt JS (1991) Tumor necrosis factor-alpha mRNA and protein are present in human placental and uterine cells at early and late stages of gestation. Am J Pathol 139: 327–335

Chouaib S, Welte K, Mertelsmann R, Dupont B (1985) Prostaglandin E_2 acts at two distinct pathways of T lymphocyte activation: inhibition of interleukin 2 production and down-regulation of transferrin receptor expression. J Immunol 135: 1172–1179

Clark DA, Falbo M, Rowley RB, Banwatt D, Stedronska-Clark J (1988) Active suppression of host-vs-graft reaction in pregnant mice. IX. Soluble suppressor activity obtained from allopregnant mouse decidua that blocks the cytolytic effector response to IL-2 is related to transforming growth factor-β. J Immunol 141: 3833–3840

Crainie M, Guibert L, Wegmann TG (1990) Expression of novel cytokine transcripts in the murine placenta. Biol Reprod 45: 999–1005

Cunha CR, Chung LWK, Shannon JM, Taguchi O, Fujii IT (1983) Hormone-induced morphogenesis and growth: role of mesenchymal-epithelial interactions. Rec Prog Horm Res 39: 559–598

Daiter E, Pampfer S. Yeung Y-G, Barad D, Stanley ER, Pollard JW (1992) Expression of colony stimulating factor-1 (CSF-1) in the human uterus and placenta. J Clin Endocrinol Metab 14: 850–858

Dayer JM, Beutler B, Cerami A (1985) Cachectin/tumor necrosis factor (TNF) stimulates collagenase and PGE_2 production by human synovial cells and dermal fibroblasts. J Exp Med 162: 2163–2168

De M, Wood GW (1990) Influence of oestrogen and progesterone on macrophage distribution in the mouse uterus. J Endocrinol 126: 417–424

Dijkstra CD, Dopp EA, Joling P, Draal G (1985) The heterogeneity of mononuclear phagocytes in lymphoid organs: distinct macrophage subpopulations in the rat recognized by monoclonal antibodies ED1, ED2 and ED3. Immunology 54: 589–599

Dinarello CA (1988) Biology of interleukin-1. FASEB J 2: 108–115

Elias JA, Gustilo K, Baeder W, Freundlich B (1987) Synergistic stimulation of fibroblast prostaglandin production by recombinant interleukin 1 and tumor necrosis factor. J Immunol 138: 3812–3816

Evans DB, Bunning RA, Russell RG (1989) The effects of recombinant human granulocyte-macrophages colony-stimulating factor (rhGM-CSF) on human osteoblast-like cells. Biochem Biophys Res Commun 160: 588–595

Finn CA, Pope M (1986) Control of leucocyte infiltration into the decidualized mouse uterus. J Endocrinol 110: 93–96

Flynn A, Finke JH, Hilfiker ML (1982)Placental mononuclear phagocytes as a source of interleukin-1. Science 218: 475–477

Fowler AK, Reed CD, Giron DJ (1980) Identification of an interferon in murine placentas. Nature 286: 266–267

Fox H (1978) The development and structure of the placenta. Major Prob Pathol 7: 1–37

Frolik CA, Dart LL, Meyers CA, Smith DM, Sporn MB (1983) Purification and initial characterization of a type β transforming growth factor from human placenta. Proc Natl Acad Sci USA 80: 3676–3680

Goldstein J, Braverman M, Salafia C, Buckley P (1988) The phenotype of human placental macrophages and its variation with gestational age. Am J Pathol 133: 648–659

Hamilton JA, Stanley ER, Burgess AW, Shadduck RK (1980) Stimulation of macrophage plasminogen activator by colony stimulating factors. J Cell Physiol 103: 435–445

Hancock RJ, Faruki S (1986) Assessment of immune responses to H-Y antigen in naturally inseminated and sperm-injected mice using cell-mediated cytotoxicity assays. J Reprod Immunol 9: 187–194

Haskill S, Johnson C, Eierman D, Becker S, Warren K (1988) Adherence induces selective mRNA expression of monocyte mediators and proto-oncogenes. J Immunol 140: 1690–1694

Head JR, Gaede SD (1986) Ia antigen expression in the rat uterus. J Reprod Immunol 9: 137–141

Heine UI, Munoz EF, Flanders KC et al. (1987) Role of transforming growth factor-β in the development of the mouse embryo. J Cell Biol 105: 2861–2876

Hill DJ (1989) Growth factors and their cellular actions. J Reprod Fertil 85: 723–734

Hu X-L, Yang Y, Hunt JS (1992) Differential distribution of interleukin-1α and interleukin-1β proteins in human placentas. J Reprod Immunol (in press)

Hume DA, Robinson AP, McPherson GG, Gordon S (1983) The mononuclear phagocyte system of the mouse defined by immunohistochemical localization of antigen F4/80. J Exp Med 158:1522–1536

Hunt JS (1989a) Cytokine networks in the uteroplacental unit: macrophages as pivotal regulatory cells. J Reprod Immunol 16: 1–17

Hunt JS (1989b) Macrophages in human uteroplacental tissues. Am J Reprod Immunol 21: 119–122

Hunt JS (1990) Current topic: the role of macrophages in the uterine response to pregnancy. Placenta 11 : 467–475

Hunt JS (1991) Prostaglandins, immunoregulation, and macrophage function. In: Coulam C, McIntyre JA, Faulk WP (eds) Immunologic obstetrics. W.W. Norton, New York (in press)

Hunt JS, King CR Jr, Wood GW (1984a) Evaluation of human chorionic trophoblasts and placental macrophages as stimulators of maternal lymphocyte proliferation in vitro. J Reprod Immunol 6 : 377–391

Hunt JS, Manning LS, Wood GW (1984b) Macrophages in murine uterus are immunosuppressive. Cell Immunol 85 : 499–510

Hunt JS, Manning LS, Mitchell D, Selanders JR, Wood GW (1985) Localization and characterization of macrophages in murine uterus. J Leukocyte Biol 38 : 255–265

Hunt JS, Soares MJ, Lei M-G, Smith RN, Wheaton D, Atherton RA, Morrison DC (1989) Products of lipopolysaccharide-activated macrophages (tumor necrosis factor-α, transforming growth factor-β) but not lipopolysaccharide modify DNA synthesis by rat trophoblast cells exhibiting the 80 kDa lipolysaccharide-binding protein. J Immunol 143 : 1606–1613

Hunt JS, Atherton RA, Pace JL (1990) Differential responses of rat trophoblast cells and embryonic fibroblasts to cytokines that regulate proliferation and class I MHC antigen expression. J Immunol 145 : 184–189

Jaattela M, Kuusela P, Saksela E (1988) Demonstration of tumor necrosis factor in human amniotic fluids and supernatants of placental and decidual tissues. Lab Invest 58 : 48–52

Kabawat SE, Mostoufi-Zadeh M, Driscoll SG, Bhan AK (1985) Implantation site in normal pregnancy. Am J Pathol 118 : 76–84

Kameda T, Matsuzaki N, Sawai K et al. (1990) Production of interleukin-6 by normal human trophoblast. Placenta 11 : 205–213

Kato K, Yokoi T, Takano N, Kanegane H, Yachie A, Miyawaki T, Taniguchi N (1990) Detection by in situ hybridization and phenotypic characterization of cells expressing IL-6 mRNA in human stimulated blood. J Immunol 144:1317–1322

Kaushansky K, Broudy VC, Harlan JM, Adamson JW (1988) Tumour necrosis factor-α and tumour necrosis factor-β (lymphotoxin) stimulate the production of granulocyte-macrophage colony-stimulating factor, macrophage colony stimulating factor, and IL-1 in vivo. J Immunol 141 : 3410–3415

Keshav S, Lawson L, Chung LP, Stein M, Perry VH, Gordon S (1990) Tumour necrosis factor mRNA localized to Paneth cells of normal murine intestinal epithelium by in situ hybridization. J Exp Med 171 : 327–332

Khong TY (1987) Immunohistologic study of the leukocytic infiltrate in maternal uterine tissues in normal and preeclamptic pregnancies at term. Am J Reprod Immunol 15 : 1–8

King A, Wellings V, Gardner L, Loke YW (1989) Immunocytochemical characterization of the unusual large granular lymphocytes in human endometrium throughout the menstrual cycle. Hum Immunol 24 : 195–205

Klein JO, Remington JS (1990) Current concepts of infections of the fetus and newborn infant. In: Remington JS, Klein JO (eds) Infectious diseases of the fetus and newborn infant, W.B. Saunders, Philadelphia

Kojima S, Tadenuma H, Inada Y, Saito Y (1989) Enhancement of plasminogen activator activity in cultured endothelial cells by granulocyte colony-stimulating factor. J Cell Physiol 138 : 192–196

Laguens G, Goni JM Jr, Laguens M, Goni JM, Laguens R (1990) Demonstration and characterization of HLA-DR positive cells in the stroma of human endometrium. J Reprod Immunol 18 : 179–186

Lala PK, Kennedy TG, Parhar RS (1988) Suppression of lymphocyte alloreactivity by early gestational human decidua. II. Characterization of the suppressor mechanisms. Cell Immunol 116 : 411–422

Lande IJ (1986) Systemic immunity developing from intrauterine antigen exposure in the nonpregnant rat. J Reprod Immunol 9 : 57–66

Le J, Vilcek J (1987) Biology of disease. Tumor necrosis factor and interleukin 1: cytokines with multiple overlapping biological activities. Lab Invest 56 : 234–248

Lessin DL, Hunt JS, King CR Jr, Wood GW (1988) Antigen expression by cells near the maternal-fetal interface. Am J Reprod Immunol Microbiol 16 : 1–7

Lewis SH, Reynolds-Kohler C, Fox HE, Nelson JA (1990) HIV-1 in trophoblastic and villous Hofbauer cells and haematological precursors in eight-week fetuses. Lancet 335 : 565–568

Loke YW, Eremin O, Ashby J, Day S (1982) Characterization of the phagocytic cells isolated from the human placenta. J Reticuloendothel Soc 31 : 317–324

Main EK, Strizki J, Schochet P (1987) Placental production of immunoregulatory factors: trophoblast is a source of interleukin-1. Trophoblast Res 2 : 149–160

Martin L, Finn CA, Trinder G (1973) DNA synthesis in the endometrium of progesterone treated mice. J Endocrinol 56: 303–307

Matthews CJ, Searle RF (1987) The role of prostaglandins in the immunosuppressive effects of supernatants from adherent cells of murine decidual tissue. J Reprod Immunol 12: 109–124

Matthews CJ, Adams AM, Searle RF (1985) Detection of macrophages and the characterization of Fc receptor bearing cells in the mouse decidua, placenta and yolk sac using the macrophage-specific monoclonal antibody F4/80. J Reprod Immunol 7: 315–323

McGregor JA, French JI, Lawellin D, Todd JK (1988) Preterm birth and infection: pathogenic possibilities. Am J Reprod Immunol Microbiol 16: 123–132

Meagher R, Amsden A, Hunt J, Soares M, Sogor L, Smith RN (1990) Placental CSF-like activity. Exp Hematol 18: 448–451

Ming WJ, Bersani L, Mantovani A (1987) Tumour necrosis factor-alpha is chemotactic for monocytes and polymorphonuclear leukocytes. J Immunol 138: 1469–1474

Moskalewski S, Ptak W, Strzyżewska J (1974) Macrophages in mouse placenta. I. Morphological and functional identification. J Reticuloendothel Soc 16: 9–14

Moskalewski S, Ptak W, Czarnik Z (1975) Demonstration of cells with IgG receptor in human placenta. Biol Neonate 26: 268–272

Motro B, Ahuva I, Sachs L, Keshet E (1990) Pattern of interleukin-6 gene expression in vivo suggests a role for this cytokine in angiogenesis. Proc Natl Acad Sci USA 87: 3092–3096

Mues B, Langer D, Zwadlo G, Sorg C (1989) Phenotypic characterization of macrophages in human term placenta. Immunology 67: 303–307

Muller R, Slamon DJ, Adamson ED, Tremblay JM, Muller D, Cline MJ, Verma IM (1983) Transcription of c-*onc* genes c-*ras*^{Ki} and c-*fms* during mouse development. Mol Cell Biol 3: 1062–1069

Nash AD, Uren S, Hawes CS, Boyle W (1989) Application of a novel immunization protocol to the production of monoclonal antibodies specific for macrophages in human placenta. Immunology 68: 332–340

Nathan CF (1987) Secretory products of macrophages. J Clin Invest 79: 319–326

Nelson BJ, Ralph P, Green SJ, Nacy CA (1991) Differential susceptibility of activated macrophage cytotoxic effector reactions to the suppressive effects of transforming growth factor-β1. J Immunol 146: 1849–1857

Nicol T (1935) The female reproductive system in the guinea pig: intravitam staining; fat production; influence of hormones. Trans R Soc Edinb 58: 449–483

Okazaki T, Casey ML, Okita JR, MacDonald PC, Johnston JM (1981) Initiation of human parturition. XII. Biosynthesis and metabolism of prostaglandins in human fetal membranes and uterine decidua. Am J Obstet Gynecol 139: 373–381

Oksenberg JR, Mor-Yosef S, Ezra Y, Brautbar C (1988) Antigen presenting cells in human decidual tissue: III. Role of accessory cells in activation of suppressor cells. Am J Reprod Immunol Microbiol 16: 151–158

Parhar RS, Kennedy TG, Lala PK (1988) Suppression of lymphocyte alloreactivity by early gestational human decidua. I. Characterization of suppressor molecules. Cell Immunol 116: 392–410

Parr EL,Young LHY, Parr MB, Young JD-E (1990) Granulated metrial gland cells of pregnant mouse uterus are natural killer-like cells that contain perforin and serine esterases. J Immunol 145: 2365–2372

Pollard JW (1990) Regulation of polypeptide growth factor synthesis and growth factor-related gene expression in the rat and mouse uterus before and after implantation. J Reprod Fertil 88: 721–731

Pollard JW, Bartocci A, Arceci R, Orlofsky A, Ladner MB, Stanley ER (1987) Apparent role of the macrophage growth factor, CSF-1, in placental development. Nature 330: 484–487

Pollard JW, Hunt JS, Wiktor-Jedrzejczak W, Stanley ER (1991a) A pregnancy defect in the osteopetrotic (*op/op*) mouse demonstrates the requirement for CSF-1 in female fertility. Dev Biol 148: 273–283

Pollard JW, Pampfer S, Daiter E, Barad D, Areci RJ (1991b) Colony stimulating factor-1 in the mouse and human uteroplacental unit. In: Schonberg DW (ed) Growth factors in reproduction. Plenum, New York, p 219–229

Rambaldi A, Young DC, Griffin JD (1987) Expression of the M-CSF (CSF-1) gene by human monocytes. Blood 69: 1409–1413

Redline RW, Lu CY (1988) Specific defects in the anti-listerial immune response in discrete regions of the murine uterus and placenta account for susceptibility to infection. J Immunol 140: 3947–3955

Redline RW, Lu CY (1989) Localization of fetal major histocompatibility complex antigens and maternal leukocytes in murine placenta. Lab Invest 61: 27–36

Redline RW, McKay DB, Vazquez MA, Papaioannou VE, Lu CY (1990) Macrophage functions are regulated by the substratum of murine decidual stromal cells. J Clin Invest 85: 1951–1958
Regenstreif LJ, Rossant J (1989) Expression of the c-*fms* proto-oncogene and of the cytokine, CSF-1 during mouse embryogenesis. Dev Biol 133: 284–294
Renz H, Gong J-H, Schmidt A, Nain M, Gemsa D (1988) Release of tumor necrosis factor-α from macrophages. Enhancement and suppression are dose-dependently regulated by prostaglandin E_2 and cyclic nucleotides. J Immunol 141: 2388–2393
Rizzino A (1988) Transforming growth factor -β: multiple effects on cell differentiation and extracellular matrices. Dev Biol 130: 411–422
Robbins DS, Shirazi Y, Drysdale BE, Leiberman A, Shin HS, Shin ML (1987) Production of cytotoxic factor for oligodendrocytes by stimulated astrocytes. J Immunol 139: 2593–2597
Robertson SA, Seamark RF (1990) Granulocyte macrophage colony stimulating factor (GM-CSF) in the murine reproductive tract: stimulation by seminal factors. Reprod Fertil Dev 2: 359–368
Romero R, Emamian M, Wan M, Quintero R, Hobbins JC, Mitchell M (1987) Prostaglandin concentrations in amniotic fluid of women with intra-amniotic infections and preterm labor. Am J Obstet Gynecol 157: 1461–1467
Romero R, Brody DT, Oyarzun E, Mazor M, Wu YK, Hobbins JC, Durum SK (1989a) Infection and labor. III. Interleukin-1: a signal for the onset of parturition. Am J Obstet Gynecol 160: 1117–1123
Romero R, Manogue KR, Mitchell MD, Wu YK, Oyarzun E, Hobbins JC, Cerami A (1989b) Infection and labor. IV. Cachectin-tumor necrosis factor in the amniotic fluid of women with intraamniotic infection and preterm labor. Am J Obstet Gynecol 161: 336–341
Romero R, Avila C, Santhanam U, Sehgal PB (1990) Amniotic fluid and interleukin-6 in preterm labor: association with infection. J Clin Invest 85: 1392–1400
Russell SW, Pace JL (1987) The effects of interferons on macrophages and their precursors. Vet Immunol Immunopathol 15: 129–165
Simmen FA, Simmen RC (1991) Peptide growth factors and proto-oncogenes in mammalian conceptus development. Biol Reprod 44: 1–5
Sminia T, Jeurissen SHM (1986) The macrophage population of the gastro-intestinal tract of the rat. Immunobiology 172: 72–80
Sokol S, Wong GG, Melton DA (1990) A mouse macrophage factor induces head structures and organizes a body axis in *Xenopus*. Science 249: 561–564
Sporn MB, Roberts AB (1989) Transforming growth factor-β. Multiple actions and potential clinical applications. JAMA 262: 938–941
Starkey PM, Turley I, Gordon S (1987) The mouse macrophage-specific glycoprotein defined by monoclonal antibody F4/80: characterization, biosynthesis and demonstration of a rat analogue. Immunology 60: 117–122
Sugarman BJ, Aggarwal BB, Hass PE, Figari IS, Palladino MA, Shepard HM (1985) Recombinant human tumor necrosis factor-α: effects on proliferation of normal and transformed cells in vitro. Science 230: 943–945
Tabibzadeh SS (1990) Proliferative activity of lymphoid cells in human endometrium throughout the menstrual cycle. J Clin Endocrinol Metabol 70: 437–443
Tabibzadeh SS, Santhanam U, Sehgal PB, May LT (1988) Cytokine-induced production of IFN-β/IL-6 by freshly explanted human endometrial stromal cells. J Immunol 142: 3134–3139
Tachi C, Tachi S (1989) Role of macrophages in the maternal recognition of pregnancy. J Reprod Fertil [Supp] 37: 63–68
Tachi C, Tachi S, Knyszynski A, Linder HR (1981) Possible involvement of macrophages in embryo-maternal relationships during ovum implantation in the rat. J Exp Zool 217: 81–92
Takacs L, Kovacs EJ, Smith MR, Young HA, Durum SK (1988) Detection of IL-1α and IL-1β gene expression by in situ hybridization. J Immunol 141: 3081–3095
Tamada H, McMaster MT, Flanders KC, Andrews GK, Dey SK (1990) Cell type-specific expression of TGF-β1 in the mouse uterus during the periimplantation period. Mol Endocrinol 4: 965–972
Tamatani T, Tsunoda H, Iwasaki H, Kaneko M, Hashimoto T, Onozaki K (1988) Existence of both IL-1α and β in normal amniotic fluid: unique high molecular weight form of IL-1β. Immunology 65: 337–342
Taniguchi T (1988) Regulation of cytokine gene expression. Annu Rev Immunol 6: 439–464
Tawfik OW, Hunt JS, Wood GW (1986a) Implication of prostaglandin E_2 in soluble factor-mediated immune suppression by murine decidual cells. Am J Reprod Immunol Microbiol 12: 111–117
Tawfik OW, Hunt JS, Wood GW (1986b) Partial characterization of uterine cells responsible for suppression of murine maternal anti-fetal immune responses. J Reprod Immunol 9: 213–224

Toth FD, Juhl C, Norskov-Lauritsen N, Mosborg-Petersen P, Ebbesen P (1990) Interferon production by cultured human trophoblasts induced with double stranded polyribonucleotide. J Reprod Immunol 17: 217–227

Tushinski RJ, Oliver IT, Guilbert LJ, Tynan PW, Warner JR, Stanley ER (1982) Survival of mononuclear phagocytes depends on a lineage-specific growth factor that the differentiated cells selectively destroy. Cell 28: 71–81

Ulich TR, Guo K, del Castillo J (1989) Endotoxin-induced cytokine gene expression in vivo. Am J Pathol 134: 11–14

Umeda T, Hara T, Niijima T (1983) Cytotoxic effect of tumor necrosis factor on human lymphocytes and specific binding of the factor to the target cells. Cell Mol Biol 29: 349–352

Uzumaki H, Okabe T, Sasaki N et al. (1989) Identification and characterization of receptors for granulocyte colony-stimulating factor on human placenta and trophoblastic cells. Proc Natl Acad Sci USA 86: 9323–9326

Van Damme J, Van Beeumen J, Decock B, Van Snick J, De Ley M, Billiau F (1988) Separation and comparison of two monokines with LAF activity (interleukin-1β and hybridoma growth factor): identification of leukocyte-derived HGF as interleukin-6. J Immunol 140: 1534–1541

Van Rood.JJ, Eernisse JG, Van Leeuwen A (1958) Leucocyte antibodies in sera from pregnant women. Nature 181: 1735–1736

Van Snick J (1990) Interleukin-6: an overview. Annu Rev Immunol 8: 253–278

Wahl SM, Hunt DA, Wakefield LM, McCartney-Francis N, Wahl LM, Roberts AB, Sporn MR (1987) Transforming growth factor type β induces monocyte chemotaxis and growth factor production. Proc Natl Acad Sci USA 84: 5788–5792

Wahl SM, Hunt DA, Wong HL et al. (1988) Transforming growth factor-β is a potent immunosuppressive agent that inhibits IL-1-dependent lymphocyte proliferation. J Immunol 140: 3026–3032

Wahl SM, McCartney-Francis N, Mergenhagen SE (1989) Inflammatory and immunomodulatory roles of TGF-β. Immunol Today 10: 258–261

Wang JM, Colella S, Allavena P, Mantovani A (1987) Chemotactic activity of human recombinant granulocyte-macrophage colony stimulating factor. Immunology 60: 439–444

Wang JM, Griffin JD, Rambaldi A, Chen ZG, Mantovani A (1988) Induction of monocyte migration by recombinant macrophage colony-stimulating factor. J Immunol 141: 575–579

Wiktor-Jedrzejczak W, Ahmed A, Szczylik C, Skelly RR (1982) Hematological characterization of congenital osteopetrosis in *op/op* mouse. J Exp Med 156: 1516–1527

Wiktor-Jedrzejczak W, Bartocci A, Ferrnate AW, Ahmed-Ansari A, Snell KW, Pollard JW, Stanley ER (1990) Total absence of colony stimulating factor-1 in the macrophage-deficient osteopetrotic (*op/op*) mouse. Proc Natl Acad Sci USA 87: 4828–4832

Wiktor-Jedrzejczak W, Urbanowska E, Aukerman SL et al. (1991) Correction by CSF-1 of defects in the osteopetrotic *op/op* mouse suggest local, developmental and humoral requirements for this growth factor. Exp Hematol 19:1049–1054

Wood GW (1980) Immunohistological identification of macrophages in murine placentae, yolk sac membranes and pregnant uteri. Placenta 1: 309–318

Wood GW, King CR Jr (1982) Trapping antigen-antibody complexes within the human placenta. Cell Immunol 69: 347–362

Wood GW, Reynard J, Krishnan E, Racela L (1978) Immunobiology of the human placenta. II. Localization of macrophages, in vivo bound IgG and C3. Cell Immunol 35: 205–216

Yang CP, DePinho SG, Greenberger LM, Arceci RJ, Horwitz SB (1989) Progesterone interacts with P-glycoprotein in multidrug-resistant cells and in the endometrium of gravid uterus. J Biol Chem 264: 782–788

Yelavarthi KK, Chen H-L, Yang Y, Cowley BD, Fishback JL, Hunt JS (1991) Tumor necrosis factor-alpha mRNA and protein in rat uterine and placental cells. J Immunol 146: 3840–3848

Yoshida H, Hayashi S-I, Kunisada T et al. (1990) The murine mutation osteopetrosis is in the coding region of the macrophage colony stimulating factor gene. Nature 345: 442–444

Zheng Y, Zhou Z-Z, Lyttle CR, Teuscher C (1988) Immunohistochemical characterization of the esterogen-stimulated leukocyte influx in the immature rat uterus. J Leukocyte Biol 44: 27–32

Urokinase-Catalyzed Plasminogen Activation at the Monocyte/Macrophage Cell Surface: A Localized and Regulated Proteolytic System

J.-D. VASSALLI[1], A. WOHLWEND[1], and D. BELIN[2]

1 Introduction

In the adult organism, monocytes and macrophages are among the few cell types that can migrate within and between body compartments. To do so, they must have the capacity to clear for themselves a path through the macromolecular barriers of basement membranes and other extracellular matrices. This requires the controlled and localized degradation of matrix proteins by extracellular proteases. Mononuclear phagocytes can produce a number of such enzymes, including collagenolytic, elastinolytic, and gelatinolytic hydrolases (TAKEMURA and WERB 1984). Because they can, directly or indirectly, catalyze the degradation of most components of extracellular matrices, plasminogen activators (PAs) are thought to play a key role in the proteolytic events that accompany the migration of a wide variety of cell types, during ontogeny as well as in pathologic circumstances. Monocytes and macrophages can produce PAs, and the regulation of their PA-dependent proteolytic activity has been a focus of attention in recent years. The findings of a number of investigators converge to suggest that the expression of PA activity is a tightly controlled phenotypic property of human and murine mononuclear phagocytes, and that multiple mechanisms act concurrently to achieve the exquisitely focused and

[1] Institute of Histology and Embryology, and
[2] Department of Pathology, University of Geneva Medical School, CH-1211 Geneva 4, Switzerland

Current Topics in Microbiology and Immunology, Vol. 181

Fig. 1. In vitro modulation of the PA activity of mouse macrophages. Peritoneal macrophages from thioglycollate-induced exudates were plated at low density, treated for 16 h with different cytokines (M-CSF, 100 U/ml; IFN-γ, 20 U/ml; TNF-α; 10 ng/ml, IL-1, 100 U/ml) and overlaid with a substrate mixture containing plasminogen and casein (VASSALLI et al. 1985). The picture was taken under dark ground illumination: plaques represent zones of substrate lysis around individual cells

regulated generation of plasmin precisely where and when it is needed to allow cell migration in the context of inflammatory reactions.

Hormones and cytokines play a particularly important role in the regulation of PA activity in many cell types, including fibroblasts, endothelial cells, ovarian granulosa cells, Sertoli cells, and mammary epithelial cells (DANØ et al. 1985; SAKSELA 1985; MOSCATELLI and RIFKIN 1988; SAKSELA and RIFKIN 1988). Similarly, the PA activity of cultured mouse peritoneal macrophages can be readily altered as a function of the cytokine balance in the macrophage environment (Fig. 1): Macrophages embedded in a layer of casein degrade the substrate in their immediate vicinity through the generation of plasmin from its inactive precursor plasminogen; this proteolytic activity is enhanced in the presence of the macrophage-activating cytokine interferon-γ (IFN-γ), and decreased in the presence of the macrophage growth factor M-CSF. It is evident that such regulation can dramatically alter the extent of extracellular substrate degradation and this suggests that a clear and complete understanding of the physiologic and pharmacologic control of the macrophage PA system could be of great help in the therapeutic management of inflammatory reactions and their associated tissue destruction. The mechanisms involved in controlling plasmin generation by monocytes/macrophages have been, at least in part, elucidated, and this review will summarize the roles of the PAs themselves, of PA inhibitors, and of a plasma membrane binding site specific for the urokinase-type PA.

2 Plasminogen Activators

Plasminogen activators are serine proteases of tryptic specificity. Their major macromolecular substrate is the zymogen plasminogen; other proteins, such as fibrinogen (WEITZ et al. 1988) and fibronectin (GOLD et al. 1989), can be directly cleaved by PAs, although this occurs only at very high enzyme to substrate ratios, and thus may not be of physiologic relevance except under particular circumstances. Plasminogen is abundant in plasma and in most extracellular fluids and constitutes a reservoir of broad spectrum proteolytic activity that can be recruited by PA-catalyzed conversion of the single-chain zymogen to the two-chain tryptic protease plasmin. Since plasmin appears to be required for the activation of a metalloprotease cascade that leads to the generation of active collagenase (WERB et al. 1977), for instance, PAs could play a key role in catalyzing extracellular proteolysis.

In all mammalian species explored to date, two PAs have been identified: urokinase-type PA (uPA) and tissue-type PA(tPA). They are the products of distinct genes and differ in certain aspects of their catalytic and binding properties. The single-chain form of uPA is a zymogen with no (or very little) activity (pro-uPA) (PETERSEN et al. 1988); both uPA and pro-uPA bind to a cell surface receptor that localizes plasmin generation to the immediate cell environment (see below). By contrast, tPA is active both as a single- and as a two-chain enzyme (although there are differences in the catalytic properties of the two forms of tPA); tPA binds to components of extracellular matrices, in particular fibrin (HOYLAERTS et al. 1982) but also fibronectin and laminin (SALONEN et al. 1984, 1985). The available evidence suggests that tPA-catalyzed proteolysis is preferentially involved in the maintenance of fluidity of the extracellular milieu, while uPA plays a role in the cell surface proteolysis necessary for cell migration. However, it is clear that this tentative model is a simplification. Indeed, in at least one case, the same cell type produces a different PA in two different species (CANIPARI et al. 1987). In addition, abundant uPA is produced by cells that are not in a process of migration (epithelial cells of the nephron and of the male genital tract, for instance) (LARSSON et al. 1984; HUARTE et al. 1987). Finally, plasma membrane binding sites for tPA have also been identified (HAJJAR et al. 1987; BARNATHAN et al. 1988). It should also be noted here that PAs have not been associated with the traffic of lymphocytes, which are endowed with remarkable migratory properties.

3 Monocytes and Macrophages Produce Plasminogen Activators

In the early 1970s, studies on the production of proteases by cells in culture led to the identification of the PA-plasmin system as a widespread mechanism used by many different cell types to catalyze extracellular proteolysis (REICH 1978). UNKELESS and co-workers (1974) described the production of a PA by mouse peritoneal macrophages and human monocytes, and demonstrated that production of this enzyme could vary dramatically as a function of the state of the cells: Macrophages from thioglycollate-elicited peritoneal exudates produced high levels of PA activity as compared to resident noninflammatory macrophages; also, certain populations of "in vivo primed" macrophages [e.g., after injection of lipopolysaccharide (LPS)] could be triggered in vitro to produce high levels of enzyme, for instance by phagocytosis of latex particles (GORDON et al. 1974).

The idea that PA production by macrophages could be a marker for one stage in the life cycle of these cells, i.e., in the early phases of their participation in inflammatory reactions, received support from further studies on the modulation of enzyme production in vitro and in vivo. A striking observation in this context was the inhibition of PA production by inflammatory macrophages under the influence of anti-inflammatory glucocorticoids (VASSALLI et al. 1976); similarly, increased enzyme production by macrophages exposed to products of activated T lymphocytes suggested that PAs could be a marker of macrophage "activation" (KLIMETZEK and SORG 1977; NOGUEIRA et al. 1977; VASSALLI and REICH 1977; GORDON et al. 1978). Thus, high levels of PA production appear to be a hallmark of murine inflammatory peritoneal macrophages. At this time only a few studies have attempted to determine whether this is true for other macrophage populations; in view of the difficulties inherent to the assay of PAs in tissue extracts, in particular because of the presence of PA inhibitors, assays that do not rely on the catalytic evaluation of PA amounts (i.e., immunoassays or determination of mRNA levels) should be the tools for such a study.

The types of PAs produced by mature human and murine monocytes/macrophages have been identified using immunologic techniques and nucleic acid hybridization. Most studies that have investigated this issue have shown that these cells usually produce exclusively uPA (VASSALLI et al. 1984; SAKSELA et al. 1985; COLLART et al. 1987). The enzyme appears to be secreted as a single-chain zymogen, the physiologic activation of which is still poorly understood; in vitro, it can be achieved by plasmin or plasma kallikrein (WUN et al. 1982a; ICHINOSE et al. 1986). The culture medium of certain leukemic cell lines has been reported to contain two-chain active enzyme, a result which suggests that these cells can produce a pro-uPA activator (STEPHENS et al. 1988). It is also possible that, in the presence of plasminogen, the intrinsic activity of single-chain tPA could initiate a uPA-dependent cascade (VIHKO et al. 1989).

Interestingly, tPA production has been observed in cultures of granulocyte/macrophage progenitors, and the switch to uPA production was proposed to be an index of cell differentiation (WILSON and FRANCIS 1987); in this context, the poor response to chemotherapy of patients with acute myeloblastic leukemia whose cells secrete tPA (WILSON et al. 1983) supports the notion that tPA is a marker of early monocyte precursors. Recently, new light has been shed on this issue by the observation that human peripheral blood monocytes, in addition to uPA, also produce tPA in response to LPS or interleukin-4 (IL-4) (HART et al. 1989a, b). It is intriguing that, under certain conditions, these cells can produce both enzymes; the identification of their respective roles is a challenge for future work. At this time, bone marrow-derived and peritoneal mouse macrophages have not been found to produce tPA, but circulating mouse monocytes have not been explored.

Given that uPA production is a marker of the activation state of at least certain macrophage populations, what are the molecular mechanisms that account for the large differences in enzyme activity between noninflammatory (resident) and exudate peritoneal macrophages? Nuclear run-on experiments that assay the transcriptional activity of the uPA gene show that this is markedly higher in thioglycollate-elicited than in resident macrophages (Fig. 2; see also COLLART et al. 1986). A comparable difference in the steady state levels of uPA mRNA between the two populations (Fig. 3) confirms that the modulation of uPA activity is, at least in part, due to changes in uPA gene transcription. Analyses of uPA gene transcription and mRNA levels in bone marrow-derived macrophages indicate that these cells are, in terms of uPA gene expression, similar to inflammatory exudate cells (Figs. 2, 3). In the context of such comparative studies between different populations of mouse macrophages, it is important to note that quite marked differences in absolute levels of uPA mRNA have been noted between different preparations of "inflammatory" (COLLART et al. 1987) or "resident" macrophages. This may be due to the difficulty inherent in achieving an inflammatory response of similar intensity from one experiment to the next, and in obtaining cells from animals completely

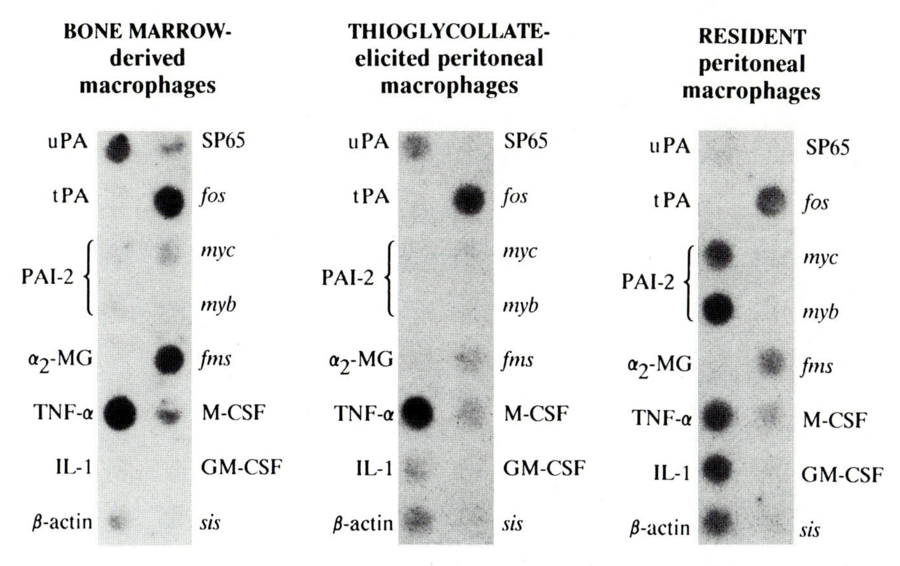

Fig. 2. Run-on transcription in isolated nuclei. Nuclei were prepared from mouse bone marrow-derived macrophages (WOHLWEND et al. 1987b) or thioglycollate-elicited or resident peritoneal macrophages (COLLART et al. 1986, 1987). Run-on transcription was performed as described by COLLART et al. (1987). The two PAI-2 DNAs used are from the 5′ (lower dot, nucleotides 1–815) and the 3′ (upper dot, nucleotides 810–1245) regions of the human cDNA (SCHLEUNING et al. 1987)

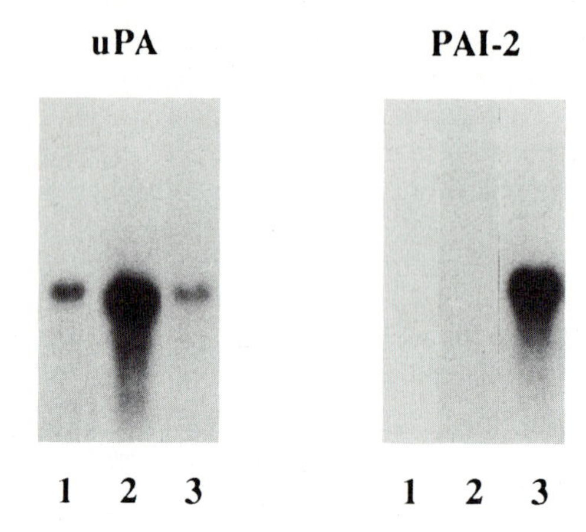

1) BONE MARROW-derived macrophages

2) THIOGLYCOLLATE-elicited peritoneal macrophages

3) RESIDENT peritoneal macrophages

Fig. 3. Northern blot analysis of total cellular RNA. Total RNA was prepared from mouse bone marrow-derived macrophages (Wohlwend et al. 1987b), thioglycollate-elicited or resident peritoneal macrophages (Collart et al. 1987). Murine uPA (Belin et al. 1985; Collart et al. 1987) and human PAI-2 (Schleuning et al. 1987; Belin et al. 1989) cRNA probes were prepared and used as described by Busso et al. (1986)

devoid of a peritoneal reaction. Macrophage uPA levels are very sensitive indices of the "activation" state of the cells and may vary from one preparation to the next; despite this variation, inflammatory cells obtained following intraperitoneal injection of thioglycollate broth always have at least ten fold higher levels (10–50 molecules per cell) of uPA mRNA than their resident counterparts obtained from uninjected animals.

In vitro pharmacologic studies have reinforced the notion that changes in uPA gene transcription play an important part in the modulation of uPA production. Those agents that had previously been shown to decrease macrophage uPA activity, such as glucocorticoids or compounds that raise intracellular cAMP levels, caused a decrease in uPA transcription and mRNA levels; similarly, agents that increase uPA production, such as lectins or IFN-γ (Vassalli et al. 1977; Vassalli and Reich 1977), appear to do so through an effect on transcription of the gene (Collart et al. 1987). A comparison with the changes in the transcription rates of other genes for macrophage-secreted proteins [e.g., tumor necrosis factor-α (TNF-α) and IL-1] in response to these agents suggests that the control of uPA production is quite distinct and hence supports the hypothesis that uPA production is a marker for a specific stage of macrophage activation. Interestingly, changes in uPA transcription were

Table 1. Agents that modulate in vitro the PA activity of mononuclear phagocytes[a]

	PA	uPA		PAI-2	
	Activity	Enzyme	mRNA	Serpin	mRNA
IFN-γ	+	+	+	=	=
Con A	+	+	+	=	n.d.
PMA	+	+	+	+	+
Endocytosis	+	+	n.d.	=	n.d.
1,25-VIT D_3	+	+	+	−	−
LPS	−	−	−	+	+
M-CSF	−	=	=	+	+
cAMP	−	−	−	+	+
Glucocorticoids	−	−	−	−	−
Retinoids	−	−	−	n.d.	n.d.
IL-4	+	tPA+	tPA mRNA+	n.d.	n.d.

+, increase; =, no effect; −, decrease; n.d., not determined

[a] Agents for which only overall changes in PA activity have been reported are not included. For references, see the text.

found to be preceded by opposite changes in c-*fos* mRNA levels, suggesting a possible role for the c-*fos* gene product in the modulation of uPA gene expression (COLLART et al. 1987). In addition, protein synthesis inhibitors have been found to cause a rapid and transient induction of uPA mRNA synthesis (COLLART et al. 1986), suggesting the existence of short-lived repressors of uPA gene transcription.

To summarize the data that relate to the control of macrophage PA activity, we can say that, in most cases where this has been studied, there is a good correlation between changes in PA production and in uPA gene transcription or uPA mRNA levels (Table 1). A striking exception, however, is the effect of the macrophage growth factor M-CSF: while M-CSF decreases macrophage PA activity (Fig. 1), it does not affect the steady state level of uPA mRNA in these cells, at least at early times (Fig. 4). A probable solution to this apparent paradox will be discussed below: M-CSF induces the production of a PA-specific inhibitor (PAI) by mouse peritoneal macrophages. Further studies of the effects of M-CSF will be of interest, since stimulation of PA activity in cultures of peritoneal and bone marrow-derived macrophages has also been observed (HAMILTON et al. 1980; HUME and GORDON 1984). Finally, it should also be noted that some agents (i.e., LPS and cAMP) cause both a decrease in uPA transcription and an increase in PAI production, thus causing a profound decrease in uPA activity through combined effects. The mechanisms of increased uPA production in response to phagocytosis (GORDON et al. 1974; SCHNYDER and BAGGIOLINI 1978), endocytosis (FALCONE and FERENC 1988; FALCONE 1989), or 1,25-dihydroxyvitamin D_3 (GYETKO et al. 1988) and of decreased uPA production in the presence of retinoids or auranofin (VASSALLI et al. 1976; LISON et al. 1989) have not been investigated; changes in mRNA translation or stability could account for some of these effects.

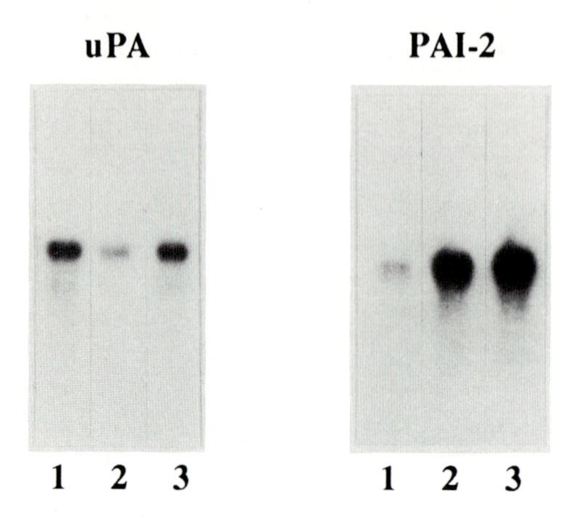

1) **Control**
2) **LPS** (4h ; 1 μg/ml)
3) **M-CSF** (4h ; 20 U/ml)

Fig. 4. Northern blot analysis of total cellular RNA. Thioglycollate-elicited murine peritoneal macrophages were incubated as indicated. Total RNA was prepared and analyzed as for Fig. 3

4 Monocytes and Macrophages Produce A Plasminogen Activator-Specific Inhibitor

Control of protease activity by macromolecular inhibitors is a physiologically important mechanism to avoid the damage that can be caused by excessive proteolysis. A demonstration of this is provided by the progressive destruction of lung elastic tissue in patients deficient in α_1-protease inhibitor, an elastase inhibitor (HUBER and CARRELL 1989). Antiproteases amount to some 10% of total plasma proteins, and production in tissues may also contribute to the local proteolytic balance.

Monocytes and macrophages have been shown to produce different antiproteases, including α_1-protease inhibitor and the broad spectrum inhibitor α_2-macroglobulin (GANTER et al. 1989). A search for antiproteases directed against PAs has revealed the production by these cells of a PA-specific inhibitor: addition of uPA to medium from cultures of mouse peritoneal macrophages or human peripheral blood monocytes resulted in an inhibition of enzyme activity (KLIMETZEK and SORG 1979; CHAPMAN et al. 1982; GOLDER and STEPHENS 1983; VASSALLI et al. 1984; CHAPMAN and STONE 1985a; KOPITAR et al. 1985; WOHLWEND et al. 1987b). A number of cell lines from the mononuclear phagocyte lineage were also found to produce such an inhibitor. Biochemical purification from cultures of human U937 cells resulted in the isolation of a PA-specific antiprotease (KRUITHOF et al. 1986), similar to a uPA inhibitor that had originally been identified in human placenta (KAWANO et al. 1968). This "placental-type" PAI was named PAI-2, and was shown to be a member of the family of *ser*ine *p*roteases *in*hibitor*s*, the serpins (CARRELL and TRAVIS 1985). Two other serpins with specificity for arginine-proteases and high affinity for PAs have also been identified: PAI-1, originally identified in cultures of endothelial cells (LOSKUTOFF et al. 1983), and protease nexin 1 (PN-1), a fibroblast and glial cell product that also inhibits thrombin (BAKER et al. 1980; GLOOR et al. 1986). PAI-2 appears to be the predominant PA inhibitor produced by mononuclear phagocytes; PAI-1 has been detected in two human monocytic cell lines (WOHLWEND et al. 1987a; LUND et al. 1988). PAI-2 inhibits most efficiently uPA (second order rate constant $5 \times 10^7 M^{-1} s^{-1}$, and also reacts with two-chain tPA ($10^5 M^{-1} s^{-1}$); it does not react with pro-uPA, nor with single-chain tPA, plasmin, or thrombin (KRUITHOF et al. 1986). Besides mononuclear phagocytes, PAI-2 has been reported in human placenta (syncytiotrophoblasts) (ÅSTEDT et al. 1986; FEINBERG et al. 1989), keratinocytes (HASHIMOTO et al. 1989), and endothelial cells (WEBB et al. 1987).

Cloning and sequencing of the human (SCHLEUNING et al. 1987; WEBB et al. 1987; YE et al. 1987; ANTALIS et al. 1988) and murine (BELIN et al. 1989) PAI-2 cDNAs and of the human PAI-2 gene (YE et al. 1989) confirmed that this inhibitor belongs to the serpin gene family, which also includes, in addition to PAI-1 and PN-1, α_1-protease inhibitor, α_2-antiplasmin, antithrombin-III, and other proteins with no known antiprotease function, such as ovalbumin, angiotensinogen, and cortisol-binding globulin (see HUBER and CARRELL 1989 for a recent review). Gene sequence comparisons reveal quite extensive divergence between members of the serpin family, and the similarity between mammalian PAI-2 and avian ovalbumin suggests that the two genes are closely related (YE et al. 1987, 1989).

Like other serpins, PAI-2 forms 1:1 complexes with its target proteases; these complexes are not dissociated in the presence of detergents such as SDS. Using preparations of radiolabeled enzyme, e.g., ^{125}I-uPA, the presence of PAI-2 can be demonstrated by gel electrophoresis: free radiolabeled enzyme can be separated from PAI-2-complexed enzyme, and, given excess enzyme, this also allows quantitation of PAI-2. Using this electrophoretic assay, two forms of PAI-2 were identified in cultures of human and murine monocytes/macrophages (WOHLWEND et al. 1987a, b): one form, of M_r 40000, accumulates in the cell, where it appears to be stored in the cytosol; another form, of M_r 55000–60000, is glycosylated and preferentially secreted. These two forms of PAI-2 are functionally and immunologically indistinguishable, and enzymatic removal of the polysaccharide portion of secreted PAI-2 yields a protein that comigrates with the cytosolic inhibitor. Thus these two forms differ only in the extent of their glycosylation (WOHLWEND et al. 1987a, b; GENTON et al. 1987; YE et al. 1988). Of newly synthesized PAI-2 molecules, approximately one-half remain in the cytosol, where they are stable for many hours, while the other half enter the secretory pathway and rapidly leave the cell. Cytosolic PAI-2 represents an abundant store of the inhibitor: the amount of nonglycosylated

PAI-2 in the cell is comparable to the amount of glycosylated inhibitor that can be secreted over a 24-h period. Whether the cytosolic inhibitor can be released through a post-translational translocation process that does not involve cell death, or whether it is only released under conditions of cell suffering or apoptosis, is not known. Interestingly, monocytes/macrophages also contain in their cytosol other proteins that are believed to act following their release in the extracellular milieu, i.e., the cytokine IL-1 (SINGER et al. 1988) and elastase inhibitors (REMOLD-O'DONNELL et al. 1989; POTEMPA et al. 1988); it is possible that their mechanisms of release are similar. Alternatively, cytosolic PAI-2 may have an intracellular function, although neither serine proteases nor enzymes of specificity comparable to that of the PAs have been identified in the cytosol.

Interestingly, the two forms of PAI-2 are encoded by a single mRNA (BELIN et al. 1989). Indeed, only one PAI-2 mRNA can be detected by Northern blot hybridization and by RNase protection, and transfection of a PAI-2 cDNA leads to the synthesis of both forms of the protein. In vitro translation of the mRNA transcript of a PAI-2 cDNA in the presence of microsomal membranes yields the two topologically distinct forms of the inhibitor: a membrane-enclosed and a "cytosolic" product. In this context, it is noteworthy that translation of the two forms of PAI-2 initiates at the same AUG, and that the secreted form is released without removal of the putative signal peptide (YE et al. 1988). The latter observation is reminiscent of ovalbumin secretion, which also occurs without removal of the N-terminal region of the protein; it may be relevant that ovalbumin is the closest serpin relative of PAI-2.

Taken together, the available information on PAI-2 indicates that the inhibitor is secreted through a process of cotranslational, but facultative, translocation. To our knowledge such a situation has not been described previously. Other proteins that are distributed bi-topologically to the cytosol and the extracellular milieu, such as yeast invertase (CARLSON et al. 1983) and mammalian gelsolin (KWIATKOWSKI et al. 1988), are translated from distinct mRNAs, one of which encodes a signal peptide. Another mechanism to achieve bitopological distribution is illustrated by secreted and nuclear rat prostatic probasins, which result from alternate translational initiation on a bifunctional mRNA (SPENCE et al. 1989). The unusual mechanism of PAI-2 partition may be related to structural features of its N-terminal region: this region (amino acids 1–22) contains, in addition to a hydrophobic stretch, two negatively charged residues near the N-terminus, and two asparagines within the hydrophobic core. It is striking that these features, which have not been found in other secreted proteins, are conserved between human and murine PAI-2 (BELIN et al. 1989); it thus seems unlikely that PAI-2 is simply a secreted protein which is translocated with poor efficiency, and a possible role for the bitopological distribution of this powerful protease inhibitor will be discussed below.

The production of PAI-2 varies dramatically between different mouse macrophage populations (WOHLWEND et al. 1987b). While PAI-2 is essentially undetectable in cultures of bone marrow-derived macrophages, it is abundant in resident peritoneal macrophages. Peritoneal exudate macrophages induced by injection of thioglycollate broth contain very little PAI-2; the inhibitor detected in the latter cultures is probably that produced by a preexisting subpopulation of "resident" macrophages. These differences in PAI-2 production can be accounted for by remarkably different rates of transcription of the PAI-2 gene (Fig. 2) and steady state levels of PAI-2 mRNA (Fig. 3); in fact, of those genes whose transcription rates we have compared, PAI-2 is that which differs the most between the resident and the inflammatory exudate macrophage populations (Fig. 2).

In vitro studies have identified some of the agents which may be responsible for controlling macrophage PAI-2 production (Table 1). Culture of peritoneal macrophages in the presence of LPS (CHAPMAN et al. 1982), of the phorbol ester phorbol myristate acetate (PMA), of agents which raise intracellular cAMP levels, or of M-CSF (WOHLWEND et al. 1987b) results in increased PAI-2 production, and in correspondingly increased steady state levels of PAI-2 mRNA (Fig. 4). It remains to be determined whether these changes in PAI-2 mRNA levels are due to changes in transcription of the gene or in message stability; PAI-2 mRNA contains in its 3′ untranslated region structural determinants (AU-rich sequences) which have been shown to be responsible for instability of other mRNAs (SHAW and KAMEN 1986). Interestingly, in the context of the overall regulation of the macrophage PA system, IFN-γ, which increases uPA production, does not affect PAI-2 while glucocorticoids, which decrease the production of uPA, also decrease that of PAI-2 (WOHLWEND et al. 1987b).

Less is known about modulation of PAI-2 production in human monocytes/macrophages. PMA (GENTON et al. 1987; SCHLEUNING et al. 1987; WEBB et al. 1987; WOHLWEND et al. 1987a; YE et al. 1987), LPS (CHAPMAN and STONE 1985b; SCHWARTZ et al. 1988), muramyl dipeptide (STEPHENS et al. 1985), and infection by dengue virus (KRISHNAMURTHI et al. 1989) increase PAI-2 levels in monocytes or related cell lines, while 1,25-dihydroxyvitamin D_3 decreases PAI-2 production by U937 cells (GYETKO et al. 1988). Mononuclear phagocytes cultured from blood, peritoneal cavity, bone marrow, or alveolar lavage, but not those prepared from the colonic mucosa, have been shown to produce PAI-2 (CHAPMAN and STONE 1985b; STEPHENS et al. 1985).

While the available evidence does not allow a complete understanding of the role of PAI-2 in the biology of mononuclear phagocytes, it is reasonable to speculate that it may be required to control uPA-catalyzed proteolysis at certain stages of the inflammatory reaction. Monocyte uPA is thought to be necessary for migration of these cells from the blood into the tissues, and it may perhaps also play a part in removal of fibrin clots. However, fibrin also constitutes a transitory scaffold that helps the reconstitution of an appropriate extracellular matrix following tissue damage, and premature or excessive fibrin degradation would clearly be detrimental to the healing process. Once in the tissue, perhaps under the influence of locally produced cytokines such as M-CSF, monocytes/macrophages could start producing PAI-2 and thereby contribute to the delicate protease–antiprotease balance within the extracellular matrix. In this view, release of the large stores of cytosolic PAI-2 could play an important part in limiting plasmin-mediated tissue destruction, even if the local conditions lead to some macrophage lethality.

5 A Plasma Membrane Binding Site for Urokinase-type Plasminogen Activator

The elucidation of the primary structure of uPA (GÜNZLER et al. 1982) revealed an unexpected characteristic of this protein: the presence, in its N-terminal region, of a domain with substantial homology to epidermal growth factor (EGF) and transforming growth factor α (TGFα). This suggested that this growth factor-like domain might dictate the binding of uPA to a receptor site on the surface of certain cells, and that, if this was the case, uPA could function as a cell surface-associated enzyme. Exploration of this hypothesis led to the discovery of a binding site for the M_r 55 000 form of uPA on human monocytes and monocyte-like U937 cells (VASSALLI et al. 1985; STOPPELLI et al. 1985). A K_d of approximately 10^{-10} M, i.e., in the range of the concentration of pro-uPA

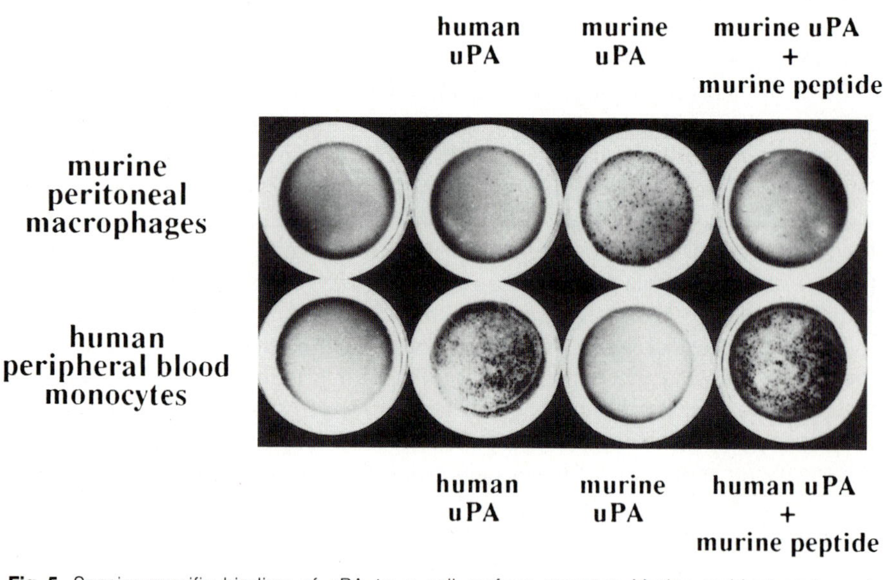

Fig. 5. Species-specific binding of uPA to a cell surface receptor. Murine resident peritoneal macrophages (WOHLWEND et al. 1987b) and human peripheral blood monocytes (VASSALLI et al. 1985) were prepared and plated at low cell density as described. Where indicated, cells were preincubated in the presence of a synthetic peptide corresponding to part of the growth factor-like domain of murine uPA (BELIN et al. 1985; APPELLA et al. 1987) (10^{-6} M, 15 min). They were then incubated for 45 min in the presence of equivalent catalytic amounts of human or murine uPA (0.6 Ploug unit/ml, approximately 10^{-10} M), washed, and analyzed for PA activity as for Fig. 1

in blood plasma (TISSOT et al. 1982; WUN et al. 1982b), and a number of 60000 sites per cell were calculated for U937 cells. This receptor was found to be specific for uPA, in that other proteins with similar domains, including tPA and EGF, did not compete for uPA binding (VASSALLI et al. 1985).

The uPA receptor can bind both the active enzyme and its zymogen pro-uPA (CUBELLIS et al. 1986). Binding occurs through the noncatalytic A chain of the enzyme (VASSALLI et al. 1985; STOPPELLI et al. 1985), and, in confirmation of the above-mentioned hypothesis, involves the growth factor-like domain of the molecule (APPELLA et al. 1987). Bound pro-uPA can be activated by plasmin (CUBELLIS et al. 1986), and the bound enzyme is catalytically active (VASSALLI et al. 1985) (Fig. 5). Thus, while pro-uPA is a secreted protein, its subsequent binding to a high affinity cell surface receptor localizes the enzyme to the plasma membrane; this probably accounts for previous reports describing membrane-associated forms of uPA (SOLOMON et al. 1980; CHAPMAN et al. 1982; LEMAIRE et al. 1983). Binding of uPA clearly contributes to limit the activity of the uPA-plasmin system to the close environment of the cell. It is clear that such a proteolytic system could operate to catalyze the focal lysis of extracellular substrates, in an optimal configuration to facilitate cell migration. In this context, it is interesting to note that immunochemical studies on human fibroblasts and fibrosarcoma cell cultures have demonstrated the presence of uPA at sites of attachment of the cells to the substratum, and its codistribution with the cytoskeletal component vinculin (PÖLLÄNEN et al. 1988; HÉBERT and BAKER 1988).

The biochemical characterization of the uPA receptor has shown that it behaves as an integral membrane protein (ESTREICHER et al. 1989) and that it comprises at least one carbohydrate-containing M_r 55000 polypeptide chain (NIELSEN et al. 1988). Binding, detergent partitioning, and chemical cross-linking studies have revealed the presence of uPA receptors with similar properties on human cells other than mononuclear phagocytes, for instance fibroblasts, polymorphonuclear leukocytes, or endothelial cells (BAJPAI and BAKER 1985; MILES and PLOW 1987; MILES et al. 1988). Studies in other species have been hampered by the species specificity of uPA binding (ESTREICHER et al 1989) and the limited availability of purified homologous uPAs. However, utilizing a catalytic assay, the presence of a receptor for murine uPA on mouse peritoneal macrophages and the species specificity of the interaction can be demonstrated (Fig. 5): addition of homologous uPA to human and murine monocytes/macrophages results in their acquisition of cell-associated PA activity, while the heterologous enzyme does not bind. The specificity is further demonstrated by the use of a synthetic peptide corresponding to a part of the growth factor-like domain of mouse uPA [mouse Ala^{20}-uPA(13-33)] (APPELLA et al. 1987). This peptide markedly inhibits the binding of murine uPA to murine macrophages, but it does not affect the binding of human uPA to human cells. The observed species specificity of binding can probably be explained by the structural differences between human and murine uPAs within the growth factor-like domains of the molecules (ESTREICHER et al. 1989).

In addition to localizing plasminogen activation to the immediate vicinity of the cell surface, binding of uPA (or pro-uPA) to its plasma membrane receptor could change the catalytic specificity or efficiency of the enzyme, or convert the single-chain protein to an active enzyme. Such studies are not easy to perform, since they require that the activity of soluble and immobilized molecules be compared in quantitative terms. Despite these difficulties, a 16-fold acceleration of the activation of cell-bound plasminogen (see below) by cell-bound pro-uPA was observed in cultures of monocyte-like U937 cells (ELLIS et al. 1989). This could be accounted for by an increase in the rate of feedback activation of pro-uPA by cell-bound plasmin. Since such a potentiation was not observed in the presence of 6-aminohexanoic acid, which prevents cellular binding of plasminogen, or of the amino-terminal fragment of uPA, which prevents binding of pro-uPA, it appears that binding does not alter the activity or specificity of the individual molecules, but rather acts by increasing their rate of reaction, probably through a receptor-mediated concentration effect on the cell surface. This interesting study illustrates how the assembly of the components of the PA-plasmin system on the plasma membrane, through binding to their respective receptors, could dramatically favor proteolysis in the close cellular environment.

It has previously been suggested that receptor-bound uPA may be protected from rapid inhibition by antiproteases (BLASI et al. 1987). However, recent studies have shown that the rates of uPA inactivation by PAI-1 (CUBELLIS et al. 1989) or PAI-2 (KIRCHHEIMER and REMOLD 1989; ESTREICHER et al., submitted for publication)[1] (for footnote see p. 79) are not markedly different whether the enzyme is bound or free in solution. Thus, the receptor does not shield uPA from its specific inhibitors, and the controlled production and secretion of PAI-2 can play an important part in modulating plasminogen activation at the cell surface. Nevertheless, it will be of interest to compare the rates of plasmin formation in the presence and absence of PAIs, using the U937 cell system with cell-bound and soluble zymogens, as described above.

The affinity, density, and distribution of the uPA receptor are all subject to modulation. Differentiation of U937 monocyte-like cells in response to PMA leads to a marked increase in receptor number and a decrease in binding affinity (STOPPELLI et al. 1985; PICONE et al. 1989). A comparable modulation has been reported for HeLa cells exposed to PMA or to the growth factor EGF (ESTREICHER et al. 1989). The mechanisms responsible for these changes have not been elucidated, although a change in the extent of receptor glycosylation appears to accompany the change in affinity (PICONE et al. 1989). The biologic relevance of these changes is not clear. The K_d value for the high affinity state of the receptor is close to the concentration of pro-uPA in plasma; at this concentration, the net result of the PMA effect would be only a small increase in receptor-bound uPA. Higher concentrations of uPA may prevail in the close environment of uPA-producing cells in tissues; under these conditions, a large increase in the amount of membrane-bound uPA could be achieved. Other studies have reported an increase in uPA receptor number, with no change in binding affinity, for U937 cells (LU et al. 1988) and peripheral blood monocytes

(KIRCHHEIMER et al. 1988) exposed to IFN-γ or to IFN-γ and TNF-α, respectively. Finally, the polarization of uPA receptor distribution on monocytes and U937 cells placed in a chemotactic gradient (ESTREICHER et al., submitted for publication)[1] suggests an additional dimension in the modulation of receptor expression and provides further evidence in favor of the hypothesis that the receptor can serve to focus plasmin generation to the leading edge of migrating cells.

6 Binding of Plasminogen to the Cell Surface

The discovery of plasminogen-binding sites on cells that also bear uPA receptors, including human monocytes and U937 cells, has added a new dimension to the concept of a cell surface system of PA-catalyzed extracellular proteolysis (HAJJAR et al. 1986; PLOW et al. 1986; MILES et al. 1988). The K_d value for plasminogen binding is approximately 10^{-6} M, i.e., close to its plasma concentration (2×10^{-6} M), and a large fraction of the 200000 plasminogen receptors on circulating monocytes should thus be occupied. Plasminogen binding can be inhibited by 6-aminohexanoic acid, indicating that the lysine-binding sites present on the plasminogen kringles may be involved. The nature of the plasminogen receptor has not been elucidated, but recent data suggest a possible role for gangliosides (MILES et al. 1989).

Studies summarized above (ELLIS et al. 1989) have shown that coexpression of uPA and plasminogen receptors on the same cells facilitates pro-uPA activation, thus significantly accelerating the generation of plasmin. Most importantly perhaps, like fibrin-bound plasmin, cell-bound plasmin appears to be protected from inhibition by α_2-antiplasmin, a highly effective inhibitor of soluble plasmin. Although the precise mechanism of this protection is not known, it may rely on the fact that the lysine-binding sites of cell surface-bound plasmin are occupied; indeed, the interaction of soluble plasmin with α_2-antiplasmin involves both the lysine-binding sites on the non-catalytic part of the enzyme and the active site. In any event, it is clear that the combined effects of accelerated plasmin generation and prolonged plasmin action through resistence to inhibition should result in a highly effective cell surface-bound proteolytic system (STEPHENS et al. 1989). Future studies of plasminogen-binding sites on other cells of the mononuclear phagocyte lineage, such as murine peritoneal macrophages, should be of considerable interest in further evaluating the role of this cell surface catalytic cascade in extracellular proteolysis.

[1] *Note added in proof*: The paper by Estreicher et al. that had been quoted as "submitted for publication" has been published. Estreicher A, Muhlhauser J, Carpentier J-L, Orci L, Vassalli J-D (1990) The receptor for urokinase-type plasminogen activator polarizes expression of the protease to the leading edge of migrating monocytes and promotes degradation of enzyme inhibitor complexes. J Cell Biol 111: 783–792

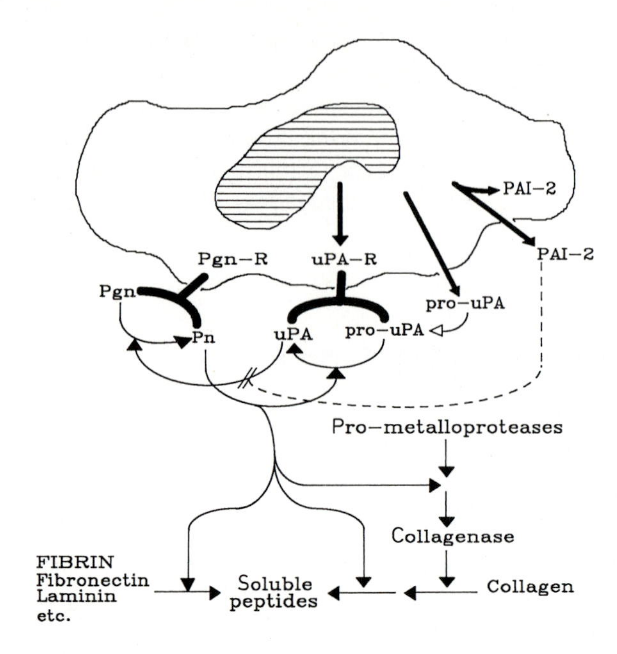

Fig. 6. Schematic representation of the PA-plasmin system on the cell surface of monocytes/macrophages. *Pgn*, plasminogen; *Pn*, plasmin; *Pgn-R*, plasminogen-binding site; *uPA-R*, uPA receptor

7 A Powerful Cell Surface Proteolytic System

The large body of information that has been summarized here converges to indicate that uPA could be a key determinant of a very powerful and highly regulated proteolytic system at the surface of mononuclear phagocytes, at certain stages of the inflammatory process. Plasmin generated at the level of the monocyte/macrophage plasma membrane could catalyze the degradation of fibrin, and also, either directly or indirectly through the activation of zymogens to the metalloproteases, that of collagen, elastin, and other components of extracellular matrices such as laminin and fibronectin (Fig. 6) (CHAPMAN et al. 1984). It is striking that uPA production has been associated with a number of situations where cell migration is required (trophoblast cells during embryo implantation, keratinocytes in reepithelializing wounds, endothelial cells during angiogenesis, malignant cells) (SAPPINO et al. 1989; GRØNDAHL-HANSEN et al. 1988; PEPPER et al. 1987; MIGNATTI et al. 1989; DANØ et al. 1985). Moreover, a direct role for uPA activity has been demonstrated in extracellular matrix invasion by tumor and endothelial cells (MIGNATTI et al. 1986, 1989; OSSOWSKI 1988a), and receptor-bound uPA has been shown to enhance tumor cell invasiveness (OSSOWSKI 1988b). Hence, although this has not as yet been directly tested, it is reasonable to envision that the PA-plasmin system could be essential for the migration of monocytes and macrophages.

Acknowledgments. The work from our laboratory was supported by grants from the Fonds National Suisse de la Recherche Scientifique, the Sir Jules Thorn Charitable Trust, and the

Commission Fédérale des Maladies Rhumatismales. We thank all of our colleagues who have participated in certain aspects of this work, and in particular D. Baccino, N. Busso, M.A. Collart, J.-M. Dayer, A. Estreicher, J. Huarte, M. Pepper, and P. Vassalli. The technical assistance of C. Combépine, P. Gubler, G. Marchi, V. Monney, and F. Silva and the photographic work of J.-P. Gerber are gratefully acknowledged.

Literature surveyed up to January 1990.

References

Antalis TM, Clark MA, Barnes T et al. (1988) Cloning and expression of a cDNA coding for a human monocyte-derived plasminogen activator inhibitor. Proc Natl Acad Sci USA 85: 985–989

Appella E, Robinson EA, Ullrich SJ, Stoppelli MP, Corti A, Cassani G, Blasi F (1987) The receptor-binding sequence of urokinase: a biological function for the growth-factor module of proteases. J Biol Chem 262: 4437–4440

Åstedt B, Hägerstrand I, Lecander I, (1986) Cellular localisation in placenta of placental type plasminogen activator inhibitor. Thromb Haemost 56: 63–65

Bajpai A, Baker JB (1985) Cryptic urokinase binding sites on human foreskin fibroblasts. Biochem Biophys Res Commun 133: 475–482

Baker JB, Low DA, Simmer RL, Cummingham DD (1980) Proteasenexin: a cellular component that links thrombin and plasminogen activator and mediates their binding to cells. Cell 21: 37–45

Barnathan ES, Kuo A, Van der Keyl H, McCrae, KR, Larsen GR, Cines DB (1988) Tissue-type plasminogen activator binding to human endothelial cells. J Biol Chem 263: 7792–7799

Belin D, Vassalli, J-D, Combépine C, et al. (1985) Cloning, nucleotide sequencing and expression of cDNAs encoding mouse urokinase-type plasminogen activator. Eur J Biochem 148: 225–232

Belin D, Wohlwend A, Schleuning W-D, Kruithof EKO, Vassalli J-D (1989) Facultative polypeptide translocation allows a single mRNA to encode the secreted and cytosolic forms of plasminogen activators inhibitor 2. EMBO J 8: 3287–3294

Blasi F, Vassalli J-D, Danø K (1987) Urokinase-type plasminogen activator: proenzyme, receptor, and inhibitors. J Cell Biol 104: 801–804

Busso N, Belin D, Failly-Crépin C, Vassalli J-D (1986) Plasminogen activators and their inhibitors in a human mammary cell line (HBL-100). J Biol Chem 261: 9309–9315

Canipari R, O'Connell ML, Meyer G, Strickland S (1987) Mouse ovarian granulosa cells produce urokinase-type plasminogen activator, whereas the corresponding rat cells produce tissue-type plasminogen activator. J Cell Biol 105: 977–981

Carrell R, Travis J (1985) α_1-Antitrypsin and the serpins: variation and countervariation. Trends Biochem Sci 10: 20–24

Carlson M, Taussig R, Kustu S, Botstein D (1983) The secreted form of invertase in *Saccharomyces cerevisiae* is synthesized from mRNA encoding a signal sequence. Mol Cell Biol 3: 439–447

Chapman HA Jr, Stone OL (1985a) Characterization of macrophage-derived plasminogen activator inhibitor (similarities with placental urokinase inhibitor). Biochem J 230: 109–116

Chapman HA Jr, Stone OL (1985b) A fibrinolytic inhibitor of human alveolar macrophages. Induction with endotoxin. Am Rev Respir Dis 132: 569–575

Chapman HA Jr, Vavrin Z, Hibbs JB Jr, (1982) Macrophage fibrinolytic activity: identification of two pathways of plasmin formation by intact cells and of a plasminogen activator inhibitor. Cell 28: 653–662

Chapman HA Jr, Stone OL, Vavrin Z (1984) Degradation of fibrin and elastin by intact human alveolar macrophages in vitro. Characterization of a plasminogen activator and its role in matrix degradation. J Clin Invest 73: 806–815

Collart MA, Belin D, Vassalli J-D, De Kossodo S, Vassalli P (1986) Gamma interferon enhances macrophage transcription of the tumor necrosis factor/cachectin, interleukin-1 and urokinase genes, which are controlled by short lived repressors. J Exp Med 164: 2113–2118

Collart MA, Belin D, Vassalli J-D, Vassalli P (1987) Modulations of functional activity in differentiated macrophages are accompanied by early and transient increase or decrease in *c-fos* gene transcription. J Immunol 139: 949–955

Cubellis MV, Nolli ML, Cassani G, Blasi F (1986) Binding of single-chain prourokinase to the urokinase receptor of human U937 cells. J Biol Chem 261: 15819–15822
Cubellis MV, Andreasen P, Ragno P, Mayer M, Danø K, Blasi F (1989) Accessibility of receptor-bound urokinase to type -1 plasminogen activator inhibitor. Proc Natl Acad Sci USA 86: 4828–4832
Danø K, Andreasen PA, Grøndahl-Hansen J, Kristensen P, Nielsen LS, Skriver L (1985) Plasminogen activators, tissue degradation, and cancer. Adv Cancer Res 44: 139–266
Ellis V, Scully MF, Kakkar VV (1989) Plasminogen activation initiated by single-chain urokinase-type plasminogen activator: potentiation by U937 cells. J Biol Chem 264: 285–2188
Estreicher A, Wohlwend A, Belin D, Schleuning W-D, Vassalli J-D (1989) Characterization of the cellular binding site for the urokinase-type plasminogen activator. J Biol Chem 264: 1180–1189
Falcone DJ (1989) Heparin stimulation of plasminogen activator secretion by macrophage-like cell line RAW264.7: role of the scavenger receptor. J Cell Physiol 140: 219–226
Falcone DJ, Ferenc MJ (1988) Acetyl-LDL stimulates macrophage-dependent plasminogen activation and degradation of extracellular matrix. J Cell Physiol 135: 387–396
Feinberg RF, Kao LC, Haimowitz JE, Queenan, JT Jr, Wun T-C, Strauss JF III, Kliman HJ (1989) Plasminogen activator inhibitor types 1 and 2 in human trophoblasts. PAI-1 is an immunocytochemical marker of invading trophoblasts. Lab Invest 61: 20–26
Ganter U, Bauer J, Majello B, Gerok W, Ciliberto G (1989) Characterization of mononuclear-phagocyte terminal maturation by mRNA phenotyping using a set of cloned cDNA probes. Eur J Biochem 185: 291–296
Genton C, Kruithof EKO, Schleuning W-D (1987) Phorbol ester induces the biosynthesis of glycosylated and nonglycosylated plasminogen activator inhibitor 2 in high excess over urokinase-type plasminogen activator in human U-937 lymphoma cells. J Cell Biol 104: 705–712.
Gloor S, Odnik K, Guenther J, Nick H, Monard D (1986) A glia-derived neurite promoting factor with protease inhibitory activity belongs to the protease nexins. Cell 47: 687–693
Gold LI, Schwimmer R, Quigley JP (1989) Human plasma fibronectin as a substrate for human urokinase. Biochem J 262: 529–534
Golder JP, Stephens RW (1983) Minactivin: a human monocyte product which specifically inactivates urokinase-type plasminogen activators. Eur J Biochem 136: 517–522
Gordon S, Cohn ZA (1978) Bacille Calmette-Guérin in the mouse. Regulation of macrophage plasminogen activator by T lymphocytes and specific antigen. J Exp Med 147: 1175–1188
Gordon S, Unkeless JC, Cohn ZA (1974) Induction of macrophage plasminogen activator by endotoxin stimulation and phagocytosis, evidence for a two-stage process. J Exp Med 140: 995–1010
Grøndahl-Hansen J, Lund LR, Ralfkiaer E, Ottevanger V, Danø K (1988) Urokinase- and tissue-type plasminogen activators in keratinocytes during woud reepithelialization in vivo. J Invest Dermatol 90: 790–795
Günzler WA, Steffens GJ, Otting F, Kim S-MA, Frankus E, Flohé L (1982) The primary structure of high molecular mass urokinase from human rine. The complete amino acid sequence of the A chain. Hoppe-Seyler's Z. Physiol Chem 363: 1155–1165
Gyetko MR, Webb AC, Sitrin RG (1988) Modulation of urokinase-type plasminogen activator and plasminogen activator inhibitor-2 expression by U-937 mononuclear phagocytes. Effects of 1α, 25-dihydroxyvitamin D_3 and phorbol ester. J Immunol 141: 2693–2698
Hajjar KA, Harpel PC, Jaffe EA, Nachman RL (1986) Binding of plasminogen to human endothelial cells. J Biol Chem 261: 11656–11662
Hajjar KA, Hamel NM, Harpel PC, Nachman RL (1987) Binding of tissue plasminogen activator to cultured human endothelial cells. J Clin Invest 80: 1712–1719
Hamilton JA, Vassalli JD, Reich E (1976) Macrophage plasminogen activator: induction by asbestos is blocked by anti-inflammatory sterioids. J Exp Med 144: 1689
Hamilton JA, Stanley ER, Burgess AW, Shadduck AA (1980) Stimulation of macrophage plasminogen activator activity by colony-stimulating factors. J Cell Physiol 103: 435–445
Hart PH, Vitti GF, Burgess DR, Singleton DK, Hamilton JA (1989a) Human monocytes can produce tissue-type plasminogen activator. J Exp Med 169: 1509–1514
Hart PH, Burgess DR, Vitti GF, Hamilton JA (1989b) Interleukin-4 stimulates human monocytes to produce tissue-type plasminogen activator. Blood 74: 1222–1225
Hashimoto K, Wun TC, Baird J, Lazarus GS, Jensen PJ (1989) Characterization of keratinocyte plasminogen activator inhibitors and demonstration of the prevention of pemphigus IgG-induced acantholysis by a purified plasminogen activator inhibitor. J Invest Dermatol 92: 310–314

Hébert CA, Baker JB (1988) Linkage of extracellular plasminogen activator to the fibroblast cytoskeleton: colocalization of cell surface urokinase with vinculin. J Cell Biol 106: 1241–1247

Holmberg L, Lecander I, Persson B, Astedt B (1978) An inhibitor from placenta specifically binds urokinase and inhibits plasminogen activator released from ovarian carcinoma in tissue culture. Biochim Biophys Acta 544: 128

Hoylaerts M, Rijken DC, Lijnen HR, Collen D (1982) Kinetics of the activation of plasminogen by human tissue plasminogen activator. Role of fibrin. J Biol Chem 257: 2912–2919

Huarte J, Belin D, Bosco A-P, Sappino A-P, Vassalli J-D (1987) Plasminogen activator and mouse supermatozoa: urokinase synthesis in the male genital tract and binding of the enzyme to the sperm cell surface. J Cell Biol 104: 1281–1289

Huber R, Carrell RW, (1989) Implications of the three-dimensional structure of α_1-antitrypsin for structure and function of serpins. Biochemistry 28: 8951–8966

Hume DA, Gordon S (1984) The correlation between plasminogen activator activity and thymidine incorporation in mouse bone marrow-derived macrophages. Exp Cell Res 150: 347–355

Ichinose A, Fujikawa K, Suyama T (1986) The activation of pro-urokinase by plasma kallikrein and its inactivation by thrombin. J Biol Chem 261: 3486–3489

Kawano T, Morimoto K, Uemura Y (1968) Urokinase inhibitor in human placenta. Nature 217: 253–255

Kirchheimer JC, Remold HG (1989) Functional characteristics of receptor-bound urokinase on human monocytes: catalytic efficiency and susceptibility to inactivation by plasminogen activator inhibitors. Blood 74: 1396–1402

Kirchheimer JC, Nong Y-H, Remold HG (1988) IFN-γ, tumor necrosis factor-α, and urokinase regulate the expression of urokinase receptors on human monocytes. J Immunol 141: 4229–4234

Klimetzek V, Sorg C (1977) Lymphokine-induced secretion of plasminogen activator by murine macrophages. Eur J Immunol 7: 185–187

Klimetzek V, Sorg C (1979) The production of fibrinolysis inhibitors as a parameter of the activation state in murine macrophages. Eur J Immunol 9: 613– 619

Kopitar M, Rozman B, Babnik J, Turk V, Mullins DE, Wun T-C (1985) Human leucocyte urokinase inhibitor. Purification, characterization and comparative studies against different plasminogen activators. Thromb Haemost 54: 750

Krishnamurti C, Wahl LM, Alving BM (1989) Stimulation of plasminogen activator inhibitor activity in human monocytes infected with dengue virus. Am J Trop Med Hyg 40: 102–107

Kruithof EKO, Vassalli J-D, Schleuning W-D, Mettaliano RJ, Bachmann F (1986) Purification and characterization of a plasminogen activator inhibitor from the histiocytic lymphoma cell line U-937. J Biol Chem 261: 11207–11213

Kwiatkowski DJ, Mehl R, Yin HL (1988) Genomic organization and biosynthesis of secreted and cytoplasmic forms of gelsolin. J Cell Biol 106: 375–384

Larsson LI, Skriver LS, Nielsen LS, Grondahl-Hansen J, Kristensen P, Danø K (1984) Distribution of urokinase-type plasminogen activator immunoreactivity in the mouse. J Cell Biol 98: 894–903

Lemaire G, Drapier JC, Petit JF (1983) Importance, localization, and functional properties of the cell-associated form of plasminogen activator in mouse peritoneal macrophages. Biochim Biophys Acta 755: 332–343

Lison D, Knoops B, Collette C, Lauwerys R (1989) Comparison of the effects of auranofin and retinoic acid on plasminogen activator activity of peritoneal macrophages and Lewis lung carcinoma cells. Biochem Pharmacol 38: 2107–2112

Loskutoff DJ, Van Mourik JA, Erickson LA, Lawrence D (1983) Detection of an unusually stable fibrinolytic inhibitor produced by bovine endothelial cells. Proc Natl Acad Sci USA 80: 2956–2960

Lu H, Misrhahi MC, Krief P et al. (1988) Parallel induction of fibrinolysis and receptors for plasminogen and urokinase by interferon gamma on U937 cells. Biochem Biophys Res Commun 155: 418–422

Lund LR, Georg B, Nielsen LS, Mayer M, Danø K, Andreasen PA (1988) Plasminogen activator inhibitor type 1: cell-specific and differentiation-induced expression and regulation in human cell lines, as determined by enzyme-linked immunosorbent assay. Mol Cell Endocrinol 60: 43–53

Mignatti P, Robbins E, Rifkin DB (1986) Tumor invasion through the human amniotic membrane: requirement for a proteinase cascade. Cell 47: 487–498

Mignatti P, Tsuboi R, Robbins E, Rifkin DB (1989) In vitro angiogenesis on the human amniotic membrane: requirements for basic fibroblast growth factor-induced proteases. J Cell Biol 108: 671–682

Miles LA, Plow EF (1987) Receptor mediated binding of the fibrinolytic components, plasminogen and urokinase, to peripheral blood cells. Thromb Haemost 58: 936–942
Miles LA, Levin EG, Plescia J, Collen D, Plow EF (1988) Plasminogen receptors urokinase receptors, and their modulation on human endothelial cells. Blood 72: 628–635
Miles LA, Dahlberg CM, Levin EG, Plow EF (1989) Gangliosides interact directly with plasminogen and urokinase and may mediate binding of these fibrinolytic components to cells. Biochemistry 28: 9337–9343
Moscatelli D, Rifkin DB (1988) Membrane and matrix localization of proteinases: a common theme in tumor cell invasion and angiogenesis. Biochim Biophys Acta 948: 67–85
Nielsen LS, Kellerman GM, Behrendt N, Picone R, Danø K, Blasi F (1988) A 55,000-60,000 Mr receptor protein for urokinase-type plasminogen activator: identification in human tumor cell lines and partial purification. J Biol Chem 263: 2358–2363
Nogueira N, Gordon S, Cohn ZA (1977) *Tryanosoma cruzi*: the immunological induction of macrophage plasminogen activator requires thymus-derived lymphocytes. J Exp Med 146: 172–183
Ossowski L (1988a) Plasminogen activator dependent pathways in the dissemination of human tumor cells in the chick embryo. Cell 52: 321–328
Ossowski L (1988b) In vivo invasion of modified chorioallantoic membrane by tumer cells: the role of cell surface-bound urokinase. J Cell Biol 107: 2437–2445
Pepper MS, Vassalli J-D, Montesano R, Orci L (1987) Urokinase-type plasminogen activator is induced in migrating capillary endothelial cells. J Cell Biol 105: 2535–2541
Petersen LC, Lund LR, Nielsen LS, Danø K, Skriver L (1988) One-chain urokinase-type plasminogen activator from human sarcoma cells is a proenzyme with little or no intrinsic activity. J Biol Chem 263: 11189–11195
Picone R, Kajtaniak EL, Nielsen LS et al. (1989) Regulation of urokinase receptors in monocytelike U937 cells by phorbol ester phorbol myristate acetate. J Cell Biol 108: 693–702
Plow EF, Freaney DE, Plescia J, Miles LA (1986) The plasminogen system and cell surfaces: evidence for plasminogen and urokinase receptors on the same cell type. J Cell Biol 103: 2411–2420
Pöllänen J, Hedman K, Nielsen LS, Danø K, Vaheri A (1988) Ultrastructural localization of plasma membrane-associated urokinse-type plasminogen activator at focal contacts. J Cell Biol 106: 87–95
Potempa J, Dubin A, Watorek W, Travis J (1988) An elastase inhibitor from equine leukocyte cytosol belongs to the serpin superfamily. Further characterization and amino acid sequence of the reactive center. J Biol Chem 263: 7364–7369
Reich E (1978) Activation of plasminogen; a general mechanism for producing localized extracellular proteolysis. In: Berlin RD, Herrman M, Lepow IH (eds) Molecular basis of biological degradative process. Academic New York, pp 155–169
Remold-O'Donnell E, Nixon JC, Rose RM (1989) Elastase inhibitor. Characterization of the human elastase inhibitor molecule associated with monocytes, macrophages, and neutrophils. J Exp Med 169: 1071–1086
Saksela O (1985) Plasminogen activation and regulation of pericellular proteolysis. Biochim Biophys Acta 823: 35–65
Saksela O, Rifkin DB (1988) Cell-associated plasminogen activation: regulation and physiological functions. Annu Rev Cell Biol 4: 93–126
Saksela O, Hovi T, Vaheri A. (1985) Urikinase-type plasminogen activator and its inhibitor secreted by cultured human monocyte-macrophages. J Cell Physiol 122: 125–132
Salonen EM, Zitting A, Vaheri A (1984) Laminin interacts with plasminogen and its tissue-type plasminogen activator. FEBS Lett 172: 29–32
Salonen EM, Saksela O, Vartio T, Vaheri A, Nielsen LS, Zeuthen J (1985) Plasminogen and tissue-type plasminogen activator bind to immobilized fibronectin. J Biol Chem 260: 12302–12307
Sappino A-P, Huarte J, Belin D, Vassalli J-D (1989) Plasminogen activators in tissue remodeling invasion: mRNA localization in mouse ovaries and implanting embryos. J Cell Biol 109: 2471–2479
Schleuning W-D, Medcalf RL, Hession C, Rothenbuhler R, Shaw A, Kruithof EKO (1987) Plasminogen activator inhibitor 2: regulation of gene transcription during phorbol ester-mediated differentiation of U-937 human histiocytic lymphoma cells. Mol Cell Biol 7: 4564–4567
Schnyder J, Baggiolini M. (1978) Role of phagocytosis in the activation of macrophages. J Exp Med 148: 1449–1457
Schwartz BS, Bradshaw JD (1989) Differential regulation of tissue factor and plasminogen activator inhibitor by human mononuclear cells. Blood 74: 1644–1650

Schwartz BS, Monroe MC, Levin EG (1988) Increased release of plasminogen activator inhibitor type 2 accompanies the human mononuclear cell tissue factor response to lipopolysaccharide. Blood 71: 734–741

Shaw G, Kamen R (1986) A conserved AU sequence from the 3′ untranslated region of GM-CSF mRNA mediates selective mRNA degradation. Cell 46: 659–667

Singer II, Scott S, Hall GL, Limjuco G, Chin J, Schmidt JA (1988) Interleukin-1β is localized in the cytoplasmic ground substance but is largely absent from the golgi apparatus and plasma membranes of stimulated human monocytes. J Exp Med 167: 389–407

Solomon JA, Chou IN, Schroder EW, Black PH (1980) Evidence for membrane association of plasminogen activator activity in mouse macrophages. Biochem Biophys Res Commun 94: 480–486

Spence AM, Sheppard PC, Davie JR et al. (1989) Regulation of a bifunctional mRNA results in synthesis of secreted and nuclear probasin. Proc Natl Acad Sci USA 86: 7843–7847

Stephens R, Alitalo R, Tapiovaara H, Vaheri A (1988) Production of an active urokinase by leukemia cells: a novel distinction from cell lines of solid tumors. Leuk Res 12: 419–422

Stephens RW, Golder JP, Fayle DRH et al. (1985) Minactivin expression in human monocyte and macrophage populations. Blood 66: 333–337

Stephens RW, Pöllännen J, Tapiovaara H et al. (1989) Activation of pro-urokinase and plasminogen on human sarcoma cells: a proteolytic system with surface-bound reactions. J Cell Biol 108: 1987–1995

Stoppelli MP, Corti A, Soffientini A, Cassani G, Blasi F, Assoian RK (1985) Differentiation-enhanced binding of the amino-terminal fragment of human urokinase plasminogen activator to a specific receptor on U937 monocytes. Proc Natl Acad Sci USA 82: 4939–4943

Takemura R, Werb Z (1984) Secretory products of macrophages and their physiological functions. Am J Physiol 246: C1–C9

Tissot J-D, Schneider P, Hauert J, Ruegg M, Kruithof EKO, Bachmann F (1982) Isolation from plasma of a plasminogen activator indentical to urinary high molecular weight urokinase. J Clin Invest 70: 1320–1323

Unkeless JC, Gordon S, Reich E (1974) Secretion of plasminogen activator by stimulated macrophages. J Exp Med 139: 834–850

Vassalli J-D, Reich E (1977) Macrophage plasminogen activator: induction by products of activated lymphoid cells. J Exp Med 145: 429–437

Vassalli J-D, Hamilton J, Reich E (1976) Macrophage plasminogen activator: modulation of enzyme production by anti-inflammatory steroids, mitotic inhibitors, and cyclic nucleotides. Cell 8: 271–281

Vassalli J-D, Hamilton J, Reich E (1977) Macrophage plasminogen activator: induction by concanavalin A and phorbol myristate acetate. Cell 11: 695–705

Vassalli J-D, Dayer J-M, Wohlwend A, Belin D (1984) Concomitant secretion of prourokinase and of a plasminogen activator-specific inhibitor by cultured human monocytes-macrophages. J Exp Med 159: 1653–1668

Vassalli J-D, Baccino D, Belin D (1985) A cellular binding site for the Mr 55,000 form of the human plasminogen activator, urokinase. J Cell Biol 100: 86–92

Vihko KK, Penttilä TL, Parvinen M, Belin D (1989) Regulation of urokinase- and tissue-type plasminogen activator gene expression in the rat seminiferous epithelium. Mol Endocrinol 3: 52–59

Webb AC, Collins KL, Snydr SE et al. (1987) Human monocyte arg-serpin cDNA. Sequence, chromosomal assignment, and homology to plasminogen activator-inhibitor. J Exp Med 166: 77–94

Weitz JI, Cruickshank MK, Thong B, Leslie B, Levine MN, Ginsberg J, Eckhardt T (1988) Human tissue-type plasminogen activator releases fibrinopeptides A and B from fibrinogen. J Clin Invest 82: 1700–1707

Werb Z, Mainardi CL, Vater CA, Harris ED Jr (1977) Endogenous activation of latent collagenase by rheumatoid synovial cells. Evidence for a role of plasminogen activator. N Engl J Med 296: 1017–1023

Wilson EL, Francis GE (1987) Differentiation-linked secretion of urokinase and tissue plasminogen activator by normal human hemopoietic cells. J Exp Med 165: 1609–1623

Wilson EL, Jacobs P, Dowdle EB (1983) The secretion of plasminogen activators by human myeloid leukemic cells in virtro. Blood 61: 568–574

Wohlwend A, Belin D, Vassalli J-D (1987a) Plasminogen activator-specific inhibitors produced by human monocytes-macrophages. J Exp Med 165: 320–339

Wohlwend A, Belin D, Vassalli J-D (1987b) Plasminogen activator-specific inhibitors in mouse macrophages: in vivo and in vitro modulation of their synthesis and secretion. J Immunol 139: 1278–1284

Wun T-C, Ossowski L, Reich E (1982a) A proenzyme form of human urokinase. J Biol Chem 257: 7262–7268

Wun T-C, Schleuning W-D, Reich E (1982b) Isolation and characterization of urokinase from human plasma. J Biol Chem 257: 3276–3283

Ye RD, Wun TC, Sadler JE (1987) cDNA cloning and expression in *Escherichia coli* of a plasminogen activator inhibitor from human placenta. J Biol Chem 262: 3718–3725

Ye RD, Wun TC, Sadler JE (1988) Mammalian protein secretion without signal peptide removal. Biosynthesis of plasminogen activator inhibitor-2 in U-937 cells. J Biol Chem 263: 4869–4875

Ye RD, Ahern SM, Le Beau MM, Lebo RV, Sadler JE (1989) Structure of the gene for human plasminogen activator inhibitor-2. The nearest mammalian homologue of chicken ovalbumin. J Biol Chem 264: 5495–5502

Macrophage-Derived Growth Factors

D. A. RAPPOLEE* and Z. WERB

1 Introduction

In the early decades of the twentieth century biologists sought to grow cells in culture. Clotted blood was found to contain molecules that accomplished this purpose (CARREL 1912), but only later did biochemists seek to purify these molecules. By the middle of the century, biochemists and biologists sought to explain neonatal eye opening in mice in molecular terms (COHEN 1987; LEVI-MONTALCINI 1987). Each of these goals ultimately led to the isolation of single species of molecules called growth factors by using in vitro or in vivo bioassays for growth and a biochemical algorithm for isolation. Epidermal growth factor

* Current address: Departments of Obstetrics and Gynecology, and Cell, Molecular and Structural Biology, Northwestern University, Chicago, IL 60611, USA
Laboratory of Radiobiology and Environmental Health, University of California, San Francisco, CA 94143-0750, USA

Current Topics in Microbiology and Immunology, Vol. 181

(EGF), nerve growth factor (NGF), platelet-derived growth factor (PDGF), transforming growth factor-β (TGF-β), interleukin-1 (IL-1), and macrophage colony-stimulating factor (M-CSF, or CSF-1) were isolated and directly sequenced or molecularly cloned (based on partial sequences) by these means in the 1970s and early 1980s. The production of transformed foci of cells by introduction of fragments of cloned transcripts or genes from tumors also produced a subclass of oncogenes that turned out to be growth factors [c-*sis*, or PDGF-B chain, and Kaposi's sarcoma-fibroblast growth factor (kFGF, or FGF-4)]. Most recently, the formation of tumors in vivo after random integration of a highly active viral promoter upstream of cellular genes has produced the int-1 and int-2 (also known as FGF-3) growth factors. Finally, after the founding member of a growth factor family is identified with a bioassay, low-stringency cDNA library screens and polymerase chain reaction can be used to complete the family (JAKOWLEW et al. 1988; HEBERT et al. 1990). All growth factors are operationally isolated and defined by their ability to cause growth, but may also act as nonmitogenic inflammatory factors.

Growth factors have a number of hallmarks. First, they are generally secreted and therefore act on nearby cells in a paracrine or autocrine fashion (ROSS and VOGEL 1978). Some growth factors may exit cells slowly (e.g. IL-1, M-CSF) or on cell death (basic and acidic forms of FGF), but all enter the extracellular milieu. Other growth factors have a membrane-bound form (e.g., IL-1, M-CSF, TGF-α, and EGF), but these growth factors also act locally between cells. The second hallmark of growth factors is that they tend to act in localized areas either within a cell (ROSS and VOGEL 1978; ROSS et al. 1986; DEUEL 1987) or within a few cell distances [as seen with NGF in pancreatic innervation (EDWARDS et al. 1989)]. Third, growth factors are generally not stored inside the cell (exceptions are PDGF and TGF-β in the platelet) but are highly inducible at the level of transcription (IL-1), translation [TGF-β, tumor necrosis factor (TNF)-α, IL-1] and post-translational activation (TGF-β). Fourth, since these molecules are extremely powerful, they are tightly regulated at the various levels of production. They are unstable at the transcriptional level because of an AUUUA motif in the 3′ untranslated area of many growth factor transcripts [c-*sis*/PDGF, IL-1, IL-2, and granulocyte-macrophage colony-stimulating factor (GM-CSF) (SHAW and KAMEN 1986)], and at post-translational levels. PDGF and TGF-β are both inactivated by a macrophage product, α_2-macroglobulin (DANIELPOUR and SPORN 1990; ROBERTS and SPORN 1990). In addition, the mediators of the growth factors (e.g., prostaglandins induced by PDGF or IL-1) have been demonstrated to negatively regulate the transcription of the inducing growth factor (KUNKEL et al. 1986; DANIEL et al. 1987). Fifth, growth factors are pleiotropic, acting as secretagogues, chemoattractants, and differentiation factors. Finally, growth factors act on the target cell through a transmembrane receptor. Some growth factors allosterically activate protein kinase activity on the cytosolic end of the receptor and trigger an amplified cascade of events within the target cell that lead to the pleiotropic events described. In some cases the growth factor receptor does not have intrinsic kinase activity but

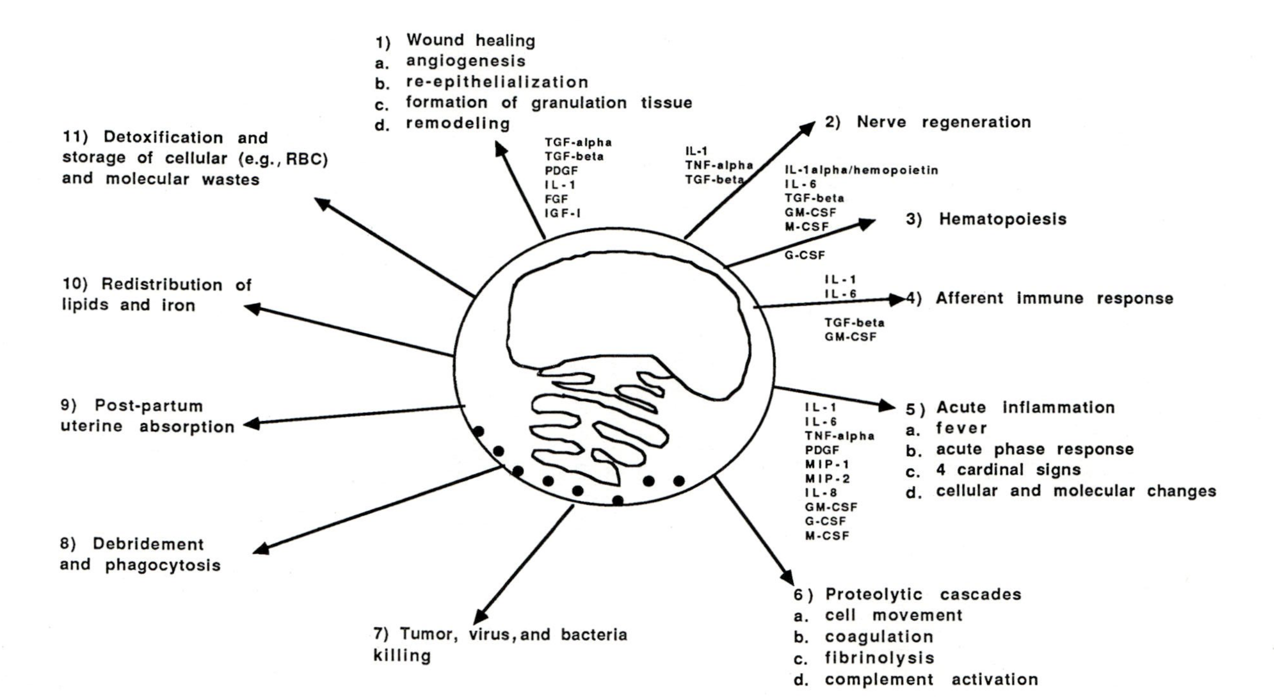

Fig. 1. Growth factors correlated with macrophage functions. Functions 1–5 are discussed in detail in the text

activates other kinases within the cell (e.g., TNF-α, IL-1, IL-6, GM-CSF, IL-2 receptors).

The number of growth factors and growth factor families has grown enormously within the last decade (Tables 1–13). The central interest of intercellular control of growth in abnormal circumstances, such as tumor formation, and in normal embryogenesis and postnatal growth as well as wound healing has led to the discovery of several major families of growth factors: PDGF (which currently has four members), FGF (seven members), EGF (five members), TGF-β (five members in one subgroup, eight in another), insulin-like growth factors (four members), and immediate-response-gene growth factors KC/JE/PF-4 (15 members). There are a number of single-member growth factor "families" (Table 13), but for each founder growth factor, molecular biologic techniques are likely to identify several homologous family members, as they have for several of the growth factor families (JAKOWLEW et al. 1988; HEBERT et al. 1990). It is not clear whether these homologous factors are strong growth inducers or whether they induce other effects, such as inflammation. The current list of 80 growth factors may expand to several hundred in the next decade.

A case in point is the increase in the number of growth factors known to be synthesized by the macrophage. In the early 1980s IL-1 was the first growth factor to be isolated and cloned from macrophages. IL-1 is now known to be produced by many cells in smaller amounts than in macrophages. It is also pleiotropic in its functions (Tables 1–3). Since then, approximately 30 growth factors have been detected in macrophages (Table 4). Only about 15 growth factors have been found that are not synthesized by macrophages. Undoubtedly, these numbers will also increase rapidly in coming years (Tables 5–13).

A challenge for those who study the differentiation, production, and function of macrophages is to understand these processes in terms of the production and function of growth factors. In this chapter we will examine the current knowledge of macrophage function in terms of growth factor production and function. We will first examine the macrophage-derived growth factors as structurally or functionally related families. We will then examine negative modulation of growth factors. Finally, we will examine several specific models that are partially understood: inflammation, wound healing, nerve regeneration, the immune response, and hematopoiesis (Fig. 1).

2 Growth Factors Produced by Macrophages

2.1 Growth Factors Functionally Related to IL-1

Interleukin-1α, IL-1β, TNF-α, IL-6, and leukemia inhibitory factor/differentiation-inhibiting activity (LIF/DIA) are macrophage-derived growth factors with overlapping effects (Table 5). They are functionally but not structurally related. IL-1α and IL-1β were the first growth factors purified, cloned, and sequenced from macrophages (AURON et al. 1984). IL-1α and IL-1β arise from two distinct genes in a variety of cell types and act on two distinct receptors expressed in many cell types (Table 1). The IL-1 receptors exist in high-affinity and low-affinity forms (K_d of 5–10 pM and 200–400 pM, respectively) (DINARELLO 1989), which bind to both IL-1α and IL-1β. It is not clear what functional differences these receptors have. One of these receptors has been cloned and, like the IL-2 and IL-6 receptors, has no intrinsic enzymatic activity but associates with other plasmalemmal proteins to transduce signals. The IL-1 receptor is in the immunoglobulin superfamily (as are the PDGF α/β, M-CSF, and FGF receptors). Its number varies from 200 to 20000 per cell, which is low compared with the 10^5 copies of PDGF and EGF receptor per fibroblast but is similar to the number of GM-CSF receptors per cell (METCALF 1985).

Interleukin-1 has the longest molecular history of macrophage-derived growth factors and a correspondingly large body of knowledge about its endocrine, paracrine and autocrine effects (Tables 2, 3). IL-6 has biologic responses that broadly overlap those of IL-1; its receptor is also widely distributed, has been cloned and resembles the IL-1 receptor in that it lacks enzymatic

Table 1. Regulation of IL-1 (modified from RAPPOLEE and WERB 1988)

Agents stimulating IL-1 production				
IL-1	LPS	LTD_4	Zymosan	Adherence
TGF-β1	LTC_4	Muramyl dipeptide	IFN-α[a]	PMA
TNF-α	C5a	Silica	IFN-γ[a]	Urate crystals
Producer cell types				
Macrophages	Kupffer cells	Keratinocytes	Microglial cells	B cells
Endothelial cells	Fibroblasts	Mesangial cells	Astrocytes	Natural killer cells

[a] GERRARD et al. (1987)
IFN, interferon; LTC, leukotriene C; PMA, phorbol 12-myristate 13-acetate

Table 2. Effects of IL-1 on immune and hematopoietic cells (modified from RAPPOLEE and WERB 1983)

T cells		
Cytotoxic T cell generation	Chemoattractant	Comitogen
IL-2 receptor induction	MEL-14	IL-2 ligand induction
LFA-1 induction		
B cells		
Chemoattractant	Comitogen	Maturation inducer
LFA-1 induction	MEL-14	
Natural killer cells		
Cytotoxicity inducer	IL-2 receptor inducer	
Macrophages		
Cytotoxicity inducer	Reactive oxygen inducer	Chemoattractant
PGE_2 inducer	IFN-β_2 inducer	IL-1 inducer
TNF-α inducer	MIP-2 inducer	Thromboxane β_2 inducer
MIP-1 inducer	IL-8 inducer	JE inducer
IL-1RA inducer		

IL-1RA, IL-1 receptor antagonist

Table 3. Effects of IL-1 on connective tissue and other cells (modified from Rappolee and Werb 1988)

Fibroblast and synovial cells		
PGE_2 inducer	Collagenase inducer	IFN-$\beta 1/\beta 2$ inducer
PI inducer	Hyaluronate inducer	Class II MHC inducer
GM-CSF inducer	IL-1 inducer	
Stromelysin inducer	Proliferation	
PDGF-A chain/mitosis		
Bone		
Resorption		
Cartilage		
Proteoglycan synthesis suppressor		Metalloproteinase inducer
Endothelial cells		
Procoagulant activity inducer	PA suppressor	IL-1 inducer
PAF, PGF_2, PGI_2 inducer	PAI inducer	Adherence of T and B cells, PMNs, macrophages
Antiocoagulant activity suppressor	GM-CSF, G-CSF, M-CSF inducer	
Systemic changes		
Drowsiness	Shock	Glucocorticoid induction
Acute-phase response inducer (C3, factor B, haptoglobin, fibrinogen)	Iron decrease (due to lactoferrin secretion)	Fever (reset hypothalamic set point)
Schwann cells		
NGF inducer		
Basophils and mast cells		
Histamine release[a]		
Megakaryocytes		
Proliferation[b]		

[a] Subramanian and Bray (1987)
[b] Williams and Morrissey (1989)
PA, plasminogen activator; PAI, plasminogen activator inhibitor; PG, prostaglandin; PI, phosphatidyl inositol turnover

activity and is a member of the immunoglobulin superfamily (Sims et al. 1988; Kishimoto 1989; Bauer et al. 1989; Beagley et al. 1989). LIF/DIA was cloned recently as both an embryonic stem cell differentiation-inhibiting factor and a leukemia-inhibiting factor that causes differentiation. It is similar to TNF-α, IL-1, and IL-6 in its stimulation of acute-phase reactants in liver (Baumann and Wong 1989). Its major biologic activites most closely resemble those of IL-6 (Moreau et al. 1988; Abe et al. 1989).

Tumor necrosis factor-α is another macrophage-derived growth factor with a distinct, recently cloned, receptor, which is widely distributed and mediates biologic responses that broadly overlap those of IL-1 (Old 1985; Beutler and Cerami 1988; Dinarello 1989; Smith et al. 1990; Schall et al. 1990). It induces fever, the acute-phase response, T and B cell activation, fibroblast proliferation, collagen synthesis, and many other effects. Several of these effects require a higher concentration of TNF-α than IL-1 (one to two orders of magnitude). IL-1

Table 4. Macrophage-derived growth factors and regulatory molecules

Growth factor	Reference	Growth Factor	Reference
Polypeptide production by macrophages			
IL-1α	a	———	
IL-1β	a	MIP-1α	a
IL-1 receptor	a	MIP-1β	a
IL-1 receptor antagonist	a	JE	Koerner et al. 1987; Introna et al. 1987
———			
PDGF-A	a	Mig (monokine induced by IFN-γ)	Farber 1990
PDGF-B/c-*sis*	a		
PDGF-related	a	IL-6	a
Vascular permeability factor	a	M-CSF	a
		KC	Koerner et al. 1987; Introna et al. 1987
———		IP-10	a
TGF-β1	a	IP-8	a
TGF-β2	a	MIP-2	a
Activin	Erämaa et al. 1990		
———		———	
Miscellaneous		TGF-α	Rappolee et al. 1988
		bFGF	Baird et al. 1985
TNF-α	a	IGF-I	Rappolee et al. 1988
GM-CSF	a	Defensins	Ganz et al. 1989
G-CSF	a	Thymosin	a
LIF/DIA	a	Bombesin	Wiedermann et al. 1986
Erythropoietin	Rich et al. 1982; Paul et al. 1984	ACTH	Smith et al. 1986
		Fibronectin	Alitalo et al. 1980
Polypeptide growth factors not produced by macrophages			
NGF	Rappolee, unpublished data	IL-10	a
EGF	Rappolee et al. 1988	Neuroleukin	a
IL-2	a	IGF-II	Rappolee, unpublished data
IL-3	a	Insulin	Rappolee, unpublished data
IL-4	a	TNF-β	a
IL-5	a	IFN-γ	a
IL-7	a		
Other regulatory products produced by macrophages			
PGE_2	a	Respiratory burst products	a
Acidic isoferritin	Broxmeyer et al. 1985		
Nitric oxide	Stuehr and Nathan 1989	Nitrates	a

[a] References from Rappolee and Werb (1988) or in Tables 5–11. Structurally related genes are grouped between horizontal lines

ACTH, adrenocorticotropic hormone; IFN, interferon

is distinct from TNF-α in that it affects stem cells in the bone marrow (hemopoietin activity), whereas TNF-α suppresses colony formation (Dinarello 1989). On the other hand, TNF-α is more powerful in inducing vascular shock, possibly by its greater effect in inducing capillary leak syndrome. Biologic responses, such as enhanced motility of endothelial cells in sprouting capillaries, may

Table 5. Growth factors functionally related to IL-1

Growth factor	Synthesis by macrophages	Induction	Specific properties	mRNA		Protein		Reference
				Size (kb)	AUUUA instability sequence	Precursor amino acid number	Mature peptide (kDa)	
IL-1α	+	LPS	Cell surface form	2.0	+	271	27	Lomedico et al. 1984; March et al. 1985
IL-1β	+	LPS	Cell surface form	1.3	+	269	33	March et al. 1985; Gray et al. 1986
IL-1 receptor antagonist	+	LPS GM-CSF		1.9	–	177	25	Eisenberg et al. 1990; Carter et al. 1990
TNF-α	+	LPS	Cell surface form	2.0	+	157	17	Pennica et al. 1985; Wang et al. 1985; Sesson et al. 1987
TNF-β	–			155	+	177	18	Semon et al. 1987
IL-6/IFN-β2	+	Adherence		1.3	+	211	21	Hirano et al. 1986; Zilberstein et al. 1986; Van Snick et al. 1988; Tanabe et al. 1988
LIF/DIA	+	LPS		1.8/4.0	+	202	45	Moreau et al. 1988; Gearing et al. 1988; Yamamori et al. 1989

increase the sensitivity of the TNF-α receptor (GERLACH et al. 1989). TNF-α also has a larger array of targets for inducing DNA fragmentation and concomitant cell death (BEUTLER and CERAMI 1988). TNF-α and IL-1 have similar effects on the major cells in the inflammatory response, inducing respiratory oxidative burst, chemotaxis, and adhesion in polymorphonuclear leukocytes (PMNs); production of granulocyte colony-stimulating factor (G-CSF), IL-6, and GM-CSF and procoagulants and adhesion in endothelial cells; production of multiple growth factors [G-CSF, GM-CSF, M-CSF, IL-1, TNF-α, IL-6, macrophage inflammatory protein (MIP)-1, MIP-2, and IL-8], chemotaxis, and adhesion in macrophages; and proliferation, growth factor, and extracellular matrix molecule expression in fibroblasts (CAVENDER et al. 1986; MUNKER et al. 1986; BEUTLER and CERAMI 1988; DINARELLO 1989; see also references in Tables 2 and 3). TNF-α induces IL-1 and IL-6 in vitro and in vivo (BROUCKAERT et al. 1989; DINARELLO 1989; FONG et al. 1989).

The control of adhesion molecules on interacting blood cells and endothelial cells by TNF-α and IL-1 is becoming clearer. The constitutive expression of MEL-14, an adhesion molecule for normal recirculating leukocytes and lymphocytes that is required for diapedesis by peritoneal exudate cells, is down-regulated by TNF-α and IL-1 rapidly (within minutes), whereas Mac-1/gp155/90 inflammatory adhesion molecules and intercellular cell adhesion molecules (ICAM) in PMNs and macrophages and ICAM and endothelial leukocyte adhesion molecules (ELAM-1) in endothelial cells are up-regulated more slowly (4 h) (GAMBLE et al. 1985; SCHLEIMER and RUTLEDGE 1986; NAWROTH et al. 1986; DOHERTY et al. 1987; BEUTLER and CERAMI 1988; KISHIMOTO et al. 1989). It is speculated that the shedding of MEL-14 may prevent activated leukocytes from entering normal lymphoid tissue, or it may be a required step in diapedesis as leukocytes disconnect from their initial binding to the activated endothelial cells of the vessel wall (BEVILACQUA et al. 1986, 1989; BRETT et al. 1989). The interaction molecule for ELAM-1 on leukocytes is not known (POBER and COTRAN 1990). ICAM on endothelial cells and Mac-1/gp155/90 on leukocytes are also up-regulated within 4 h but decay hours after ELAM-1, perhaps mediating immediate and long-range adhesion. This coordinate temporal expression parallels the autocrine effects of growth factors on stimulation of macrophages.

2.2 Immediate-Response-Gene Growth Factors

The founders of the two groups of inflammatory response genes (group 1: KC, MIP-2, IL-8, and PF-4; group 2: JE and MIP-1α,β) were originally cloned as response genes to PDGF (JE and KC) and interferon-γ (IP-10) (STILES 1983; LUSTER et al. 1985; DEUEL 1987; ROLLINS et al. 1988; KAWAHARA and DEUEL 1989; OQUENDO et al. 1989; STOECKLE and BARKER 1990). It is interesting to note that the PDGF-inducible genes JE and KC were recently found to be much more highly induced by IL-1 (HALL et al. 1989). IL-8 (also called neutrophil activity protein-1/monocyte-derived neutrophil chemotactic factor/T cell chemotactic

Table 6. Immediate-response-gene growth factor family

Growth factor	Synthesis by macrophages	Induction	Specific properties	mRNA		Protein		Reference
				Size (kb)	AUUUA instability sequence	Precursor amino acid number	Mature peptide (kDa)	
JE/monocyte chemoattractant protein-1	+	LPS	Heparin binding	4.5	+	148	16.3	ROLLINS et al. 1988; KAWAHARA and DEUEL 1989; YOSHIMURA et al. 1989
MIP-1α/LD78	+	LPS	Heparin binding	0.8	+	92	8.0	DAVATELIS et al. 1989; KWON and WEISSMAN 1989
MIP-1β	+	LPS	Heparin binding	0.65	+	109	8.0	SHERRY et al. 1988; BROWN et al. 1989
TCA-3				0.65	–	92		BURD et al. 1987
M-CSF	+	Adherence	Transmembrane	2.3/3.8/4.5	+	118 × 2	70.0	KAWASAKI et al. 1985; RAJAVASHISTH et al. 1987
IL-6	+	Adherence		1.3	+	211	21.0	YAMASAKI et al. 1988; VAN SNICK et al. 1988
IL-2	–			1.1	+	153	15–17	TANIGUCHI et al. 1983; YOKOTA et al. 1985

IFN-α (4–10 members)	+	LPS		1.0/2.8/5.5	+	189/190	20, 29–35, 35–40	Shaw et al. 1983; Epstein et al. 1990
KC/Gro/MGSA	+	LPS		0.9	+	106	10.2	Richmond et al. 1988; Oquendo et al. 1989
PF-4	+	LPS	Heparin binding	0.7/2.1/5.0	+	105	10.0	Deuel et al. 1977; Doi et al. 1987
β-TG/CTAP-III			Heparin binding			81/85	8.0	Begg et al. 1978
IP-10	+	IFN-γ		1.2	+	98	20.0	Luster et al. 1985; Vanguri and Farber 1989
IL-8/NAP-1/ MDNCF/310-C	+	LPS	Heparin binding	1.0	+	72	8.0	Schmid and Weissman 1987; Matsushima et al. 1988
MIP-2	+	LPS	Heparin binding				6.0	Wolpe et al. 1989
Mig monokine induced by γ-IFN	+	IFN-γ		1.6	+	126	14.5	Farber 1990

LD, leukocyte-derived; MGSA, melanocyte growth-stimulating activity; TCA, T cell activator; β-TG, β-thromboglobulin factor), MIP-2, and MIP-1α,β have been purified on the basis of biologic assays for inflammation and cloned from N-terminal sequences of the purified protein (DAVATELIS et al. 1988; SHERRY et al. 1988; MUKAIDA et al. 1989; STRIETER et al. 1989; WOLPE et al. 1989). The two groups are related by sequence homology, intron/exon conservation, and biologic function (Table 6).

Many of the factors reach high mRNA and protein concentrations quickly after exposure of macrophages and other cells to inflammatory stimuli (DEUEL 1987; see also Fig. 2). Therefore, these factors are available from multiple cell sources early in the inflammatory response or after trauma. They may also have roles in wound healing, nerve regeneration, delayed-type hypersensitivity, and other macrophage-mediated pathophysiologic events. As with the more highly characterized macrophage-derived growth factors (IL-1, TNF-α, and IL-6), these factors have common targets and effects: PMN chemoattraction (PF-4, IL-8, MIP-2, MIP-1α,β), pyrogenesis (MIP-1α,β), and macrophage chemoattraction (JE).

MIP-1 α,β. MIP-1α and MIP-1β were cloned from protein sequences obtained from a lipopolysaccharide (LPS)-stimulated macrophage cell line. They are 69 amino acid residue heparin-binding polypeptides that induce neutrophil chemotaxis and cytotoxicity. MIP mRNA is induced 1 h after LPS stimulation and continues at high levels for 16 h before decreasing at 24 h. MIP-1α is a pyrogen and, 1 h after injection, induces a fast, monophasic fever that is not prostaglandin dependent and is therefore distinct from fever induced by TNF-α, IL-1, interferon-α, and perhaps IL-6 (DAVATELIS et al. 1989). MIP-1α

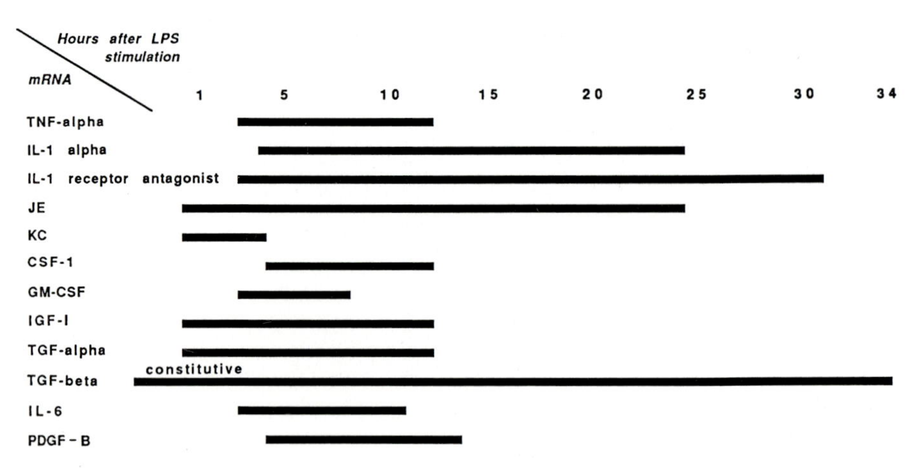

Fig. 2. The kinetics of RNA induction after addition of LPS. Note that various species and tissues were the sources of macrophages, and various types of LPS were used in the experiments. The interleukin receptor antagonist was stimulated by adhesion of macrophages to IgG. References: TNF-α, IL-1α, SCALES et al. 1989; IL-1 receptor antagonist, EISENBERG et al. 1990; JE, KC, INTRONA et al. 1987; CSF-1, BECKER et al. 1989; LEE et al. 1990; GM-CSF, THORENS et al. 1987; LEE et al. 1990; IGF-I, NAGAOKA et al. 1990b; TGF-α, RAPPOLEE et al. 1988; TGE-β, ASSOIAN et al. 1987; IL-6, NAVARRO et al. 1989; NORTHEMANN et al. 1989; PDGF-B, NAGAOKA et al. 1990a

also regulates the differentiation of macrophage precursors, but MIP-1β does not (GRAHAM et al. 1990).

MIP-2. MIP-2, a 6-kDa heparin-binding polypeptide, is induced by LPS in macrophages. It is a chemoattractant for PMNs but does not activate them for a respiratory burst. It is most closely related to the KC gene and is in the family of genes related to PF-4 (WOLPE et al. 1989). MIP-1 and MIP-2 also modulate in vitro granulopoiesis and monocytopoiesis (BROXMEYER et al. 1989).

IL-8. IL-8 is an 8-kDa heparin-binding polypeptide induced in LPS-stimulated macrophages. It is a chemoattractant and respiratory burst activator in neutrophils and causes leukotriene production by neutrophils but also appears to be important in the attraction of T cells to the sites of delayed-type hypersensitivity (MATSUSHIMA et al. 1989; SCHRÖDER 1989). IL-8 is not, however, a mitogen or comitogen for thymocytes. IL-1 and TNF-α induce IL-8 in macrophages, fibroblasts, and endothelial cells (MATSUSHIMA et al. 1988). Interestingly, IL-8 is not a chemoattractant for monocytes and does not amplify the macrophage response by increasing ingression of macrophages (MATSUSHIMA et al. 1989; LARSEN et al. 1989). It is closely related to the platelet α-granule protein β-thromboglobulin/CTAP-III.

PF-4. PF-4 is a small polypeptide (76 amino acid residues) that is induced by PDGF in fibroblasts and delivered to wounds in large quantites by degranulation of platelet α-granules. Serum contains only nanogram quantities of PDGF, but up to 20 μg/ml of PF-4. It is a neutrophil chemoattractant and is thought to modulate megakaryopoiesis, and it is anti-angiogenic (DEUEL et al. 1981; DEUEL 1987; OQUENDO et al. 1989; MAIONE et al. 1990).

JE. JE is a 148 amino acid polypeptide induced by interferon-γ and IL-1 in macrophages. Its function is not yet fully characterized, but its importance is indicated by the speed of its induction (2 h) and high copy number (3000 copies per fibroblast, which is comparable to TNF-α and IL-1 transcript copy number in macrophages) (ROLLINS et al. 1988; KAWAHARA and DEUEL 1989; PRPIC et al. 1989; HALL et al. 1989). The transcript for JE accumulates for longer time periods in macrophages than in fibroblasts (see Fig. 2). JE is a chemoattractant for macrophages (YOSHIMURA and LEONARD 1990).

2.3 Transforming Growth Factor-β

The TGF-β family of growth factors, which affect macrophage function at many levels and whose function is controlled by macrophages at many levels, is pleiotropic (Table 7). TGF-β1 is one of the first major growth factors delivered to wounds by platelets. Most of the published data refer to TGF-β1, but other TGF-βs have similar effects (ROBERTS and SPORN 1990). TGF-β is highly chemoattractive for macrophages; its ED_{50} is 40–400 fM for macrophage chemotaxis in vitro. It is autoinductive for macrophages so that motile macrophages may synthesize TGF-β transcript as they enter the inflammatory locus. TGF-β also induces other growth factor transcripts in macrophages: PDGF-B, IL-1, TGF-α,

Table 7. TGF-β family

Growth factor	Synthesis by macrophages	Induction	Specific properties	mRNA		Protein		Reference
				Size (kb)	AUUUA instability sequence	Precursor amino acid number	Mature peptide (kDa)	
TGF-β1	+	LPS	Heparin binding, proteolytic activation required	2.5	+	390/112 × 2[a]	25	Derynck et al. 1985, 1986, 1987
TGF-β2	+			6.5/5.1/4.2		412/112 × 2		de Martin et al. 1987; Madišen et al. 1988; Miller et al. 1989
TGF-β3				3.0	+	412/112 × 2		ten Dijke et al. 1988; Derynck et al. 1988; Miller et al. 1989
TGF-β4				1.7		304/114 × 2		Jakowlew et al. 1988
TGF-β5				3.0		382/112 × 2		Kondaiah et al. 1990
Vg-1				2.8	+	387/114 × 2		Rebagliati et al. 1985
Vgr-1				3.5	−	132		Lyons et al. 1989
BMP-2A						396(110, 80)	30	Wozney et al. 1988
BMP-2B						408	30	Wozney et al. 1988
BMP-3				2.0		472	30	Wozney et al. 1988
MIS				2.0	+	575	72	Cate et al. 1986
Inhibin α				1.8(4.5, 7.0)		368/134 × 2	32	Forage et al. 1986; Mason et al. 1986
Activin βA	+		Heparin binding	2.8 (6.0, 4.0, 1.7)		407/115 × 2	32	Mason et al. 1986; Ying 1988; Takahashi et al. 1990
Op-1				2.4		431	30	Özkaynak et al. 1990
Activin βB				~1.8		407/115 × 2	22	Mason et al. 1989

[a] Dimeric molecules

BMP, bone morphogenetic factor; MIS, müllerian inhibiting substance

TNF-α, and FGF (WAHL et al. 1987; CHANTRY et al. 1989; ROBERTS and SPORN 1990). It has been argued that IL-1 is induced by TGF-β only at the transcriptional level and that a second signal may be required for translation (ROBERTS and SPORN 1990). Once TGF-β is secreted, it must be activated by cleavage of the N-terminal fragment to liberate an active C-terminal fragment. TGF-β can be activated by one of two macrophage-mediated steps: activation of plasminogen by macrophage-derived plasminogen activator or acidification of the local environment by lysosomal leakage (MASSAGUE 1987; FAVA et al. 1989; ROBERTS and SPORN 1990). Macrophages may also control the effects of TGF-β by releasing α_2-macroglobulin, which inactivates it (HOVI et al. 1977; ROBERTS and SPORN 1990). The effects of TGF-β are important during the resolution of the wound. TGF-β decreases both T cell-mediated cellular immunity and the production of hydrogen peroxide by macrophages (TSUNAWAKI et al. 1988; ROBERTS and SPORN 1990). The injection of TGF-β into dermis causes formation of granulation tissue and neovascularization (ROBERTS et al. 1986; PIERCE et al. 1989). Although these effects may be secondary to the chemoattraction for macrophages, TGF-β also has chemoattractive and synthetic effects on periwound fibroblasts. It causes fibrosis by up-regulating transcription and accumulation of the extracelluar matrix components collagen (types I, III, IV, V) and fibronectin (MASSAGUE 1987; KHALIL et al. 1989; ROBERTS and SPORN 1990). It also down-regulates transcription of extracellular matrix-degrading proteinases and up-regulates the transcription of their inhibitors, such as tissue inhibitor of metalloproteinases (TIMP) (EDWARDS et al. 1987). TGF-β also increases the expression of integrins, specifically the α and β units of fibronection receptor (ROBERTS and SPORN 1990). This may increase the adhesive characteristics of cells for basal lamina, where TGF-β itself is found sequestered in basal lamina at times corresponding to peak TGF-β expression by macrophages in lung disease (KHALIL et al. 1989). Finally, TGF-β causes the immunoglobulin class switch of B cells preferentially to IgA while suppressing IgG (COFFMAN et al. 1989).

Another member of the TGF-β superfamily, activin/erythrocyte differentiation factor, has been found to be synthesized by macrophages and may be important in the regulation of erythropoiesis by macrophages (ERÄMAA et al. 1990).

2.4 Platelet-Derived Growth Factor

Platelet-derived growth factor is found in serum at a concentration of 20 ng/ml but is not found in plasma (< 1 ng/ml). It stimulates a variety of cells through receptors with a K_d of $1-100 \times 10^{-10}$. There are two isoforms of PDGF, A and B (isoform B is c-*sis*)(Table 8), which are composed of dimers of the related A and/or B chains of PDGF, and two receptors for these isoforms with overlapping biologic effects in wound healing and inflammation (DEUEL 1987). Macrophages produce both isoforms but with different kinetics (MARTINET et al. 1986; RAPPOLEE

Table 8. PDGF family

Growth factor	Synthesis by macrophages	Induction	Specific properties	mRNA		Protein		Reference
				Size (kb)	AUUUA instability sequence	Precursor amino acid number	Mature peptide (kDa)	
PDGF-A	+	LPS	Heparin, collagen binding	1.3/1.9/2.1	+	211	31	BETSHOLTZ et al. 1986
PDGF-B/c-*sis*	+	LPS	Heparin, collagen binding	3.6,[a] 4.2	+	209	24	JOSEPHS et al. 1984
PDGF-r	+						36	PENCEV and GROTENDORST 1988
Vascular permeability factor	+			3.8		189/215	25/40	KECK et al. 1989; LEUNG et al. 1989

[a] Macrophage-specific size (SHIMOKADO et al. 1985)

and WERB 1989). The receptors number from 200000 in fibroblasts to 50000 in smooth muscle cells.

The PDGF isoforms have varied effects, causing immediate degranulation of fibroblasts, chemoattraction of neutrophils and monocytes (at 1 ng/ml and 20 ng/ml, respectively), membrane ruffling, actin reorganization, and mitosis. Early in hemostasis, PDGF delivered by platelets may act as a powerful vasoconstrictor and also attracts and activates leukocytes for microbicidal action (ROSS and VOGEL 1978; STILES 1983; BERK et al. 1986; DEUEL 1987). It also rapidly induces a transient increase in several immediate-response genes, including JE (induced to 3000 copies per cell), KC (induced to 700 copies per cell), and IL-6. A second immediate and transient response is that of nuclear *trans*-activating factors, c-*fos* (induced within 15 min by a cycloheximide-insensitive mechanism or superinduced by cycloheximide) and c-*myc* (induced to five to ten copies per cell) (DERYNCK 1988; ROLLINS et al. 1988; KAWAHARA and DEUEL 1989; OQUENDO et al. 1989). In the immediate phase of inflammation, PDGF liberates fibroblast enzymes and increases the potential of local cells to mount a second program of growth factor and cytokine expression whose inflammatory effects are poorly understood. Since PDGF also induces expression of prostaglandin E_2(PGE_2), it may also limit its own production (via prostaglandins) and action on target cells (through the action of IL-6) (DANIEL et al. 1987). Macrophage-derived α_2-macroglobulin is also known to sequester and inactivate PDGF (HOVI et al. 1977; ROSS et al. 1986).

Platelet-derived growth factor may also have pleiotropic actions in the later stages of wound healing and inflammation. It induces expression of both proteinases, such as collagenase, and extracellular matrix proteins, such as types I, III, IV, and V collagen (CHUA et al. 1985; BAUER et al. 1985; ROSS et al. 1986). The actions of these induced molecules may mediate movement of cells, diapedesis, remodeling, or mitosis. In various molecular phases of wound healing, PDGF can act much later than TGF-β when injected into dermis (PIERCE et al. 1989). However, PDGF also induces TGF-β, and some of its effects may be mediated by this growth factor. Since both PDGF and TGF-β bind to various matrix components, their effects may be residual to the expression of their corresponding mRNA and protein by local cells.

2.5 Transforming Growth Factor-α

Transforming growth factor-α is in the EGF family of growth factors (Table 9). It is transcribed by macrophages in response to lipopolysaccharides, lipids, and, under certain conditions, adhesion (RAPPOLEE et al. 1988; MADTES et al. 1988). TGF-α and EGF share a receptor (the proto-oncogene form of the *erb-B* gene) and bind with an identical K_d of 10^{-9} M. Macrophages derived from wound cylinders transcribe and translate TGF-α. Both macrophages and megakaryocytes (which generate platelets) transcribe TGF-α and liberate a protein that binds the EGF receptor (RAPPOLEE et al. 1988; MADTES et al. 1988).

Table 9. EGF family

Growth factor	Synthesis by macrophages	Induction	Specific properties	mRNA		Protein		Reference
				Size (kb)	AUUUA instability sequence	Precursor amino acid number	Mature peptide (kDa)	
EGF	–		Transmembrane precursor	4.8	–	1218	6	Scott et al. 1983; Gray et al. 1983; Bell et al. 1986b
TGF-α	+	LPS	Transmembrane precursor	4.5	+	160	6	Derynck et al. 1984; Lee et al. 1990
Amphiregulin				1.4		162	9	Shoyab et al. 1989
BMP-1				2.8/4.0		700+	50	Wozney et al. 1988
NTERA/EGF				2.0		188		Ciccodicola et al. 1989
Schwannoma-derived growth factor				1.7/3.0/4.5		243	31–35	Kimura et al. 1990

BMP, bone morphogenetic factor; NTERA, neuronal teratoma

Neither macrophages (RAPPOLEE et al. 1988) nor megakaryocytes (RAPPOLEE, unpublished data) transcribe EGF mRNA. This suggests that TGF-α, which was originally thought to be an EGF isoform peculiar to transformed cells, may act in pathophysiologic events (DERYNCK et al. 1984). The amount of TGF-α polypeptide secreted by macrophages is low compared with that secreted by eosinophils and epithelial cells, but when acting in a local environment its concentration may be near the K_d of its receptor (RAPPOLEE et al. 1988; DERYNCK 1990). It is also synthesized as a 159 amino acid residue transmembrane precursor of the mature secreted 50 amino acid polypeptide. The transmembrane molecule has been shown to mediate biologic effects on target cells bearing the EGF receptor (BRACHMANN et al. 1989). This indicates that macrophage-derived TGF-α may act as a secreted molecule or as a transmembrane "precursor." The relative activities of the various membrane-bound and secreted molecules are not understood.

Macrophages do not bind TGF-α (RAPPOLEE, unpublished data). This suggests that, unlike TGF-β, PDGF, IL-1, M-CSF, TNF-α, and MIP-1, TGF-α must work "downstream" on other cell types in a paracrine manner and does not act in an autocrine manner or recruit or influence new monocytes. According to current information, TGF-α is the only macrophage-derived growth factor that mediates all three parts of dermal wound healing: reepithelialization, formation of granulation tissue, and induction of neovascularization (SCHREIBER et al. 1986; ROBERTS et al. 1986; SCHULTZ et al. 1987). It also induces interferon-γ in lymphocytes (ABDULLAH et al. 1989). Like PDGF and FGF, TGF-α induces collagenases and stromelysin and the synthesis of collagens in fibroblasts (MATRISIAN et al. 1985; EDWARDS et al. 1987). It also induces interferons in fibroblasts (LEE and WEINSTEIN 1978).

2.6 Fibroblast Growth Factors

The FGF family currently consists of seven sequenced members (Table 10), only one of which has been identified in macrophages (BAIRD et al. 1985). Basic FGF (bFGF) has no signal sequence and is "secreted" by stimulated P388D1 macrophages in vitro but not by stimulated primary macrophages in vitro (although these cells have bFGF in the cytosol). It has been hypothesized that bFGF may be liberated on cell death, and it is possible that P388D1 macrophages have a higher cell turnover rate in vitro (RAPPOLEE et al. 1988). At least two other members of the FGF family have signal sequences, but it is not known whether they are synthesized by macrophages. There are three receptors in the FGF receptor family, but they have not been well characterized. Two of them are related to the only FGF receptor that has been cloned by biochemical means (BURGESS and MACIAG 1989). All of the receptors have intrinsic tyrosine kinase activity and are more closely related to the M-CSF/PDGF receptor than to the insulin/EGF group. Biochemically, it is known that there are two FGF receptors on many cell types: a high-affinity receptor with low copy number

Table 10. FGF family

Growth factor	Synthesis by macrophages	Induction	Gene structure (intron/exon)	Specific properties	mRNA		Protein		Reference
					Size (kb)	AUUUA instability sequence	Precursor amino acid number	Mature peptide (kDa)	
FGF-1/ acidic FGF			3/2	Heparin binding; no signal peptide	4.8	+	155	17	Jaye et al. 1986
FGF-2/ basic FGF	+	LPS, thioglycollate	3/2	Heparin binding; no signal peptide	2.2/4.6/6.0	+	146–155	16	Abraham et al. 1986a,b; Shimasaki et al. 1988
FGF-3/ int-2			3/2	Heparin binding; partial signal peptide	1.4/1.7/2.6/2.9	+	245	27	Moore et al. 1986
FGF-4/ kFGF			3/2	Heparin binding; full signal peptide	1.2/3.5	+	245		Delli-Bovi et al. 1987; Brookes et al. 1989; Hebert et al. 1990
FGF-5			3/2	Heparin binding; no signal peptide	1.6/4.0		267		Zhan et al. 1988; Hebert et al. 1990
FGF-6			3/2		2.8/3.9/4.7		267		Marics et al. 1989
FGF-7/ keratinocyte growth factor			3/2		2.4		206	22.5	Finch et al. 1989

Table 11. Hematopoietic growth factors[a]

Growth factor	Synthesis by macrophages	Induction	Gene structure (intron/exon)	Specific properties	mRNA		Protein		Reference
					Size (kb)	AUUUA instability sequence	Precursor amino acid number	Mature peptide (kDa)	
IL-1α/ hemopoietin	+	LPS		Membrane form	2.0	+	271	27	LOMEDICO et al. 1984; MARCH et al. 1985; KURT-JONES et al. 1985a,b
IL-6/IFN-β2/ B-cell stim. factor II	+	Adherence	5/4		1.3	+	211	22–29	NORDAN and POTTER 1986; HIRANO et al. 1986; ZILBERSTEIN et al. 1986; VAN SNICK et al. 1988
G-CSF	+	LPS	5/4		1.4/1.6	+	204/208	25(20 human)	TSUCHIYA et al. 1986
GM-CSF	+	Adherence	4/3	Binds GAGs	0.8/1.2	+	141	18–25	MIYATAKE et al. 1985; CANTRELL et al. 1985; GORDON et al. 1987
M-CSF/CSF-1	+	Adherence		Transmembrane form	1.4/2.3/3.8/4.5	+	118 × 2 (dimer)	70	KAWASAKI et al. 1985; RAJAVASHISTH et al. 1987
Erythropoietin	+	LPS	5/4		1.0	+	193	18.4	LIN et al. 1985; SHOEMAKER and MITSOCK 1986
IL-3	–	Antigen	5/4		1.9	+	166	15–30	FUNG et al. 1984; YANG et al. 1986

[a] Note that these factors represent a partial list of the best-characterized molecules. Activin, PF-4, MIP-1α, and others that have less well characterized effects are not included

GAG, glycosaminoglycan

($K_d = 50–500$ pM, $5–50 \times 10^3$/cell) and a low-affinity receptor with heparin-like qualities with high copy number ($K_d = 10$ nM, $5–20 \times 10^5$/cell). Heparin binds and potentiates FGF when both are in solution, and it has been speculated that the cell surface heparin-like receptor may focus the FGF on the high-affinity receptor (BURGESS and MACIAG 1989).

Fibroblast growth factor may function late in inflammation in wound healing and remodeling. It is highly angiogenic, a chemoattractant and mitogen for endothelial cells, and a mitogen for smooth cells. It also has numerous immediate effects on fibroblasts as a secretagogue and chemoattractant and is a mitogen for 3T3 fibroblasts. It may have some of the same effects on fibroblasts in regulating proliferation fibrosis as do TGF-β, TGF-α, and PDGF (GROSS et al. 1983; ABRAHAM et al. 1986a; EDWARDS et al. 1987).

2.7 Colony-Stimulating Factors

Of the major colony-stimulating factors (CSFs) currently characterized, macrophages synthesize M-CSF, G-CSF, GM-CSF, IL-1, and IL-6 (Table 11) but do not synthesize multi-CSF (IL-3) or IL-5 (NICOLA 1989). CSFs mediate survival, proliferation, functional modulation (chemotaxis, degranulation, activation, adhesion, cytotoxicity, mRNA phenotype changes), and differentiation on various populations of precursor and mature blood cells (GRABSTEIN et al. 1986; HORIGUCHI et al. 1987; BECKER et al. 1987; DONAHUE et al. 1988; RAPPOLEE and WERB 1989; ALVARO-GRACIA et al. 1989; BUSSOLINO et al. 1989; HOANG et al. 1989). A number of rules have emerged in classifying the activities of CSFs:

1. The ED_{50} for activating mature cells or causing mitosis in precursors is higher than that for maintaining survival (although the K_d of all the CSFs for their cognate ligands is low—between 10 pM and 1000 pM). Bone marrow precursors in vitro do not survive longer than 24 h unless a specific CSF is present.
2. At low, limiting concentrations, CSFs are specific for a restricted lineage (M-CSF for macrophages and G-CSF for granulocytes; IL-3 and GM-CSF have their highest effects on macrophages).
3. CSFs often act synergistically; IL-3 synergizes with GM-CSF or IL-6 in maintaining proliferation of committed stem cells.

Fig. 3. a The three tiers of development of hematopoietic stem cells. **b** The effect of four macrophage-derived growth factors that affect the balance of hematopoietic proliferation and differentiation. The eight mature cell types produced from a common totipotent stem cell are red blood cells (*RBC*), polymorphonuclear leukocytes (*PMN*), macrophages (*MAC*), eosinophils (*EOS*), megakaryocytes (*MEGA*), mast cells (*MAST*), B lymphocytes (*B CELL*), and T lymphocytes (*T CELL*). *Heavy lines* indicate a mitogenic effect at low concentrations of growth factor; *medium lines* indicate a mitogenic effect only at high concentrations of growth factor; *light lines* indicate no effect. *Dots* indicate a modulation of function in the mature cell. (Modified from METCALF et al. 1985; NICOLA 1989)

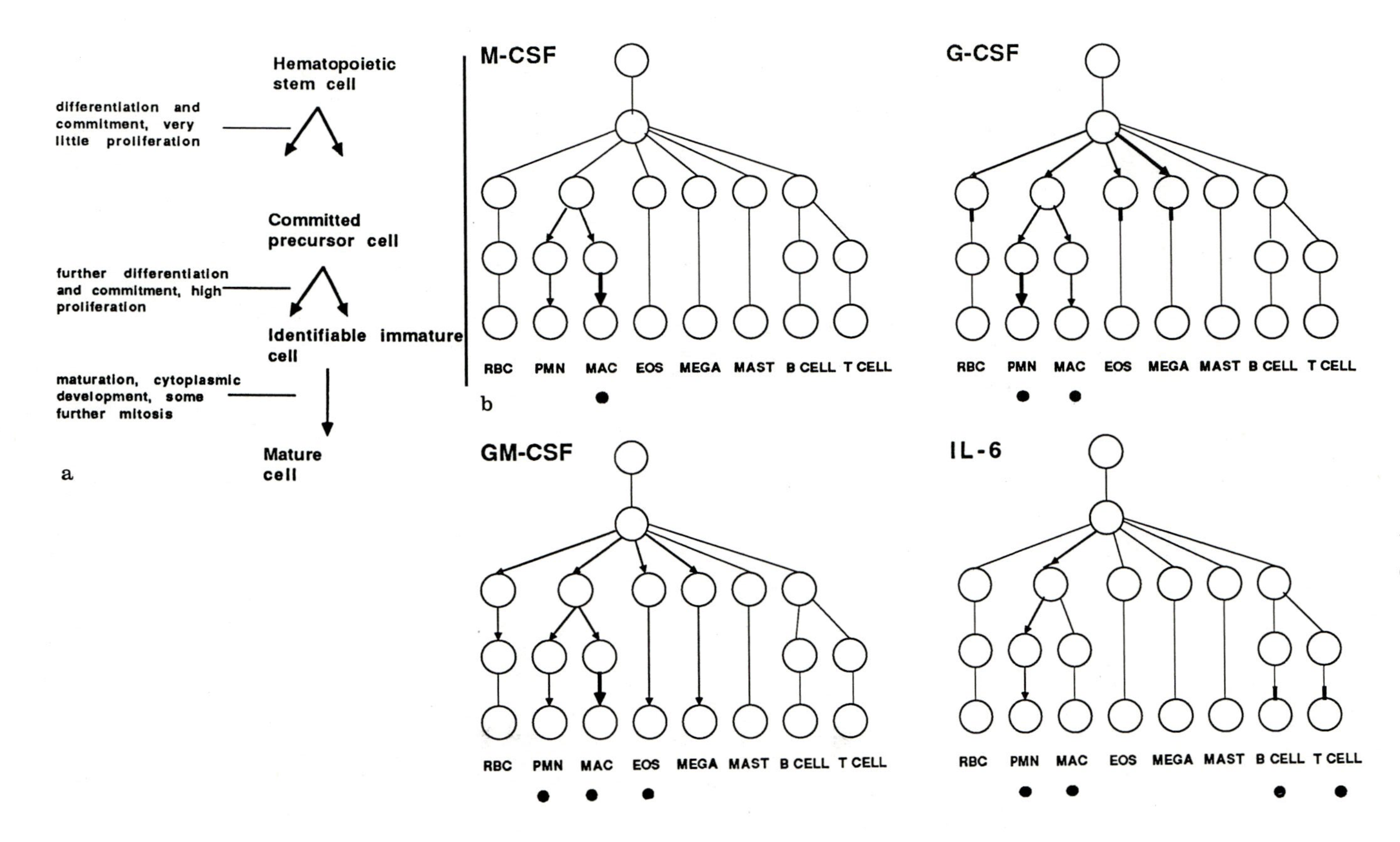
Hematopoietic stem cell
differentiation and commitment, very little proliferation
Committed precursor cell
further differentiation and commitment, high proliferation
Identifiable immature cell
maturation, cytoplasmic development, some further mitosis
Mature cell
a
b
M-CSF
G-CSF
GM-CSF
IL-6
RBC PMN MAC EOS MEGA MAST B CELL T CELL
RBC PMN MAC EOS MEGA MAST B CELL T CELL
RBC PMN MAC EOS MEGA MAST B CELL T CELL
RBC PMN MAC EOS MEGA MAST B CELL T CELL

4. In the lineage of cells in blood cell formation, growth factors act in the following order during maturation: unknown CSFs for the earliest stem cells → IL-1α/hemopoietin → IL-3 → IL-6/GM-CSF → G-CSF → M-CSF.
5. Synergism of CSFs operates at many levels, including induction of receptor and ligand expression, expansion of precursor populations, and modulation of receptors.
6. The history of a cell determines how it responds to a given CSF (METCALF 1985; NICOLA 1989). For example, granulocyte-macrophage precursor cell lines respond to G-CSF to become granulocytes or macrophages or mixed granulocytes and macrophages, depending on which factor these precursors have previously been exposed to (NICOLA 1989).

The concentrations of M-CSF and IL-6 in the blood rise after injection of LPS or in response to bacterial infection, but GM-CSF and IL-3 concentrations do not rise in response to the same stimulants. However, injection of recombinant CSFs has resulted in localized activation of macrophages and PMNs and some proliferation of progenitors, but little rise in numbers of blood cells because of endocrine activation or proliferation of stem cells in the bone marrow. The local production of CSFs at sites of inflammation is clearly important in regulating the functions of endothelial cells, fibroblasts, and invading blood cells through a large number of molecular mechanisms (see Sects. 4 and 8). CSFs are mitogenic for endothelial cells (BUSSOLINO et al. 1989) and white blood cells (METCALF 1985; DONAHUE et al. 1988) and modulate the molecular phenotype of white blood cells (HORIGUCHI et al. 1987; RALPH and NAKOINZ 1987; WEISBART et al. 1988; NATHAN 1989; ZUCKERMAN and SUPRENANT 1989). The endocrine, paracrine, and autocrine activities of macrophage-derived CSFs are not defined (METCALF 1985; NICOLA 1989), although macrophage products can modulate granulocytopoiesis and monocytopoiesis in vitro (CHERVENICK and LOBUGLIO 1972; BROXMEYER et al. 1989). In vitro heparan sulfate-bound GM-CSF (and IL-3) activate hematopoiesis, suggesting a localization of these factors in a paracrine manner by bone marrow fibroblasts and macrophages (ROBERTS et al. 1988; Fig. 3). Macrophages are unique among blood-derived cells in their capacity to undergo a further mitotic division at sites of activity of the most differentiated cells. This may result in an increased number of local macrophages. The level of plasma M-CSF is regulated by the population size of macrophages because macrophages quickly remove M-CSF injected into the bloodstream (BARTOCCI et al. 1987).

2.8 Insulin-Like Growth Factors

Macrophages are induced by phorbol 12-myristate 13-acetate to transcribe insulin-like growth factor (IGF)-I (Table 12). In addition, macrophages have a preformed pool of IGF-I that is secreted on stimulation (NAGAOKA et al. 1990b). Since wound-derived macrophages express IGF-I mRNA, it is inferred that IGF-I is translated by macrophages and has a function in the wound (RAPPOLEE et al.

Table 12. Insulin-like growth factor family

Growth factor	Synthesis by macrophages	Induction	Gene structure (intron/exon)	mRNA		Protein		Reference
				Size (kb)	AUUUA instability sequence	Precursor amino acid number	Mature peptide (kDa)	
IGF-I	+	LPS	5/4	0.9/5.3/7.7	–	130(70)	10	RINDERKNECHT and HUMBEL 1978; BELL et al. 1986a
IGF-II	–		8/7	2.0/4.9/6.0	–	180(67)	10	STEMPIEN et al. 1986
Insulin (2 members)			3/2	0.6/0.7/2.4		110(51)	7	MUGLIA and LOCKER 1983; WENTWORTH et al. 1986
Relaxin (2 members)				1.2/1.6/2.5/3.3		182(54)	6.3	HALEY et al. 1987

1988). IGF-I has insulin-like effects on glycogen synthesis and activation of anabolic processes in target cells. Macrophages do not synthesize either IGF-II or insulin (RAPPOLEE, unpublished results). The importance of IGF-I in macrophage function remains to be determined.

3 Negative Modulation of Macrophage-Derived Growth Factors

It has become obvious that most, if not all, macrophage-derived growth factors are either positively cross-induced or autoinductive. These inductions theoretically allow for functionally different growth factors to reach peak concentrations quickly and in a set temporal order. The control of growth factor production and function must also be down-regulated. Much exciting research has recently been reported, and there is now an understanding of the complexity of the down-regulation of growth factors.

Many macrophage-derived growth factor transcripts are unstable. IL-1α and IL-1β, PDGF-B, TNF-α, and GM-CSF share an AUUUA motif repeated in the 3′ flanking sequence that is thought to target the transcript for destruction in the cytosol (SHAW and KAMEN 1986; BRAWERMAN 1989) and have a half-life of about 1 h. Similarly, growth factor polypeptides, such as insulin, are targeted for destruction by specific peptidases in the extracellular environment (DUCKWORTH 1988). A constitutive destructive process ensures that the powerful effects of growth factors do not linger when the inductive stimuli have abated.

A second mechanism for negatively modulating the function of growth factors is to buffer or inhibit the polypeptides. IGF-I has a binding protein with a binding affinity 100 times that of the IGF-I receptor (SARA and HALL 1990). Many macrophage-derived growth factors, such as TNF-α, IL-4, IL-2, and GM-CSF, have a soluble form of the cellular receptor that may have the same affinity as the cellular form (TREIGER et al. 1986; ENGELMANN et al. 1989; NOVICK et al. 1989). These proteins bind growth factors and prevent binding to receptors. In addition, a competitive inhibitor of IL-1 has recently been cloned (CARTER et al. 1990; HANNUM et al. 1990; EISENBERG et al. 1990). This inhibitor has sequence homology with IL-1 and binds to IL-1 receptors (in the brain, T cells, and fibroblasts) with the same affinity as IL-1, but does not activate the receptor. The IL-1 inhibitor can dampen or attenuate the effects of IL-1 in some tissues (LIAO et al. 1985; AREND et al. 1990; CARTER et al. 1990). It is produced by macrophages upon stimulation that also triggers IL-1, but the inhibitor is made more slowly and lasts longer. The inhibitor transcript has no AUUUA motif and may therefore attenuate the IL-1 ligand after the initial response in the absence of the inhibitor. Macrophages also produce α_2-macroglobulin, which binds and activates PDGF and TGF-β, both macrophage-derived growth factors (ROSS et al. 1986; GRAVES and ANTONIADES 1988; MCCAFFREY et al. 1989; DANIELPOUR and SPORN 1990; ROBERTS and SPORN 1990). In addition, TGF-β can stimulate production of α_2-macroglobulin in some cells (SHI et al. 1990). Binding proteins and inhibitors may attenuate ongoing responses or limit the responses spatially.

A third class of negative modulation of growth factors is negative feedback from either the target or producer cell on production. PDGF and IL-1 induce PGE_2 in target cells, and PGE_2 suppresses the synthesis of these growth factors (KUNKEL et al. 1986; DANIEL et al. 1987; DINARELLO 1989). Growth factors also induce polypeptides that down-regulate the production of the growth factor. For example, TNF-α induces IL-1, which then inhibits the autoinduction of TNF-α and attenuates TNF-α production (EPSTEIN et al. 1990). Growth factors such as TNF-α and M-CSF also induce factors such as interferon-β, which act negatively on the target cell. When blocking antibody to interferon is added in vitro, cells responding to TNF-α and M-CSF increase their response time and effects (RESNITZKY et al. 1986; KOHASE et al. 1986). These forms of negative feedback limit the duration and magnitude of the growth factor response by inhibiting the production or effect of the factor.

Finally, more extended forms of feedback reside in the equilibrium of production of producer cells, which is governed by the growth factors they synthesize. Macrophages synthesize M-CSF, GM-CSF, interferon-α, interferon-β, and IL-1, which positively regulate macrophage production, and PGE_2 and MIP-1α, which negatively regulate macrophage production (KURLAND et al. 1978; MOORE 1984a, b; METCALF 1985; NICOLA 1989; GRAHAM et al. 1990). Mature macrophages quickly remove M-CSF from the blood and regulate M-CSF concentration in inverse proportion to the number of mature macrophages (BARTOCCI et al. 1987).

It is clear that macrophages and macrophage-derived growth factors stimulate self-limiting effects on the duration, magnitude, and location of action. Since each growth factor has unique effects, this may ensure a temporal series of specified effects. The mechanisms and significance of these self-limitations are beginning to be understood.

4 Inflammation

The role of macrophages in the early phases of acute inflammation is not clearly defined. Although resident tissue macrophages are present and able to respond to limited stimuli, their functions are unknown. Platelets and neurogenic spasms provide early hemostasis in the first seconds and minutes. Neutrophils form the mass of white cells in the first 12 h of inflammation. They are attracted by simple inflammatory stimuli. The activated macrophage is the primary cytotoxic cell in the inflammatory lesion at 24 h after stimulation (COHN 1978; NORTH 1978; RAPPOLEE and WERB 1989).

In the later phases of acute inflammation (the end of the first day after wounding), macrophages are the major leukocyte in the inflammatory locus. Migration and adhesion of macrophages are controlled by secreted factors, including growth factors. Macrophages are attracted by TGF-β, PDGF, IL-1, TNF-α, IL-2, and a large range of inflammatory debris such as *N*-formyl methionyl-capped bacterial proteins, complement split products, and fibrinopeptides (MING et al. 1987; DINARELLO 1989; RAPPOLEE and WERB 1989). Since many of the growth factors that attract macrophages are produced by macrophages, the stimulated macrophages arriving at the wound recruit other macrophages and white cells to the wound. For example, many macrophage products, such as TGF-β, IL-1α, TNF-α, platelet-activating factor, leukotriene B_4, IL-1β, and MIP-1, attract neutrophils (RAPPOLEE and WERB 1988, 1989). Adhesion or margination before diapedesis of macrophages is under the influence of growth factors. TNF-α down-regulates the constitutive expression of GMP-140-related adhesive molecules such as MEL-14 (by shedding), but up-regulates the expression of GMP-140 by immediate degranulation of endothelial cells and platelets. MEL-14 is shed from the cell surface of lymphocytes and monocytes within minutes of addition of growth factor in vitro. Both TNF-α and IL-1 induce slower changes in ELAM-1 and ICAM expression, which makes them more adhesive to macrophages and neutrophils as well as to lymphocytes. These changes in the adhesiveness of endothelial cells peak at 2–4 h and return to baseline at 24 h, even in the continued presence of TNF-α in vitro (GAMBLE et al. 1985; DOHERTY et al. 1987; BEUTLER and CERAMI 1988; KISHIMOTO et al. 1989; BEVILACQUA et al. 1989). IL-1 induces production of stromelysin in fibroblasts and macrophages, modulating the ability of these cells to move through the vascular basement membrane and interstitial extracellular matrix (FRISCH and RULEY 1987; RAPPOLEE and WERB 1989). It is interesting to note that TNF-α causes edema, which may be commensurate with increased transmigration of the macrophage vessel wall, but IL-1 causes only changes in adhesion without increasing vascular permeability (BEVILACQUA et al. 1989; BRETT et al. 1989). In addition, M-CSF and IL-4 increase the expression of two types of plasminogen activator synthesized by macrophages (WERB 1987; HART et al. 1989a, b). Other macrophage factors such as GM-CSF and G-CSF can induce endothelial cells to produce CSFs and alter their procoagulant ratios to become more adhesive to leukocytes and lymphocytes and to further wall off wounds (BUSSOLINO et al. 1989).

Macrophage growth factors such as PDGF have the capability of causing vascular spasm through their action on smooth muscle cells, although most of this activity may occur early in inflammation and be a function of platelet delivery (BERK et al. 1986). Vascular spasm can be neurogenically caused, as can macrophage influx as monocytes respond to neuropeptide substance P chemotactically and by producing IL-1, TNF-α, and IL-6 (RUFF et al. 1985; LOTZ et al. 1988). Production of macrophage TNF-α and IL-1 peaks later after LPS stimulation (at about 6 and 12 h, respectively, in vitro), suggesting that these growth factors regulate later maintenance of clot formation and inflammation. TNF-α and IL-1 may also contribute to pathologic conditions such as hemorrhagic necrosis, thrombosis, and intravascular coagulation (DURUM et al. 1985; OLD 1985; BEUTLER and CERAMI 1988). Macrophages predominate and function well in the wound after 12h for several reasons. They produce an acid environment by releasing up to 25% of their acidic lysosomes, and they survive this acid and hypoxic environment better than neutrophils. They also have a more highly functional protein production apparatus than do neutrophils and are able to respond to changes in the resolving wound. In addition, they are much more phagocytic than neutrophils (RAPPOLEE et al. 1989). Inflammatory macrophages are probably the major source of IL-1 and TNF-α, as these two growth factors can approach 1%–5% of total macrophage protein production. These growth factors have pleiotropic actions. They have endocrine effects in regulating the hypothalamic temperature set point (and therefore are pyrogenic) and the production by liver of acute-phase proteins. A recent finding suggests that IL-1 also regulates the production of glucocorticoids in the adrenal gland, a pathway that may mediate potential negative feedback on IL-1 production (BESEDOVSKY et al. 1986). In addition IL-1 autoinduces its own production as well as inducing TNF-α itself. The kinetics of these inductions are slightly different, and it has been suggested that IL-1 also down-regulates the TNF-α autoinduction (EPSTEIN et al. 1990). IL-1 also induces the production by macrophages of lactoferrin, a molecule that attracts macrophages, neutrophils, and B and T lymphocytes. Both IL-1 and TNF-α are induced by the LPS produced by bacteria.

There is a dense network of autocrine and paracrine inductions of growth factors in the cells of the wound. In the wound-derived macrophage, C5a complement split product and TGF-β induce IL-1. LPS-stimulated macrophages produce more TGF-β. IL-1 produces further IL-1 mRNA and protein in macrophages, and this induces TNF-α mRNA and protein but inhibits the later autoinduction of TNF-α. Since TNF-α induces IL-1, this leads to a self-limitation of TNF-α production through a two-growth-factor circuit (EPSTEIN et al. 1990). In addition, it ensures that TNF-α is produced early but subsides as IL-1 production increases. Macrophage TNF-α also induces the production of IL-1 by endothelial cells and fibroblasts (DINARELLO 1989). Furthermore, macrophage TNF-α and IL-1 induce the production in endothelial cells of GM-CSF, G-CSF, and M-CSF (SEELENTAG et al. 1987). The first two of these CSFs have broad functions in the wound, as their receptors are expressed on many cell types. They modulate the procoagulant activity of endothelial cells (RYAN and GECZY 1986; ZUCKERMAN and SUPRENANT 1989) and modulate other functions such as angiogenesis in these cells (FRÄTER-SCHRÖDER et al. 1987; LEIBOVICH et al. 1987). The receptor for M-CSF (also known as the *c-fms* proto-oncogene) is expressed only on macrophages (METCALF 1985; NICOLA 1989) and, therefore, the M-CSF growth factor has a much narrower cellular function. M-CSF causes the macrophage to become highly secretory (beyond and in synergism with any other macrophage stimulator) (TAKEMURA and WERB 1984a, b; WARREN and RALPH 1986; NATHAN 1987; BECKER et al. 1987). The CSFs also increase the cytotoxicity of macrophages and neutrophils by inducing the respiratory burst and production of reactive oxygen intermediates (NATHAN 1987; NICOLA 1989). M-CSF may cooperate with the interferon-γ or IL-2 produced by immune T cells to enhance the cytotoxicity of macrophages (Li et al. 1989). Antiviral effects may be mediated by IL-6/interferon-β2, which is induced by LPS and TNF-α. The endocrine effects of IL-6 on the acute-phase response of liver mimic those induced by TNF-α and IL-1 (Beutler et al. 1986; Darlington et al. 1986; Dinarello 1989). However, IL-6 induces fibronectin production by the liver, an effect not produced by the other two growth factors (Lanser and Brown 1989).

The extent of the endocrine effects of CSFs produced by macrophages is not known. The production of many other cells, such as fibroblasts, leads to the increased concentrations of M-CSF in serum characteristic of some infections. This increase may play a part in inducing further bone marrow production of macrophages and neutrophils. The M-CSF concentrations in the blood are negatively regulated by macrophages themselves (Bartocci et al. 1987).

The importance of growth factors in the early participation of macrophages in hemostasis and the early inflammatory response is great in magnitude, but not well defined. Certainly, macrophage-derived IL-1, TNF-α, PDGF, TGF-β, IL-6/interferon-β2, GM-CSF, M-CSF, and G-CSF play potentially large roles. Newly discovered macrophage growth factors such as TGF-α, IGF-I, and the immediate-response growth factors KC/JE/MIP-1/MIP-2/IL-8 also may play a part. These macrophage-derived growth factors stimulate secretion and migration and reprogram the function of inflammatory cells. They also lead to increased vascular spasm and procoagulation and adhesive interaction between endothelial cells and blood cells. They lead to enhanced microbial and phagocytic capability in neutrophils and macrophages. The importance of growth factor autoinduction, cross-induction, and negative feedback is not well understood.

5 Wound Healing

The resolving inflammatory loci provide a stage for the macrophage as central actor. In killing, debridement, and wound healing, macrophages are absolutely required (Leibovich and Ross 1975; Rappolee and Werb 1989). When macrophages are eliminated by antileukocyte serum injected locally, and monocyte production is prevented by injection of glucocorticoids, wound healing proceeds very slowly.

Macrophage production of complement and lysosomal hydrolases is synergized by M-CSF and IL-1 (Bentley et al. 1981; Takemura and Werb 1984a; Perlmutter et al. 1986; Nicola 1989). M-CSF is produced by stimulated fibroblasts, endothelial cells, and macrophages in the wound. TNF-α and IL-1 induction has been discussed above. Reactive oxygen intermediates are synthesized by interferon-γ-stimulated macrophages, but GM-CSF, IL-2, prolactin, somatotropin, M-CSF, IL-1, and TNF-α synergize in this induction (Adams and Hamilton

1984, 1989; METCALF 1985; NATHAN et al. 1985; GRABSTEIN et al. 1986; WARREN and RALPH 1986; MALKOVSKY et al. 1987; RALPH and NAKOINZ 1987; EDWARDS et al. 1988; BERNTON et al. 1988; FRAKER et al. 1989). Interferon-γ production by immune T cells is enhanced by macrophages (BENACERRAF and UNANUE 1979; LUCAS and EPSTEIN 1985). In addition, several macrophage-derived growth factors—IL-1, TNF-α, MIP-1, MIP-2, and IL-8—induce neutrophil chemotaxis and/or respiratory burst (ADAMS and HAMILTON 1984; DAHINDEN et al. 1989; DAVATELIS et al. 1989; DINARELLO 1989; LARSEN et al. 1989; WALZ et al. 1989; WOLPE et al. 1989). TNF-α kills a broader spectrum of tumor cells than does IL-1. TNF-α induces the fragmentation of DNA within tumor cells that contain the receptor for this ligand (ONOZAKI et al. 1985; URBAN et al. 1986; BEUTLER and CERAMI 1988; DINARELLO 1989). It is not clear how the macrophages control the cytotoxic molecules they secrete. Clearly, the macrophage is rather resistant to oxygen radicals. It produces catalase and superoxide dismutase, which inactivate reactive oxygen intermediates (ADAMS and HAMILTON 1984; TAKEMURA and WERB 1984b; NATHAN 1987). Macrophages also secrete several complement inhibitors, including α_2-macroglobulin, α_1-proteinase inhibitor, and C3-inhibitor, which may attenuate complement proteinases as well as restrict complement activation to its locus (TAKEMURA AND WERB 1984a, b; RAPPOLEE and WERB 1989). Production of reactive oxygen intermediates by macrophages is down-regulated by TGF-β (TSUNAWAKI et al. 1988).

Wound debridement is mediated by lysosomal hydrolases and later by neutral proteinases, which break down debris in the extracellular milieu, and by phagocytosis by macrophages and neutrophils. M-CSF synergizes with other stimulators to induce lysosomal hydrolases. Neutral proteinases, such as the serine proteinase urokinase plasminogen activator, are induced by M-CSF and TGF-α, and metalloproteinases, such as collagenase, are induced in fibroblasts, endothelial cells, and synovial cells by IL-1 and PDGF (LEE and WEINSTEIN 1978; LIN and GORDON 1979; POSTLETHWAITE et al. 1983; BAUER et al. 1985; CHUA et al. 1985; MATRISIAN et al. 1985; EDWARDS et al. 1987; SCHNYDER et al. 1987; DERYNCK 1988; DINARELLO 1989). Tissue plasminogen activator is also produced by macrophages (HART et al. 1989a, b). Uptake of the matrix debris, such as collagen fragments, induces macrophages to produce IL-1 and PGE_2. PGE_2, in turn, induces macrophages themselves to produce collagenase (FRISCH and RULEY 1987; Table 3). Collagenase and stromelysin are induced in synovial cells by IL-1 and TNF-α, and IL-1 induces collagenase in dermal fibroblasts (FRISCH and RULEY 1987; see also Table 3). Collagenase expression can also be induced in fibroblasts by several macrophage-derived growth factors, including PDGF, bFGF, and TGF-α/EGF (POSTLETHWAITE et al. 1983; CHUA et al. 1985; DAYER et al. 1985; EDWARDS et al. 1987). PDGF and TGF-α are produced by wound-derived macrophages (RAPPOLEE et al. 1988; RAPPOLEE and WERB 1990). The metalloproteinases stromelysin and collagenase are down-regulated transcriptionally by TGF-β (EDWARDS et al. 1987; ROBERTS and SPORN 1990). In addition, TGF-β, PDGF, and IL-1 induce synthesis of TIMP (EDWARDS et al. 1987; DINARELLO 1989; ROBERTS and SPORN 1990). These effects may limit the effects of the proteinases in the wound and down-regulate the production of the proteinases as the wound clears. Phagocytosis by macrophages and neutrophils is enhanced by M-CSF, GM-CSF (which induces IgA receptor expression in neutrophils), and IL-1 (WEISBART et al. 1988).

As the wound resolves, 3–7 days after trauma, dead endothelial cells (vascular beds), fibroblasts, and epidermal cells must be regenerated and extracellular matrix must be replaced. By the end of the first week after wounding, fibroblasts and endothelial cells have filled in the wound with loose connective tissue and a dense capillary network, respectively. The mass of capillaries and fibroblasts is called "granulation tissue." By the end of the second week the capillary network has thinned and fibrosis of collagen has increased (STOSSEL 1988; RAPPOLEE and WERB 1989). Ablation experiments indicate that macrophages, but not neutrophils, are required for this wound healing and angiogenesis (LEIBOVICH and ROSS 1975). Others have found that activated macrophages and wound fluid induce wound healing (POLVERINI et al. 1977; GREENBURG and HUNT 1978; BANDA et al. 1982; KOCH et al. 1986).

Macrophages secrete a number of growth factors that are known to mediate angiogenesis and an overlapping group of growth factors that induce formation of granulation tissue and reepithelialization. bFGF is synthesized by stimulated macrophages and induces fibroplasia, DNA synthesis in endothelial cells, and angiogenesis. IL-1 has limited mitogenic capacity for fibroblasts but no angiogenic property. Fibroblast mitogenesis is mediated by the ability of IL-1 to induce fibroblast PDGF-A, and a blocking antibody to PDGF-A prevents IL-1-induced mitogenesis of fibroblasts (RAINES et al. 1989). Other macrophage-derived growth factors that induce fibroblast proliferation of fibrosis in vivo or in vitro are PDGF, TGF-β, TGF-α, IGF-I, and bombesin. Angiogenesis consists of endothelial sprouting, which can account for up to 1 mm of capillary growth, and endothelial cell mitosis, which is required for further capillary lengthening (FOLKMAN 1986; RAPPOLEE and WERB 1989). TGF-β is known to cause endothelial cell chemotaxis in vitro (and sprouting in vivo) but actually inhibits endothelial cell mitosis (ROBERTS et al. 1986; HEIMARK et al. 1986; MASSAGUE 1987). TNF-α has been claimed by two investigative groups to be angiogenic. One group claims that TNF-α is directly angiogenic for endothelial cells, but because it is not mitogenic for these cells the angiogenesis is limited to capillary sprouting (LEIBOVICH et al. 1987). The second group concludes that the angiogenic effect is secondary to the chemoattractant activity of TNF-α for monocytes, which produce other angiogenic factors (FRÄTER-SCHRÖDER et al. 1987). Endothelial cells that are motile (e.g., those at sprouting capillaries) are more sensitive to TNF-α than are confluent endothelial cells in mature blood vessels (GERLACH et al. 1989). Other factors that have been shown to be mitogenic for endothelial cells in vitro or angiogenic in vivo are bFGF, TGF-α, G-CSF, and GM-CSF (THOMAS et al. 1985; SCHREIBER et al. 1986; BURGESS and MACIAG 1989; BUSSOLINO et al. 1989). It is interesting to note that hypoxia, a condition common to non-vascularized wound foci, induces macrophages to secrete a nonmitogenic angiogenic factor in vitro (KNIGHTON et al. 1983). PF-4 is an immediate-response growth factor produced by macrophages that is antiangiogenic (MAIONE et al. 1990; RAPPOLEE et al., unpublished data). It is likely that this factor is TNF-α. As well as being an inducer of granulation tissue and neovascularization, TGF-α also accelerates reepithelialization when applied in vivo (SCHULTZ et al. 1987).

It is clear that several macrophage-derived growth factors mediate the killing functions of leukocytes, the debridement functions of leukocytes, and the wound-healing functions of fibroblasts, endothelial cells, and epidermal cells.

6 Nerve Regeneration

Macrophages are prominent in peripheral nerve regeneration. Upon crushing or cutting of peripheral nerve, there is an immediate "wallerian" degeneration of the distal stump, which consists of the fragmentation of Schwann cell cytoplasm and breakdown of the distal axon. Within a few days of the trauma, monocyte-derived macrophages enter the nerve and begin to debride it (PERRY et al. 1987). During the next 2–3 weeks, the nerve is debrided and the axon regrows, and the Schwann cells undergo mitosis to populate the regenerating nerve to about 10 times their original number. If the nerve is cut and explanted to the peritoneum inside a millipore cylinder that prevents the entrance of macrophages, no fragmented axon and Schwann cell debridement and no Schwann cell mitosis occur (SCHEIDT et al. 1986). If the cylinder allows ingression of macrophages, then both debridement and Schwann cell gliosis occur (SCHEIDT et al. 1986). This suggests that both degeneration and regeneration (debridement and Schwann cell mitosis) are under the control of macrophages (HEUMANN et al. 1987; MAHLEY 1988; BAUER et al. 1989; BOYLES et al. 1989). The regenerating nerve undergoes waves of NGF and apolipoprotein E expression at times that correspond to the influx of macrophages. Macrophages synthesize apolipoprotein E after stimulation by products of injured peripheral nerves (BASU et al. 1981; BROWN and GOLDSTEIN 1983; WERB and CHIN 1983), and this synthesis is regenerative (IGNATIUS et al. 1987). If the peripheral nerve is explanted into culture, NGF is not expressed. However, IL-1 and TNF-α can replace the monocyte-derived macrophages (which do not enter the explanted nerve) and induce the transcription of NGF in cells in nerve (LINDHOLM et al. 1987, 1988; UNDERWOOD et al. 1990). Since macrophages do not synthesize NGF, macrophage-derived TNF-α and IL-1 are probably inducing NGF mRNA in the major cell of the nerve, the Schwann cell. In addition, TGF-β induces the synthesis of TIMP in the nerve, and TGF-β mRNA itself increases by more than tenfold in the crushed peripheral nerve (UNDERWOOD et al. 1990). It is tempting to speculate that TIMP protects the basal lamina in the nerve to guide the regrowing axons back to the correct target tissue.

7 The Immune Response

Macrophages are important in the afferent or generative arm of the immune response (MacKaness 1964). Macrophages endocytose and digest antigens and then present them to T helper cells in a complex with the Ia antigens on the macrophage surface (Benacerraf and Unanue 1979; Unanue and Allen 1987). This occurs in the immune tissues in lymph nodes, spleen, skin, and brain (Unanue and Allen 1987; Hickey and Kimura 1988). GM-CSF enhances, but M-CSF suppresses, Ia expression in macrophages (Willman et al. 1989). This presentation is required for the generation of immune responses to many T-dependent antigens and makes the T cells "competent" by inducing expression of IL-2 receptors. IL-1 can act as a cofactor to antigen/Ia antigen complex and stimulates T cells to produce IL-2, which stimulates progression into S phase by its receptors (Rappolee and Werb 1989). IL-6 also can act as a thymocyte comitogen.

Interleukin-1 enhances the humoral immune response by several mechanisms. First, it stimulates B cell differentiation by inducing a pre-B-cell line with only cytoplasmic μ chains to express κ light chains and subsequent surface immunoglobulin. IL-1 also enhances proliferation and the secretion of immunoglobulin in mature B cells (Durum et al. 1985; Pike and Nossal 1985; Kurt-Jones et al. 1987; Dinarello 1989). Stimulated macrophages additionally produce B cell stimulatory factor-2, also known as IL-6, and hybridoma growth factor (Gauldie et al. 1987; Van Damme et al. 1987; Tosato et al. 1988). IL-6 and IL-1 are major products of stimulated macrophages. Like IL-1, IL-6 induces B cell proliferation and immunoglobulin expression (Beagley et al. 1989). TGF-β also modulates immunoglobulin expression (Coffman et al. 1989; Sonoda et al. 1989). It is difficult to distinguish the relative contributions of IL-1 and IL-6 to B cell and T cell activation because T cells can synthesize IL-1 and B cells can synthesize IL-6, and each factor induces the expression of the other (Dinarello 1989). Another effect of IL-1 on immune cells is the induction of natural killer activity. There is more than one IL-1 gene, transcript, and protein in all species surveyed. The current hypothesis is that all IL-1 species have similar immune effects and act through a single immune cell receptor (Dinarello 1989). The development of the humoral immune response requires the secretion of IL-1 and IL-6 by macrophages, as well as expression of class II MHC molecules and processed antigen on the macrophage surface.

The development of a cellular immune response requires an interaction between macrophages and T cells and is represented by delayed-type hypersensitivity. This reaction occurs in previously sensitized individuals and requires 48 h to develop, whereas the immediate hypersensitivity of the humoral immune response subsides by 48 h. Delayed-type hypersensitivity is characterized by class II MHC-restricted interaction between the T cells and macrophages that have migrated into the interstitial site of bacterial infection. The macrophages outnumber the T cells in these lesions by more than 10 to 1, but activated T cells are required to trigger macrophage microbicidal activity (Benacerraf and

UNANUE 1979). A complex of soluble factors may mediate communication between macrophages and T cells during delayed-type hypersensitivity. Activated T cells secrete factors IL-2, GM-CSF, and interferon-γ, which attract macrophages, activate them, induce Ia expression by macrophages, and prevent their departure. In response, activated macrophages produce IL-1, TNF-α, M-CSF, and MIP-1, which further attract and activate T cells and macrophages (UNANUE and ALLEN 1987). In summary, macrophages and macrophage-derived growth factors are essential in the generation of the humoral and cellular immune responses. As previously mentioned, the cytocidal response of macrophages induced by activated T cells is an important part of the efferent immune response.

8 Hematopoiesis

Since most blood cells are short-lived, they must be replaced constantly, and production must be increased under conditions of stress such as inflammation. The increase in production occurs in the bone marrow, but in times of stress the spleen can become the primary organ of hematopoiesis.

The eight major types of blood cells arise from a common precursor stem cell in a series of three tiers of differentiation and proliferation (Fig. 3). The first tier is composed of 5-hydroxyurea-resistant nonmitotic stem cells that respond to unknown growth factors and become hemopoietin (IL-1α)-responsive and then IL-3-responsive stem cells en route to the second tier. The second tier consists of highly mitotic committed progenitor cells that respond to a broad array of CSFs in vitro by proliferating and differentiating. GM-CSF, G-CSF, M-CSF, and other factors (Fig. 3) stimulate these second-tier cells to become morphologically distinct but functionally immature end cells. At this stage the specific lineage cells may respond to primarily lineage-specific factors, such as IL-5 (eosinophils), G-CSF (neutrophils), erythropoietin (erythrocytes and megakaryocytes), and M-CSF (macrophages) (METCALF 1985).

Each growth factor has a hierarchical effect on hematopoiesis in regard to lineage specificity: M-CSF is macrophage lineage specific; GM-CSF most readily stimulates macrophages and secondarily stimulates neutrophils; IL-6 affects macrophages primarily and neutrophils secondarily; and G-CSF affects neutrophils primarily and macrophages secondarily (METCALF 1985). In any given circumstance in vivo, local cell types, interregulation of expression of CSF receptor, and factor concentration may have effects not predicted by in vitro dose-response experiments. In vitro CSFs maintain cell survival and mitosis but can also drive differentiation events. Cell lines dependent on one factor can be driven into terminal differentiation by other CSFs. This may mean that mitotically active cells can be driven to differentiate if a second factor is present (NICOLA 1989). In addition, recombinant CSFs are synergistic in vitro and in vivo. Taken together, these results suggest that expansion and differentiation can be uncoupled in such a way that the kinetics and magnitude of production of various growth factors will determine the outcome of the type and number of cells released from the bone marrow or spleen into the blood during various pathophysiologic responses. The type and number of cells released from the bone marrow will be determined by a combination of growth factors originating from blood, stromal cells, and the hematopoietic cells themselves.

Although recombinant CSFs can have hematopoietic effects when introduced in vivo or in vitro, there are small or subthreshold concentrations in the blood both normally and in times of inflammatory stimulation. Little is understood about the endocrine influences of CSFs on hematopoiesis in bone marrow. M-CSF is unique in that it can stimulate a limited proliferation of macrophages in nonhematopoietic tissues. The recent work of CROCKER and co-workers suggests that macrophages in bone marrow and spleen may be important in some form of trophic interaction during hematopoiesis (CROCKER and GORDON 1989). The production of erythrocyte differentiation factor (activin) by macrophages may be important in this interaction (ERÄMAA et al. 1990). Macrophages are the major producers of IL-1, IL-6, and G-CSF and produce large amounts of GM-CSF and M-CSF. The relative importance of macrophages or macrophage-derived growth factors is poorly understood in relation to normal or pathophysiologic homeostasis of blood cell levels.

9 Conclusions and Future Directions

Macrophages must grow, differentiate, and mediate homeostasis by conversing with other cells in their milieu. An important part of this cell communication is mediated by macrophage-derived growth factors. As the list of growth factors increases it will be important to expand the phenotype of macrophage-derived growth factors. About half of the currently known growth factors are uncharacterized with respect to macrophages. Other factors, such as soluble immune response suppressor and parathyroid hormone, may be processed but not synthesized by macrophages (AUNE and PIERCE 1981; DIMENT et al. 1989). More growth factors may be cloned directly with macrophages as a source; other extracellular matrix molecules (such as laminin) with functional growth factor domains (Table 13) will also be characterized in macrophages. Next, the transcriptional and translational controls of macrophage-derived growth factors will be defined in more detail. These can be defined in vitro by biochemical and molecular biologic methods and by examining genomic 5′ flanking sequences for possible control by *trans*-acting factors (ECONOMOU et al. 1989). Regulation of growth factor transcription in macrophages by second messengers is also being studied (ADEREM et al. 1988; PRPIC et al. 1989). These phenotypes and controls will be correlated with functions of the various distinct subpopulations of macrophages in vitro and with macrophage ontogeny. Much of the functional capability of growth factors will be defined by testing recombinant growth factors in vivo and in vitro with a knowledge of the expression of cognate receptors on the responding cells used to interpret responses. It will be important to determine the spatial expression of macrophage-derived growth factors by macrophages and other cell types by in situ analysis, reverse transcription-polymerase chain reaction, or immunocytochemistry (BAYNE et al. 1986; RAPPOLEE et al. 1988; REMICK et al. 1988). There are few good genetic models for macrophage function. The osteopetrotic mouse (*op/op*), which lacks a functional M-CSF gene (WIKTOR-JEDRZEJCZAK et al. 1990), shows promise. The construction of transgenic animals that express macrophage-derived growth factors is beginning to shed light on macrophage function, and tumor cells or macrophages that express macrophage-derived growth factors and can be injected into syngeneic mice are proving useful (LANG et al. 1987; JOHNSON et al. 1989; YOSHIDA et al. 1990). In mice, macrophage-derived growth factors can also be ablated by homologous recombination, although this is a more difficult technology (DE CHIARA et al. 1990). It may be possible to ablate macrophages or macrophage-derived growth factors by inserting heterologous promoters or macrophage-specific enhancer combinations into suicide genes or antisense growth factor constructs. In addition, understanding of macrophage-derived growth factors may allow better clinical applications through use of recombinant macrophage-derived growth factors, or expression of transformed endogenous macrophages expressing combinations of macrophage-derived growth factors.

The complex negative regulation of macrophage-derived growth factor expression is now partially understood. A large group of macrophage-derived growth factors are negatively regulated by the prostaglandins that also mediate some of their positive effects (OLD 1985; ROSS et al. 1986; DINARELLO 1989). However, one recently characterized macrophage-derived growth factor (MIP-1) does not act through prostaglandin, and its pyrogenic effects are not inhibited by indomethacin (DAVATELIS et al. 1989). Negative feedback of IL-1 is mediated by IL-1 induction of glucocorticoids (BESEDOVSKY et al. 1986). MIP-1α was recently shown to inhibit bone marrow colony formation in vitro by GM-CSF or M-CSF (GRAHAM et al. 1990). A competitive inhibitor of IL-1 with sequence homology to IL-1 was also recently cloned (EISENBERG et al. 1990). As neuro-immunologic interactions are further characterized, they may define further negative regulatory loops between the nervous system and macrophages, although only inductive effects are currently known (EDWARDS et al. 1988; LOTZ et al. 1988). Finally, the kinetics of macrophage-derived growth factor induction and attenuation may be defined, as has been partially done for TNF-α and IL-1 in macrophages.

The macrophage is an excellent model system for understanding the production and function of growth factors. Not only will it provide a satisfactory understanding of mechanisms of communication of metazoan cells, but this understanding will lead to a clinically relevant understanding of macrophage function in pathophysiology and pathology.

Table 13. Miscellaneous growth factors

Growth factor	Synthesis by macrophages	Specific properties	mRNA		Protein		Reference
			Size (kb)	AUUUA instability sequence	Precursor amino acid number	Mature peptide (kDa)	
OTP-1/IFN-α_{11}			1.1	+	172		Imakawa et al. 1987
IFN-γ	–		0.8		155	40–80	Gray and Goeddel 1983;
NGF-β	–		1.3	+	118 × 2 (dimer)	27	Ullrich et al. 1983; Scott et al. 1983
Neurotrophin-3		Related to NGF	1.4	+	120		Maisonpierre et al. 1990;
BDNF		Related to NGF	1.5	–	252	13.5	Lin et al. 1989; Leibrock et al.1989
CNTF			4.3		200	22.7	Stöckli et al. 1989
int-1 (12 members)			2.4	+	370		van Ooyen and Nusse 1984
IL-3	–		1.0	+	166	15–30	Fung et al. 1984; Yang et al. 1986
IL-4	–		0.7		153	15/50	Arai et al. 1989
IL-5	–	Related to IL-4	1.7	+	132	32–62	Tanabe et al. 1987
Proliferin		Related to prolactin	1.0		224	25	Linzer and Nathans 1984
Epithelin (2 members)						6.0	Shoyab et al. 1990
IL-7	–		1.8/2.4	–	154	15	Goodwin et al. 1989
IL-10	–		1.0/1.5		178	17.4	Moore et al. 1990
PD ECGF			1.8	+	482	45	Ishikawa et al. 1989
HGF			6.0	–	728	82	Nakamura et al. 1989
Neuroleukin	–		2.0	–	558	56	Gurney et al. 1986
Endothelin (3 members)			2.3	+	200 (21)	2.5	Yanagisawa et al. 1988; Inoue et al. 1989a, b; Bloch et al. 1989

Thymopoietin/ splenin (2 members)				49	5.0	Audhya et al. 1987
Fibronectin	+	Heparin binding	24 forms 1.2–8.5	2500 (approx)	550 (dimer)	Schwarzbauer et al. 1983; Blatti et al. 1988; Barone et al. 1989
Bombesin/GRP	+		0.8/3.8/7.8	147(27)	3.0	Wiedermann et al. 1986; Sausville et al. 1986; Lebacq-Verheyden et al. 1988
Substance P	+		1.0/1.2	130/115(11)	1.2	Krause et al. 1987
Thymosin (2 members)	+	No signal sequence	0.8/1.2/1.4	109/111	43	Goodall et al. 1986; Eschenfeldt and Berger 1986; Gondo et al. 1987
Laminin A		EGF ·			400	Panayotou et al. 1989
B1		domains	~8.0	1786	200	Sasaki et al. 1987
B2			5.5/7.6	1599	200	Pikkarainen et al. 1988

OTP, ovine trophoblast protein; NGF, nerve growth factor; BDNF, brain-derived neurotrophic factor; CNTF, ciliary neuron trophic factor; PD ECGF, platelet-derived endothelial cell growth factor; HGF, hepatocyte growth factor; GRP, gastrin-releasing peptide

Acknowledgments. We thank Rick Lyman for typing the manuscript. This work was supported by a contract from the Office of Health and Environmental Research, U.S. Department of Energy (DE-AC03-76-SF01012) and a National Research Service Award (5 T32 ES07106) from the National Institute of Environmental Health Sciences.

References

Abdullah NA, Torres BA, Basu M, Johnson HM (1989) Differential effects of epidermal growth factor, transforming growth factor-α, and vaccinia virus growth factor in the positive regulation of IFN-γ production. J Immunol 143: 113–117

Abe T, Murakami M, Sato T, Kajiki M, Ohno M, Kodaira R (1989) Macrophage differentiation inducing factor from human monocytic cells is equivalent to murine leukemia inhibitory factor. J Biol Chem 264: 8941–8945

Abraham JA, Whang JL, Tumolo A, Mergia A, Friedman J, Gospodarowicz D, Fiddes JC (1986a) Human basic fibroblast growth factor: nucleotide sequence and genomic organization. EMBO J 5: 2523–2528

Abraham JA, Mergia A, Whang JL et al. (1986b) Nucleotide sequence of a bovine clone encoding the angiogenic protein, basic fibroblast growth factor. Science 233: 545–548

Adams DO, Hamilton TA (1984) The cell biology of macrophage activation. Annu Rev Immunol 2: 283–318

Adams DO, Hamilton TA (1989) The activated macrophage and granulomatous inflammation. Curr Top Pathol 79: 151–167

Aderem AA, Albert KA, Keum MM, Wang JK, Greengard P, Cohn ZA (1988) Stimulus-dependent myristoylation of a major substrate for protein kinase C. Nature 332: 362–364

Alitalo K, Hovi T, Vaheri A (1980) Fibronectin is produced by human macrophages. J Exp Med 151: 602–613

Alvaro-Gracia JM, Zvaifler NJ, Firestein GS (1989) Cytokines in chronic inflammatory arthritis. IV. Granulocyte/macrophage colony-stimulating factor-mediated induction of class II MHC antigen on human monocytes: a possible role in rheumatoid arthritis. J Exp Med 170: 865–875

Arai N, Nomura D, Villaret D et al. (1989) Complete nucleotide sequence of the chromosomal gene for human IL-4 and its expression. J Immunol 142: 274–282

Arend WP, Welgus HG, Thompson RC, Eisenberg SP (1990) Biological properties of recombinant human monocyte-derived interleukin 1 receptor antagonist. J Clin Invest 85: 1694–1697

Assoian RK, Fleurdelys BE, Stevenson HC et al. (1987) Expression and secretion of type β transforming growth factor by activated human macrophages. Proc Natl Acad Sci USA 84: 6020–6024

Audhya T, Schlesinger DH, Goldstein G (1987) Isolation and complete amino acid sequence of human thymopoietin and splenin. Proc Natl Acad Sci USA 84: 3545–3549

Aune TM, Pierce CW (1981) Conversion of soluble immune response suppressor to macrophage-derived suppressor factor by peroxide. Proc Natl Acad Sci USA 78: 5099–5103

Auron PE, Webb AC, Rosenwasser LJ, Mucci SF, Rich A, Wolff SM, Dinarello CA (1984) Nucleotide sequence of human monocyte interleukin 1 precursor cDNA. Proc Natl Acad Sci USA 81: 7907–7911

Baird A, Mormède P, Böhlen P (1985) Immunoreactive fibroblast growth factor in cells of peritoneal exudate suggests its identity with macrophage-derived growth factor. Biochem Biophys Res Commun 126: 358–364

Banda MJ, Knighton DR, Hunt TK, Werb Z (1982) Isolation of a nonmitogenic angiogenesis factor from wound fluid. Proc Natl Acad Sci USA 79: 7773–7777

Barone MV, Henchcliffe C, Baralle FE, Paolella G (1989) Cell type specific *trans*-acting factors are involved in alternative splicing of human fibronectin pre-mRNA. EMBO J 8: 1079–1085

Bartocci A, Mastrogiannis DS, Migliorati G, Stockert RJ, Wolkoff AW, Stanley ER (1987) Macrophages specifically regulate the concentration of their own growth factor in the circulation. Proc Natl Acad Sci USA 84: 6179–6183

Basu SK, Brown MS, Ho YK, Havel RJ, Goldstein JL (1981) Mouse macrophages synthesize and secrete a protein resembling apolipoprotein E. Proc Natl Acad Sci USA 78: 7545–7549

Bauer EA, Cooper TW, Huang JS, Altman J, Deuel TF (1985) Stimulation of in vitro human skin collagenase expression by platelet-derived growth factor. Proc Natl Acad Sci USA 82: 4132–4136

Bauer J, Bauer TM, Kalb T et al. (1989) Regulation of interleukin 6 receptor expression in human monocytes and monocyte-derived macrophages. J Exp Med 170: 1537–1549
Baumann H, Wong GG (1989) Hepatocyte-stimulating factor III shares structural and functional identity with leukemia-inhibitory factor. J Immunol 143: 1163–1167
Bayne EK, Rupp EA, Limjuco G, Chin J, Schmidt JA (1986) Immunocytochemical detection of interleukin 1 within stimulated human monocytes. J Exp Med 163: 1267–1280
Beagley KW, Eldridge JH, Lee F et al. (1989) Interleukins and IgA synthesis. Human and murine interleukin 6 induce high rate IgA secretion IgA-committed B cells. J Exp Med 169: 2133–2148
Becker S, Warren MK, Haskill S (1987) Colony-stimulating factor-induced monocyte survival and differentiation into macrophages in serum-free cultures. J Immunol 139: 3703–3709
Becker S, Devlin RB, Haskill JS (1989) Differential production of tumor necrosis factor, macrophage colony stimulating factor, and interleukin 1 by human alveolar macrophages. J Leukocyte Biol 45: 353–361
Begg GS, Pepper DS, Chesterman CN, Morgan FJ (1978) Complete covalent structure of human β-thromboglobulin. Biochemistry 17: 1739–1744
Bell GI, Stempien MM, Fong NM, Rall LB (1986a) Sequences of liver cDNAs encoding two different mouse insulin-like growth factor I precursors. Nucleic Acids Res 14: 7873–7882
Bell GI, Fong NM, Stempien MM et al. (1986b) Human epidermal growth factor precursor: cDNA sequence, expression *in vitro* and gene organization. Nucleic Acids Res 14: 8427–8446
Benacerraf B, Unanue ER (1979) Cellular immunity and delayed hypersensitivity. In: Textbook of immunology. Williams and Wilkins, Baltimore MD, pp 112–122
Bentley C, Zimmer B, Hadding U (1981) The macrophages as a source of complement components. In: Pick E (ed) Lymphokines, vol 4. Academic, New York, pp 197–230
Berk BC, Alexander RW, Brock TA, Gimbrone MA Jr, Webb RC (1986) Vasoconstriction: a new activity for platelet-derived growth factor. Science 232: 87–90
Bernton EW, Meltzer MS, Holaday JW (1988) Suppression of macrophage activation and T-lymphocyte function in hypoprolactinemic mice. Science 239: 401–404
Besedovsky H, del Rey A, Sorkin E, Dinarello CA (1986) Immunoregulatory feedback between interleukin-1 and glucocorticoid hormones. Science 233: 652–654
Betsholtz C, Johnsson A, Heldin CH et al. (1986) cDNA sequence and chromosomal localization of human platelet-derived growth factor A-chain and its expression in tumour cell lines. Nature 320: 695–699
Beutler B, Cerami A (1988) Tumor necrosis, cachexia, shock, and inflammation: a common mediator. Annu Rev Biochem 57: 505–518
Beutler B, Krochin N, Milsark IW, Luedke C, Cerami A (1986) Control of cachectin (tumor necrosis factor) synthesis: mechanisms of endotoxin resistance. Science 232: 977–980
Bevilacqua MP, Pober JS, Majeau GR, Fiers W, Cotran RS, Gimbrone MA Jr (1986) Recombinant tumor necrosis factor induces procoagulant activity in cultured human vascular endothelium: characterization and comparison with the actions of interleukin 1. Proc Natl Acad Sci USA 83: 4533–4537
Bevilacqua MP, Stengelin S, Gimbrone MA Jr, Seed B (1989) Endothelial leukocyte adhesion molecule 1: an inducible receptor for neutrophils related to complement regulatory proteins and lectins. Science 243: 1160–1165
Blatti SP, Foster DN, Ranganathan G, Moses HL, Getz MJ (1988) Induction of fibronectin gene transcription and mRNA is a primary response to growth-factor stimulation of AKR-2B cells. Proc Natl Acad Sci USA 85: 1119–1123
Bloch KD, Eddy RL, Shows TB, Quertermous T (1989) cDNA cloning and chromosomal assignment of the gene encoding endothelin 3. J Biol Chem 264: 18156–18161
Boyles JK, Zoellner CD, Anderson LJ et al. (1989) A role for apolipoprotein E, apolipoprotein A-1, and low density lipoprotein receptors in cholesterol transport during regeneration and remyelination of the rat sciatic nerve. J Clin Invest 83: 1015–1031
Brachmann R, Lindquist PB, Nagashima M, Kohr W, Lipari T, Napier M, Derynck R (1989) Transmembrane TGF-alpha precursors activate EGF/TGF-alpha receptors. Cell 56: 691–700
Brawerman G (1989) mRNA decay: finding the right targets. Cell 57: 9–10
Brett J, Gerlach H, Nawroth P, Steinberg S, Godman G, Stern D (1989) Tumor necrosis factor/cachectin increases permeability of endothelial cell monolayers by a mechanism involving regulatory G proteins. J Exp Med 169: 1977–1991
Brookes S, Smith R, Thurlow J, Dickson C, Peters G (1989) The mouse homologue of *hst*/k-FGF: sequence, genome organization and location relative to *int*-2. Nucleic Acids Res 17: 4037–4045

Brouckaert P, Spriggs DR, Demetri G, Kufe DW, Fiers W (1989) Circulating interleukin 6 during a continuous infusion of tumor necrosis factor and interferon γ. J Exp Med 169: 2257–2262
Brown KD, Zurawski SM, Mosmann TR, Zurawski G (1989) A family of small inducible proteins secreted by leukocytes are members of a new superfamily that includes leukocyte and fibroblast-derived inflammatory agents, growth factors, and indicators of various activation processes. J Immunol 142: 679–687
Brown MS, Goldstein JL (1983) Lipoprotein metabolism in the macrophage: implications for cholesterol deposition in atherosclerosis. Annu Rev Biochem 52: 223–261
Broxmeyer HE, Juliano L, Waheed A, Shadduck RK (1985) Release from mouse macrophages of acidic isoferritins that suppress hematopoietic progenitor cells is induced by purified L cell colony stimulating factor and suppressed by human lactoferrin. J Immunol 135: 3224–3231
Broxmeyer HE, Sherry B, Lu L, Cooper S, Carow C, Wolpe SD, Cerami A (1989) Myelopoietic enhancing effects of murine macrophage inflammatory proteins 1 and 2 on colony formation in vitro by murine and human bone marrow granulocyte/macrophage progenitor cells. J Exp Med 170: 1583–1594
Burd PR, Freeman GJ, Wilson SD, Berman M, DeKruyff R, Billings PR, Dorf ME (1987) Cloning and characterization of a novel T cell activation gene. J Immunol 139: 3126–3131
Burgess WH, Maciag T (1989) The heparin-binding (fibroblast) growth factor family of proteins. Annu Rev Biochem 58: 575–606
Bussolino F, Wang JM, Defilippi P et al. (1989) Granulocyte- and granulocyte-macrophage-colony stimulating factors induce human endothelial cells to migrate and proliferate. Nature 337: 471–473
Cantrell MA, Anderson D, Cerretti DP et al. (1985) Cloning, sequence, and expression of a human granulocyte/macrophage colony-stimulating factor. Proc Natl Acad Sci USA 82: 6250–6254
Carrel A (1912) On the permanent life of tissues outside the organism. J Exp Med 15: 516–528
Carter DB, Deibel MR Jr, Dunn CJ et al. (1990) Purification, cloning, expression and biological characterization of an interleukin-1 receptor antagonist protein. Nature 344: 633–638
Cate RL, Mattaliano RJ, Hession C et al. (1986) Isolation of the bovine and human genes for müllerian inhibiting substance and expression of the human gene in animal cells. Cell 45: 685–698
Cavender DE, Haskard DO, Joseph B, Ziff M (1986) Interleukin 1 increases the binding of human B and T lymphocytes to endothelial cell monolayers. J Immunol 136: 203–207
Chantry D, Turner M, Abney E, Feldmann M (1989) Modulation of cytokine production by transforming growth factor-β. J Immunol 142: 4295–4300
Chervenick PA, LoBuglio AF (1972) Human blood monocytes: stimulators of granulocyte and mononuclear colony formation in vitro. Science 178: 164–166
Chua CC, Geiman DE, Keller GH, Ladda RL (1985) Induction of collagenase secretion in human fibroblast cultures by growth promoting factors. J Biol Chem 260: 5213–5216
Ciccodicola A, Dono R, Obici S, Simeone A, Zollo M, Persico MG (1989) Molecular characterization of a gene of the "EGF family" expressed in undifferentiated human NTERA2 teratocarcinoma cells. EMBO J 8: 1987–1991
Coffman RL, Lebman DA, Shrader B (1989) Transforming growth factor β specifically enhances IgA production by lipopolysaccharide-stimulated murine B lymphocytes. J Exp Med 170: 1039–1044
Cohen S (1987) Epidermal growth factor. In Vitro Cell Dev Biol 23: 239–246
Cohn ZA (1978) The activation of mononuclear phagocytes: fact, fancy, and future. J Immunol 121: 813–816
Crocker PR, Gordon S (1989) Mouse macrophage hemagglutinin (sheep erythrocyte receptor) with specificity for sialylated glycoconjugates characterized by a monoclonal antibody. J Exp Med 169: 1333–1346
Dahinden CA, Kurimoto Y, De Weck AL, Lindley I, Dewald B, Baggiolini M (1989) The neutrophil-activating peptide NAF/NAP-1 induces histamine and leukotriene release by interleukin 3-primed basophils. J Exp Med 170: 1787–1792
Daniel TO, Gibbs VC, Milfay DF, Williams LT (1987) Agents that increase cAMP accumulation block endothelial c-*sis* induction by thrombin and transforming growth factor-β. J Biol Chem 262: 11893–11896
Danielpour D, Sporn MB (1990) Differential inhibition of transforming growth factor beta 1 and beta 2 activity by alpha 2-macroglobulin. J Biol Chem 265: 6973–6977
Darlington GJ, Wilson DR, Lachman LB (1986) Monocyte-conditioned medium, interleukin-1, and tumor necrosis factor stimulate the acute phase response in human hepatoma cells in vitro. J Cell Biol 103: 787–793

Davatelis G, Tekamp-Olson P, Wolpe SD et al. (1988) Cloning and characterization of a cDNA for murine macrophage inflammatory protein (MIP), a novel monokine with inflammatory and chemokinetic properties. J Exp Med 167: 1939–1944
Davatelis G, Wolpe SD, Sherry B, Dayer J-M, Chicheportiche R, Cerami A (1989) Macrophage inflammatory protein-1: a prostaglandin-independent endogenous pyrogen. Science 243: 1066–1068
Dayer J-M, Beutler B, Cerami A (1985) Cachectin/tumor necrosis factor stimulates collagenase and prostaglandin E_2 production by human synovial cells and dermal fibroblasts. J Exp Med 162: 2163–2168
De Chiara TM, Efstratiadis A, Robertson EJ (1990) A growth-deficiency phenotype in heterozygous mice carrying an insulin-like growth factor II gene disrupted by targeting. Nature 345: 78–80
Delli-Bovi P, Curatola AM, Kern FG, Greco A, Ittmann M, Basilico C (1987) An oncogene isolated by transfection of Kaposi's sarcoma DNA encodes a growth factor that is a member of the FGF family. Cell 50: 729–737
de Martin R, Haendler B, Hofer-Warbinek R et al. (1987) Complementary DNA for human glioblastoma-derived T cell suppressor factor, a novel member of the transforming growth factor-β gene family. EMBO J 6: 3673–3677
Derynck R (1988) Transforming growth factor α. Cell 54: 593–595
Derynck R (1990) Transforming growth factor-α. Mol Reprod Dev 27: 3–9
Derynck R, Roberts AB, Winkler ME, Chen EY, Goeddel DV (1984) Human transforming growth factor-α: precursor structure and expression in *E. coli*. Cell 38: 287–297
Derynck R, Jarrett JA, Chen EY et al. (1985) Human transforming growth factor-β complementary DNA sequence and expression in normal and transformed cells. Nature 316: 701–705
Derynck R, Jarrett JA, Chen EY, Goeddel DV (1986) The murine transforming growth factor-β precursor. J Biol Chem 261: 4377–4379
Derynck R, Rhee L, Chen EY, Van Tilburg A (1987) Intron-exon structure of the human transforming growth factor-β precursor gene. Nucleic Acids Res 15: 3188–3189
Derynck R, Lindquist PB, Lee A et al. (1988) A new type of transforming growth factor-beta, TGF-β 3. EMBO J 7: 3737–3743
Deuel TF (1987) Polypeptide growth factors: roles in normal and abnormal cell growth. Annu Rev Cell Biol 3: 443–492
Deuel TF, Keim PS, Farmer M, Heinrikson RL (1977) Amino acid sequence of platelet factor 4. Proc Natl Acad Sci USA 74: 2256–2258
Deuel TF, Senior RM, Chang D, Griffin GL, Heinrikson RL, Kaiser ET (1981) Platelet factor 4 is chemotactic for neutrophils and monocytes. Proc Natl Acad Sci USA 78: 4584–4587
Diment S, Martin KJ, Stahl PD (1989) Cleavage of parathyroid hormone in macrophage endosomes illustrates a novel pathway for intracellular processing of proteins. J Biol Chem 264: 13403–13406
Dinarello CA (1989) Interleukin-1 and its biologically related cytokines. Adv Immunol 44: 153–205
Doherty DE, Haslett C, Tonnesen MG, Henson PM (1987) Human monocyte adherence: a primary effect of chemotactic factors on the monocyte to stimulate adherence to human endothelium. J Immunol 138: 1762–1771
Doi T, Greenberg SM, Rosenberg RD (1987) Structure of the rat platelet factor 4 gene: a marker for megakaryocyte differentiation. Mol Cell Biol 7: 898–904
Donahue RE, Seehra J, Metzger M et al. (1988) Human IL-3 and GM-CSF act synergistically in stimulating hematopoiesis in primates. Science 241: 1820–1823
Duckworth WC (1988) Insulin degradation: mechanisms, products, and significance. Endocrine Rev 9: 319–345
Durum SK, Schmidt JA, Oppenheim JJ (1985) Interleukin 1: an immunological perspective. Annu Rev Immunol 3: 263–287
Economou JS, Rhoades K, Essner R, McBride WH, Gasson JC, Morton DL (1989) Genetic analysis of the human tumor necrosis factor α/cachectin promoter region in a macrophage cell line. J Exp Med 170: 321–326
Edwards CK III, Ghiasuddin SM, Schepper JM, Yunger LM, Kelley KW (1988) A newly defined property of somatotropin: priming of macrophages for production of superoxide anion. Science 239: 769–771
Edwards DR, Murphy G, Reynolds JJ, Whitham SE, Docherty AJ, Angel P, Heath JK (1987) Transforming growth factor beta modulates the expression of collagenase and metalloproteinase inhibitor. EMBO J 6: 1899–1904
Edwards RH, Rutter WJ, Hanahan D (1989) Directed expression of NGF to pancreatic beta cells in transgenic mice leads to selective hyperinnervation of the islets. Cell 58: 161–170

Eisenberg SP, Evans RJ, Arend WP, Verderber E, Brewer MT, Hannum CH, Thompson RC (1990) Primary structure and functional expression from complementary DNA of a human interleukin-1 receptor antagonist. Nature 343: 341–346

Engelmann H, Aderka D, Rubenstein M, Rotman D, Wallach D (1989) A tumor necrosis factor-binding protein purified to homogeneity from human urine protects cells from tumor necrosis factor toxicity. J Biol Chem 264: 11974–11980

Engelmann H, Novick D, Wallach D (1990) Two tumor necrosis factor-binding proteins purified from human urine. Evidence for immunological cross-reactivity with cell surface tumor necrosis factor receptors. J Biol Chem 265: 1531–1536

Epstein LB, Lackides GA, Smith DM (1991) The complex relationship of tumor necrosis factor and interleukin-1β in human monocytes. In: Bonavida B, Granger G (eds) Tumor necrosis factor: structure, mechanism of action, role in disease and therapy. Karger, Basel, pp 107–113

Erämaa M, Ritvos O, Hurme M, Voutilainen R (1990) Inducible expression of erythroid differentiation factor/inhibin βA chain mRNA in cultured human monocytes (abstract). In: Growth and differentiation factors in development, UCLA symposia on molecular biology. J Cell Biochem [Suppl] 14E: 56

Eschenfeldt WH, Berger SL (1986) The human prothymosin α gene is polymorphic and induced upon growth stimulation: evidence using a cloned cDNA. Proc Natl Acad Sci USA 83: 9403–9407

Farber JM (1990) A macrophage mRNA selectively induced by γ-interferon encodes a member of the platelet factor 4 family of cytokines. Proc Natl Acad Sci USA 87: 5238–5242

Fava R, Olsen N, Keski-Oja J, Moses H, Pincus T (1989) Active and latent forms of transforming growth factor β activity in synovial effusions. J Exp Med 169: 291–296

Finch PW, Rubin JS, Miki T, Ron D, Aaronson SA (1989) Human KGF is FGF-related with properties of a paracrine effector of epithelial cell growth. Science 245: 752–755

Folkman J (1986) Angiogenesis: What makes blood vessels grow? Int U Physiol Sci Am Physiol Soc 1: 199–202

Fong Y, Tracey KJ, Moldawer LL et al. (1989) Antibodies to cachectin/tumor necrosis factor reduce interleukin 1β and interleukin 6 appearance during lethal bacteremia. J Exp Med 170: 1627–1633

Forage RG, Ring JM, Brown RW et al. (1986) Cloning and sequence analysis of cDNa species coding for the two subunits of inhibin from bovine follicular fluid. Proc Natl Acad Sci USA 83: 3091–3095

Fraker DL, Langstein HN, Norton JA (1989) Passive immunization against tumor necrosis factor partially abrogates interleukin 2 toxicity. J Exp Med 170: 1015–1020

Fräter-Schröder M, Risau W, Hallmann R, Gautschi P, Böhlen P (1987) Tumor necrosis factor type α, a potent inhibitor of endothelial cell growth in vitro, is angiogenic in vivo. Proc Natl Acad Sci USA 84: 5277–5281

Frisch SM, Ruley HE (1987) Transcription from the stromelysin promoter is induced by interleukin-1 and repressed by dexamethasone. J Biol Chem 262: 16300–16304

Fung MC, Hapel AJ, Ymer S, Cohen DR, Johnson RM, Campbell HD, Young IG (1984) Molecular cloning of cDNA for murine interleukin-3. Nature 307: 233–237

Gamble JR, Harlan JM, Klebanoff SJ, Vadas MA (1985) Stimulation of the adherence of neutrophils to umbilical vein endothelium by human recombinant tumor necrosis factor. Proc Natl Acad Sci USA 82: 8667–8671

Ganz T, Rayner JR, Valore EV, Tumolo A, Talmadge K, Fuller F (1989) The structure of the rabbit macrophage defensin genes and their organ-specific expression. J Immunol 143: 1358–1365

Gauldie J, Richards C, Harnish D, Lansdorp P, Baumann H (1987) Interferon β_2/B-cell stimulatory factor type 2 shares identity with monocyte-derived hepatocyte-stimulating factor and regulates the major acute phase protein response in liver cells. Proc Natl Acad Sci USA 84: 7251–7255

Gearing DP, King JA, Gough NM (1988) Complete sequence of murine myeloid leukaemia inhibitory factor (LIF). Nucleic Acids Res 16: 9857

Gerlach H, Lieberman H, Bach R, Godman G, Brett J, Stern D (1989) Enhanced responsiveness of endothelium in the growing/motile state to tumor necrosis factor/cachectin. J Exp Med 170: 913–931 [Published erratum appears in J Exp Med 170: 1793 (1989)]

Gerrard TL, Siegel JP, Dyer DR, Zoon K-C (1987) Differential effects of interferon-α and interferon-γ on interleukin 1 secretion by monocytes. J Immunol 138: 2535–2540

Gondo H, Kudo J, White JW, Barr C, Selvanayagam P, Saunders GF (1987) Differential expression of the human thymosin-β_4 gene in lymphocytes, macrophages, and granulocytes. J Immunol 139: 3840–3848

Goodall GJ, Dominguez F, Horecker BL (1986) Molecular cloning of cDNA for human prothymosin α. Proc Natl Acad Sci USA 83: 8926–8928

Goodwin RG, Lupton S, Schmierer A et al. (1989) Human interleukin 7: molecular cloning and growth factor activity on human murine B-lineage cells. Proc Natl Acad Sci USA 86: 302–306
Gordon MY, Riley GP, Watt SM, Greaves MF (1987) Compartmentalization of a haematopoietic growth factor (GF-CSF) by glycosaminoglycans in the bone marrow microenvironment. Nature 326: 403–405
Grabstein KH, Urdal DL, Tushinski RJ et al. (1986) Induction of macrophage tumoricidal activity by granulocyte-macrophage colony-stimulating factor. Science 232: 506–508
Graham GJ, Wright EG, Hewick R et al. (1990) Identification and characterization of an inhibitor of haemopoietic stem cell proliferation. Nature 344: 442–444
Graves DT, Antoniades HN, (1988) Characterization of a high-molecular-weight protein immunoprecipitated by platelet-derived growth factor antisera. J Cell Physiol 137: 263–271
Gray A, Dull TJ, Ullrich A (1983) Nucleotide sequence of epidermal growth factor cDNA predicts a 128,000-molecular weight protein precursor. Nature 303: 722–725
Gray PW, Goeddel DV (1983) Cloning and expression of murine immune interferon cDNA. Proc Natl Acad Sci USA 80: 5842–5846
Gray PW, Glaister D, Chen E, Goeddel DV, Pennica D (1986) Two interleukin-1 genes in the mouse: cloning and expression of the cDNA for murine interleukin 1β. J Immunol 137: 3644–3648
Greenburg GB, Hunt TK (1978) The proliferative response in vitro of vascular endothelial and smooth muscle cells exposed to wound fluids and macrophages. J Cell Physiol 97: 353–360
Gross JL, Moscatelli D, Rifkin DB (1983) Increased capillary endothelial cell protease activity in response to angiogenic stimuli in vitro. Proc Natl Acad Sci USA 80: 2623–2627
Gurney ME, Heinrich SP, Lee MR, Yin H-S (1986) Molecular cloning and expression of neuroleukin, a neurotrophic factor for spinal and sensory neurons. Science 234: 566–574
Haley J, Crawford R, Hudson P, Scanlon D, Tregear G, Shine J, Niall H (1987) Porcine relaxin. Gene structure and expression. J Biol Chem 262: 11940–11946
Hall DJ, Brownlee C, Stiles CD, (1989) Interleukin-1 is a potent regulator of JE and KC gene expression in quiescent BALB/c fibroblasts. J Cell Physiol 141: 154–159
Hannum CH, Wilcox CJ, Arend WP et al. (1990) Interleukin-1 receptor antagonist activity of a human interleukin-1 inhibitor. Nature 343: 336–340
Hart PH, Burgess DR, Vitti GF, Hamilton JA (1989a) Interleukin-4 stimulates human monocytes to produce tissue-type plasminogen activator. Blood 74: 1222–1225
Hart PH, Vitti GF, Burgess DR, Singleton DK, Hamilton JA (1989b) Human monocytes can produce tissue-type plasminogen activator. J Exp Med 169: 1509–1514
Hebert JM, Basilico C, Goldfarb M, Haub D, Martin GR (1990) Isolation of cDNAs encoding four mouse FGF family members and characterization of their expression patterns during embryogenesis. Dev Biol 138: 454–463
Heimark RL, Twardzik DR, Schwartz SM (1986) Inhibition of endothelial regeneration by type-beta transforming growth factor from platelets. Science 233: 1078–1080
Heumann R, Lindholm D, Bandtlow C et al. (1987) Differential regulation of mRNA encoding nerve growth factor and its receptor in rat sciatic nerve during development, degeneration, and regeneration: role of macrophages. Proc Natl Acad Sci USA 84: 8735–8739
Hickey WF, Kimura H (1988) Perivascular microglial cells of the CNS are bone marrow-derived and present antigen in vivo. Science 239: 290–292
Hirano T, Yasukawa K, Harada H et al. (1986) Complementary DNA for a novel human interleukin (BSF-2) that induces B lymphocytes to produce immunoglobulin. Nature 324: 73–76
Hoang T, Levy B, Onetto N, Haman A, Rodriguez-Cimadevilla JC (1989) Tumor necrosis factor α stimulates the growth of the clonogenic cells of acute myeloblastic leukemia in synergy with granulocyte/macrophage colony-stimulating factor. J Exp Med 170: 15–26
Horiguchi J, Warren MK, Kufe D (1987) Expression of the macrophage-specific colony-stimulating factor in human monocytes treated with granulocyte-macrophage colony-stimulating factor. Blood 69: 1259–1261
Hovi T, Mosher D, Vaheri A (1977) Cultured human monocytes synthesize and secrete α_2-macroglobulin. J Exp Med 145: 1580–1589
Ignatius MJ, Shooter EM, Pitas RE, Mahley RW (1987) Lipoprotein uptake by neuronal growth cones in vitro. Science 236: 959–962
Imakawa K, Anthony RV, Kazemi M, Marotti KR, Polites HG, Roberts RM (1987) Interferon-like sequence of ovine trophoplast protein secreted by embryonic trophectoderm. Nature 330: 377–379
Inoue A, Yanagisawa M, Kimura S, Kasuya Y, Miyauchi T, Goto K, Masaki T (1989a) The human endothelin family: three structurally and pharmacologically distinct isopeptides predicted by three separate genes. Proc Natl Acad Sci USA 86: 2863–2867

Inoue A, Yanagisawa M, Takuwa Y, Mitsui Y, Kobayashi M, Masaki T (1989b) The human preproendothelin-1 gene. J Biol Chem 264: 14954–14959
Introna M, Bast RC Jr, Tannenbaum CS, Hamilton TA, Adams DO (1987) The effect of LPS on expression of the early "competence" genes JE and KC in murine peritoneal macrophages. J Immunol 138: 3891–3896
Ishikawa F, Miyazono K, Hellman U et al. (1989) Identification of angiogenic activity and the cloning and expression of platelet-derived endothelial cell growth factor. Nature 338: 557–562
Jakowlew SB, Dillard PJ, Sporn MB, Roberts AB (1988) Complementary deoxyribonucleic acid cloning of a messenger ribonucleic acid encoding transforming growth factor β 4 from chicken embryo chondrocytes. Mol Endocrinol 2: 1186–1195
Jaye M, Howk R, Burgess W et al. (1986) Human endothelial cell growth factor: cloning, nucleotide sequence, and chromosome localization. Science 233: 541–545
Johnson GR, Gonda TJ, Metcalf D, Hariharan IK, Cory S (1989) A lethal myeloproliferative syndrome in mice transplanted with bone marrow cells infected with a retrovirus expressing granulocyte-macrophage colony stimulating factor. EMBO J 8: 441–448
Josephs SJ, Guo C, Ratner L, Wong-Staal F (1984) Human proto-oncogene nucleotide sequences corresponding to the transforming region of simian sarcoma virus. Science 223: 487–491
Kawahara RS, Deuel TF (1989) Platelet-derived growth factor-inducible gene *JE* is a member of a family of small inducible genes related to platelet factor 4. J Biol Chem 264: 679–682
Kawasaki ES, Ladner MB, Wang AM et al. (1985) Molecular cloning of a complementary DNA encoding human macrophage-specific colony-stimulating factor (CSF-1). Science 230: 291–296
Keck PJ, Hauser SD, Krivi G, Sanzo K, Warren T, Feder J, Connolly DT (1989) Vascular permeability factor, an endothelial cell mitogen related to PDGF. Science 246: 1309–1311
Khalil N, Bereznay O, Sporn M, Greenberg AH (1989) Macrophage production of transforming growth factor β and fibroblast collagen synthesis in chronic pulmonary inflammation. J Exp Med 170: 727–737
Kimura H, Fischer WH, Schubert D (1990) Structure, expression and function of a schwannoma-derived growth factor. Nature 348: 257–260
Kishimoto T (1989) The biology of interleukin-6. Blood 74: 1–10
Kishimoto TK, Jutila MA, Berg EL, Butcher EC (1989) Neutrophil Mac-1 and MEL-14 adhesion proteins inversely regulated by chemotactic factors. Science 245: 1238–1241
Knighton DR, Hunt TK, Scheuenstuhl H, Halliday BJ, Werb Z, Banda MJ (1983) Oxygen tension regulates the expression of angiogenesis factor by macrophages. Science 221: 1283–1285
Koch AE, Polverini PJ, Leibovich SJ (1986) Stimulation of neovascularization by human rheumatoid synovial tissue macrophages. Arthritis Rheum 29: 471–479
Koerner TJ, Hamilton TA, Introna M, Tannenbaum CS, Bast RC Jr, Adams DO (1987) The early competence genes JE and KC are differentially regulated in murine peritoneal macrophages in response to lipopolysaccharide. Biochem Biophys Res Commun 149: 969–974
Kohase M, Henriksen-De Stefano D, May LT, Vilcek J, Sehgal PB (1986) Induction of β_2-interferon by tumor necrosis factor: a homeostatic mechanism in the control of cell proliferation. Cell 45: 659–666
Kondaiah P, Sands MJ, Smith JM, Fields A, Roberts AB, Sporn MB, Melton DA (1990) Identification of a novel transforming growth factor-β (TGF-β5) mRNA in *Xenopus laevis*. J Biol Chem 265: 1089–1093
Krause JE, Chirgwin JM, Carter MS, Xu ZS, Hershey AD (1987) Three rat preprotachykinin mRNAs encode the neuropeptides substance P and neurokinin A. Proc Natl Acad Sci USA 84: 881–885
Kunkel SL, Chensue SW, Phan SH (1986) Prostaglandins as endogenous mediators of interleukin 1 production. J Immunol 136: 186–192
Kurland JI, Bockman RS, Broxmeyer HE, Moore MAS (1978) Limitation of excessive myelopoiesis by the intrinsic modulation of macrophage-derived prostaglandin E. Science 199: 552–555
Kurt-Jones EA, Beller DI, Mizel SB, Linanue ER (1985a) Identification of a membrane-associated interleukin 1 in macrophages. Proc Natl Acad Sci USA 82: 1204–1208
Kurt-Jones EA, Virgin HW IV, Unanue ER (1985b) Relationship of macrophage Ia and membrane IL1 expression to antigen presentation. J Immunol 135: 3652–3654
Kurt-Jones EA, Fiers W, Pober JS (1987) Membrane interleukin 1 induction on human endothelial cells and dermal fibroblasts. J Immunol 139: 2317–2324
Kwon BS, Weissman SM (1989) cDNA sequences of two inducible T-cell genes. Proc Natl Acad Sci USA 86: 1963–1967
Lang RA, Metcalf D, Cuthbertson RA et al. (1987) Transgenic mice expressing a hemopoietic growth factor gene (GM-CSF) develop accumulations of macrophages, blindness, and a fatal syndrome of tissue damage. Cell 51: 675–686

Lanser ME, Brown GE (1989) Stimulation of rat hepatocyte fibronectin production by monocyte-conditioned medium is due to interleukin 6. J Exp Med 170: 1781–1786
Larsen CG, Anderson AO, Appella E, Oppenheim JJ, Matsushima K (1989) The neutrophil-activating protein (NAP-1) is also chemotactic for T lymphocytes. Science 243: 1464–1466
Lebacq-Verheyden A-M, Krystal G, Sartor O, Way J, Battey JF (1988) The rat preprogastrin releasing peptide gene is transcribed from two initiation sites in the brain. Mol Endocrinol 2: 556–563
Lee L-S, Weinstein IB (1978) Epidermal growth factor, like phorbol esters, induces plasminogen activator in HeLa cells. Nature 274: 696–697
Lee M-T, Kaushansky K, Ralph P, Ladner MB (1990) Differential expression of M-CSF, G-CSF, and GM-CSF by human monocytes. J Leukocyte Biol 47: 275–282
Leibovich SJ, Ross R (1975) The role of the macrophage in wound repair. A study with hydrocortisone and antimacrophage serum. Am J Pathol 78: 71–100
Leibovich SJ, Polverini PJ, Shepard HM, Wiseman DM, Shively V, Nuseir N (1987) Macrophage-induced angiogenesis is mediated by tumour necrosis factor-α. Nature 329: 630–632
Leibrock J, Lottspeich F, Hohn A et al. (1989) Molecular cloning and expression of brain-derived neurotrophic factor. Nature 341: 149–152
Leung DW, Cachianes G, Kuang WJ, Goeddel DV, Ferrara N (1989) Vascular endothelial growth factor is a secreted angiogenic mitogen. Science 246: 1306–1309
Levi-Montalcini R (1987) The nerve growth factor 35 years later. Science 237: 1154–1162
Li H, Schwinzer R, Baccarini M, Lohmann-Matthes M-L (1989) Cooperative effects of colony-stimulating factor 1 and recombinant interleukin 2 on proliferation and induction of cytotoxicity of macrophage precursors generated from mouse bone marrow cell cultures. J Exp Med 169: 973–986
Liao Z, Haimovitz A, Chen Y, Chan J, Rosenstreich DL (1985) Characterization of a human interleukin 1 inhibitor. J Immunol 134: 3882–3886
Lin F-K, Suggs S, Lin C-H et al. (1985) Cloning and expression of the human erythropoietin gene. Proc Natl Acad Sci USA 82: 7580–7584
Lin H-S, Gordon S (1979) Secretion of plasminogen activator by bone marrow-derived mononuclear phagocytes and its enhancement by colony-stimulating factor. J Exp Med 150: 231–245
Lin L-FH, Mismer D, Lile JD, Armes LG, Butler ET III, Vannice JL, Collins F (1989) Purification, cloning, and expression of ciliary neurotrophic factor (CNTF). Science 246: 1023–1025
Lindholm D, Heumann R, Meyer M, Thoenen H (1987) Interleukin-1 regulates synthesis of nerve growth factor in non-neuronal cells of rat sciatic nerve. Nature 330: 658–659
Lindholm D, Heumann R, Hengerer B, Thoenen H (1988) Interleukin 1 increases stability and transcription of mRNA encoding nerve growth factor in cultured rat fibroblasts. J Biol Chem 263: 16348–16351
Linzer DIH, Nathans D (1984) Nucleotide sequence of a growth-related mRNA encoding a member of the prolactin-growth hormone family. Proc Natl Acad Sci USA 81: 4255–4259
Lomedico PT, Gubler U, Hellmann CP et al. (1984) Cloning and expression of murine interleukin-1 cDNA in *Escherichia coli*. Nature 312: 458–462
Lotz M, Vaughan JH, Carson DA (1988) Effect of neuropeptides on production of inflammatory cytokines by human monocytes. Science 241: 1218–1221
Lucas DO, Epstein LB (1985) Interferon and macrophages, In: Reichard SM, Filkins JP (eds) The reticuloendothelial system: a comprehensive treatise, vol 7B. Plenum, New York, pp 143–168
Luster AD, Unkeless JC, Ravetch JV (1985) γ-Interferon transcriptionally regulates an early-response gene containing homology to platelet proteins. Nature 315: 672–676
Lyons K, Graycar JL, Lee A et al. (1989) *Vgr-1*, a mammalian gene related to *Xenopus Vg-1*, is a member of the transforming growth factor β gene superfamily. Proc Natl Acad Sci USA 86: 4554–4558
MacKaness GG (1964) The immunologic basis of acquired cellular resistance. J Exp Med 120: 105–115
Madisen L, Webb NR, Rose TM et al. (1988) Transforming growth factor-β 2: cDNA cloning and sequence analysis. DNA 7: 1–8
Madtes DK, Raines EW, Sakariassen KS, Assoian RK, Sporn MB, Bell GI, Ross R (1988) Induction of transforming growth factor-α in activated human alveolar macrophages. Cell 53: 285–293
Mahley RW (1988) Apolipoprotein E: cholesterol transport protein with expanding role in cell biology. Science 240: 622–630
Maione TE, Gray GS, Petro J et al. (1990) Inhibition of angiogenesis by recombinant human platelet factor-4 and related peptides. Science 247: 77–79
Maisonpierre PC, Belluscio L, Squinto S, Ip NY, Furth ME, Lindsay RM, Yancopoulos GD (1990) Neurotrophin-3: a neurotrophic factor related to NGF and BDNF. Science 247: 1446–1451

Malkovsky M, Loveland B, North M, Asherson GL, Gao L, Ward P, Fiers W (1987) Recombinant interleukin-2 directly augments the cytotoxicity of human monocytes: Nature 325: 262–265
March CJ, Mosley B, Larsen A et al. (1985) Cloning, sequence and expression of two distinct human interleukin-1 complementary DNAs. Nature 315: 641–647
Marics I, Adelaide J, Raybaud F et al. (1989) Characterization of the HST-related FGF. 6 gene, a new member of the fibroblast growth factor gene family. Oncogene 4: 335–340
Martinet Y, Bitterman PB, Mornex JF, Grotendorst GR, Martin GR, Crystal RG (1986) Activated human monocytes express the c-*cis* proto-oncogene and release a mediator showing PDGF-like activity. Nature 319: 158–160
Mason AJ, Niall HD, Seeburg PH (1986) Structure of two human ovarian inhibins. Biochem Biophys Res Commun 135: 954–957
Mason AJ, Berkemeier LM, Schmelzer CH, Schwall RH (1989) Activin B: precursor sequences, genomic structure and in vitro activities. Mol Endocrinol 3: 1352–1358
Massague J (1987) The TGF-β family of growth and differentiation factors. Cell 49: 437–438
Matrisian LM, Glaichenhaus N, Gesnel M-C, Breathnach R (1985) Epidermal growth factor and oncogenes induce transcription of the same cellular mRNA in rat fibroblasts. EMBO J 4: 1435–1440
Matsushima K, Morishita K, Yoshimura T et al. (1988) Molecular cloning of a human monocyte-derived neutrophil chemotactic factor (MDNCF) and the induction of MDNCF mRNA by interleukin 1 and tumor necrosis factor. J Exp Med 167: 1883–1893
Matsushima K, Larsen CG, DuBois GC, Oppenheim JJ (1989) Purification and characterization of a novel monocyte chemotactic and activating factor produced by a human myelomonocytic cell line. J Exp Med 169: 1485–1490
McCaffrey TA, Falcone DJ, Brayton CF, Agarwal LA, Welt FGP, Weksler BB (1989) Transforming growth factor-β activity is potentiated by heparin via dissociation of the transforming growth factor-β/α_2-macroglobulin inactive complex. J Cell Biol 109: 441–448
Metcalf D (1985) The granulocyte-macrophage colony-stimulating factors. Science 229: 16–22
Miller DA, Lee A, Matsui Y, Chen EY, Moses HL, Derynck R (1989) Complementary DNA cloning of the murine transforming growth factor-β3 (TGFβ3) precursor and the comparative expression of TGF-β 3 and TGF-β 1 messenger RNA in murine embryos and adult tissues. Mol Endocrinol 3: 1926–1934
Ming WJ, Bersani L, Mantovani A (1987) Tumor necrosis factor is chemotactic for monocytes and polymorphonuclear leukocytes. J Immunol 138: 1469–1474
Miyatake S, Otsuka T, Yokota T, Lee F, Arai K (1985) Structure of the chromosomal gene for granulocyte-macrophage colony stimulating factor: comparison of the mouse and human genes. EMBO J 4: 2561–2568
Moore KW, Vieira P, Fiorentino DF, Trounstine ML, Khan TA, Mosmann TR (1990) Homology of cytokine synthesis inhibitory factor (IL-10) to the Epstein-Barr virus gene BCRFI. Science 248: 1230–1234
Moore RN, Larsen HS, Horohov DW, Rouse BT (1984a) Endogenous regulation of macrophage proliferative expansion by colony-stimulating factor-induced interferon. Science 223: 178–181
Moore RN, Pitruzzello FJ, Larsen HS, Rouse BT (1984b) Feedback regulation of colony-stimulating factor (CSF-1)-induced macrophage proliferation by endogenous E prostaglandins and interferon-α/β. J Immunol 133: 541–543
Moore R, Casey G, Brookes S, Dixon M, Peters G, Dickson C (1986) Sequence, topography and protein coding potential of mouse *int*-2: a putative oncogene activated by mouse mammary tumour virus. EMBO J 5: 919–924
Moreau J-F, Donaldson DD, Bennett F, Witek-Giannotti J, Clark SC, Wong GG (1988) Leukaemia inhibitory factor is identical to the myeloid growth factor human interleukin for DA cells. Nature 336: 690–692
Muglia L, Locker J (1983) Extrapancreatic insulin gene expression in the fetal rat. Proc Natl Acad Sci USA 81: 3635–3641
Mukaida N, Shiroo M, Matsushima K (1989) Genomic structure of the human monocyte-derived neutrophil chemotactic factor IL-8. J Immunol 143: 1366–1371
Munker R, Gasson J, Ogawa M, Koeffler HP (1986) Recombinant human TNF induces production of granulocyte-monocyte colony-stimulating factor. Nature 323: 79–82
Nagaoka I, Trapnell BC, Crystal RG (1990a) Upregulation of platelet-derived growth factor-A and -B gene expression in alveolar macrophages of individuals with idiopathic pulmonary fibrosis. J Clin Invest 85: 2023–2027
Nagaoka I, Trapnell BC, Crystal RG (1990b) Regulation of insulin-like growth factor I gene expression in the human macrophage-like cell line U937. J Clin Invest 85: 448–455

Nakamura T, Nishizawa T, Hagiya M et al. (1989) Molecular cloning and expression of human hepatocyte growth factor. Nature 342: 440–443
Nathan CF (1987) Secretory products of macrophages. J Clin Invest 79: 319–326
Nathan CF (1989) Respiratory burst in adherent human neutrophils: triggering by colony-stimulating factors. CSF-GM and CSF-G. Blood 73: 301–306
Nathan CF, Horowitz CR, de la Harpe J, Vadhan-Raj S, Sherwin SA, Oettgen HF, Krown SE (1985) Administration of recombinant interferon γ to cancer patients enhances monocyte secretion of hydrogen peroxide. Proc Natl Acad Sci USA 82: 8686–8690
Navarro S, Debili N, Bernaudin J-F, Vainchenker W, Doly J (1989) Regulation of the expression of IL-6 in human monocytes. J Immunol 142: 4339–4345
Nawroth PP, Handley DA, Esmon CT, Stern DM (1986) Interleukin 1 induces endothelial cell procoagulant while suppressing cell-surface anticoagulant activity. Proc Natl Acad Sci USA 83: 3460–3464
Nicola NA (1989) Hemopoietic cell growth factors and their receptors. Annu Rev Biochem 58: 45–77
Nordan RP, Potter M (1986) A macrophage-derived factor required by plasmacytomas for survival and proliferation in vitro. Science 233: 566–569
North RJ (1978) The concept of the activated macrophage. J Immunol 121: 806–809
Northemann W, Braciak TA, Hattori M, Lee F, Fey GH (1989) Structure of the rat interleukin 6 gene and its expression in macrophage-derived cells. J Biol Chem 264: 16072–16082
Novick D, Engelmann H, Wallach D, Rubenstein M (1989) Soluble cytokine receptors are present in normal human urine. J Exp Med 170: 1409–1414
Old LJ (1985) Tumor necrosis factor (TNF). Science 230: 630–632
Onozaki K, Matsushima K, Aggarwal BB, Oppenheim JJ (1985) Human interleukin 1 is a cytocidal factor for several tumor cell lines. J Immunol 135: 3962–3968
Oquendo P, Alberta J, Wen DZ, Graycar JL, Derynck R, Stiles CD (1989) The platelet-derived growth factor-inducible KC gene encodes a secretory protein related to platelet α-granule proteins. J Biol Chem 264: 4133–4137
Özkaynak E, Rueger DC, Drier EA, Corbett C, Ridge RJ, Sampath TK, Oppermann H (1990) OP-1 cDNA encodes an osteogenic protein in the TGF-β family. EMBO J 9: 2085–2093
Panayotou G, End P, Aumailley M, Timpl R, Engel J (1989) Domains of laminin with growth-factor activity. Cell 56: 93–101
Paul P, Rothmann SA, McMahon JT, Gordon AS (1984) Erythropoietin secretion by isolated rat Kupffer cells. Exp Hematol 12: 825–830
Pencev D, Grotendorst GR (1988) Human peripheral blood monocytes secrete a unique form of PDGF. Oncogene Res 3: 333–342
Pennica D, Hayflick JS, Bringman TS, Palladino MA, Goeddel DV (1985) Cloning and expression in *Escherichia coli* of the cDNA for murine tumor necrosis factor. Proc Natl Acad Sci USA 82: 6060–6064
Perlmutter DH, Goldberger G, Dinarello CA, Mizel SB, Colten HR (1986) Regulation of class III histocompatibility complex gene products by interleukin-1. Science 232: 850–852
Perry VH, Brown MC, Gordon S (1987) The macrophage response to central and peripheral nerve injury. A possible role for macrophages in regeneration. J Exp Med 165: 1218–1223
Pierce GF, Mustoe TA, Lingelbach J, Masakowski VR, Griffin GL, Senior RM, Deuel TF (1989) Platelet-derived growth factor and transforming growth factor-β enhance tissue repair activities by unique mechanisms. J Cell Biol 109: 429–440
Pike BL, Nossal GJV (1985) Interleukin 1 can act as a B-cell growth and differentiation factor. Proc Natl Acad Sci USA 82: 8153–8157
Pikkarainen T, Kallunki T, Tryggvason K (1988) Human laminin B2 chain. J Biol Chem 263: 6751–6758
Pober JS, Cotran RS (1990) Cytokines and endothelial cell biology. Physiol Rev 70: 427–451
Polverini PJ, Cotran RS, Gimbrone MA Jr, Unanue ER (1977) Activated macrophages induce vascular proliferation. Nature 269: 804–806
Postlethwaite AE, Lachman LB, Mainardi CL, Kang AH (1983) Interleukin 1 stimulation of collagenase production by cultured fibroblasts. J Exp Med 157: 801–806
Prpic V, Yu S-F, Figueiredo F et al. (1989) Role of Na^+/H^+ exchange by interferon-γ in enhanced expression of JE and I-A β genes. Science 244: 469–471
Raines EW, Dower SK, Ross R (1989) Interleukin–1 mitogenic activity for fibroblasts and smooth muscle cells is due to PDGF-AA. Science 243: 393–396
Rajavashisth TB, Eng R, Shadduck RK, Waheed A, Ben-Avram CM, Shively JE, Lusis AJ (1987) Cloning and tissue-specific expression of mouse macrophage colony- stimulating factor mRNA. Proc Natl Acad Sci USA 84: 1157–1161

Ralph P, Nakoinz I (1987) Stimulation of macrophage tumoricidal activity by the growth and differentiation factor CSF-1. Cell Immunol 105: 270–279

Rappolee DA, Werb Z (1988) Secretory products of phagocytes. Curr Opin Immunol 1: 47–55

Rappolee DA, Werb Z (1989) Macrophage secretions: a functional perspective. Bull Inst Pasteur 87: 361–394

Rappolee DA, Werb Z (1990) mRNA phenotyping for studying gene expression in small numbers of cells: platelet-derived growth factor and other growth factors in wound-derived macrophages. Am J Respir Cell Mol Biol 2: 3–10

Rappolee DA, Mark D, Banda MJ, Werb Z (1988) Wound macrophages express TGF-α and other growth factors in vivo: analysis by mRNA phenotyping. Science 241: 708–712

Rappolee DA, Wang A, Mark D, Werb Z (1989) Novel method for studying mRNA phenotypes in single or small numbers of cells. J Cell Biochem 39: 1–11

Rebagliati MR, Weeks DL, Harvey RP, Melton DA (1985) Identification and cloning of localized maternal mRNAs from *Xenopus* eggs. Cell 42: 769–777

Remick DG, Scales WE, May MA, Spengler M, Nguyen D, Kunkel SL (1988) In situ hybridization analysis of macrophage-derived tumor necrosis factor and interleukin-1 mRNA. Lab Invest 59: 809–816

Resnitzky D, Yarden A, Zipori D, Kimchi A (1986) Autocrine β-related interferon controls c-*myc* suppression and growth arrest during hematopoietic cell differentiation. Cell 46: 31–40

Rich IN, Heit W, Kubanek B (1992) Extrarenal erythropoietin production by macrophages. Blood 60: 1007–1018

Richmond A, Balentien E, Thomas HG et al. (1988) Molecular characterization and chromosomal mapping of melanoma growth stimulatory activity, a growth factor structurally related to β-thromboglobulin. EMBO J 7: 2025–2033

Rinderknecht E, Humbel RE (1978) The amino acid sequence of human insulin-like growth factor I and its structural homology with proinsulin. J Biol Chem 253: 2769–2776

Roberts AB, Sporn MB (1990) The transforming growth factor-β. In: Sporn MB, Roberts AB (eds) Peptide growth factors and their receptors. Springer, Berlin Heidelberg New York, pp 419–472

Roberts AB, Sporn MB, Assoian RK et al. (1986) Transforming growth factor type β: rapid induction of fibrosis and angiogenesis in vivo and stimulation of collagen formation in vitro. Proc Natl Acad Sci USA 83: 4167–4171

Roberts R, Gallagher J, Spooner E, Allen TD, Bloomfield F, Dexter TM (1988) Heparan sulphate bound growth factors: a mechanism for stromal cell mediated haemopoiesis. Nature 332: 376–378

Rollins BJ, Morrison ED, Stiles CD (1988) Cloning and expression of *JE*, a gene inducible by platelet-derived growth factor and whose product has cytokine-like properties. Proc Natl Acad Sci USA 85: 3738–3742

Ross R, Vogel A (1978) The platelet-derived growth factor. Cell 14: 203–210

Ross R, Raines EW, Bowen-Pope DF (1986) The biology of platelet-derived growth factor. Cell 46: 155–169

Ruff MR, Pert CB, Weber RJ, Wahl LM, Wahl SM, Paul SM (1985) Benzodiazepine receptor-mediated chemotaxis of human monocytes. Science 229: 1281–1283

Ryan J, Geczy CL (1986) Characterization and purification of mouse macrophage procoagulant-inducing factor. J Immunol 137: 2864–2870

Sara VR, Hall K (1990) Insulin-like growth factors and their binding proteins. Physiol Rev 70: 591–614

Sasaki M, Kato S, Kohno K, Martin GR, Yamada Y (1987) Sequence of the cDNA encoding the laminin B1 chain reveals a multidomain protein containing cysteine-rich repeats. Proc Natl Acad Sci USA 84: 935–939

Sausville EA, Lebacq-Verheyden A-M, Spindel ER, Cuttita F, Gazdar AF, Battey JF (1986) Expression of gastrin-releasing peptide gene in human small cell lung carcinoma. J Biol Chem 261: 2451–2457

Scales WE, Chensue SW, Otterness I, Kunkel SL (1989) Regulation of monokine gene expression: prostaglandin E_2 suppresses tumor necrosis factor but not interleukin-1α or β-mRNA and cell-associated bioactivity. J Leukocyte Biol 45: 416–421

Schall TJ, Lewis M, Koller KJ et al. (1990) Molecular cloning and expression of a receptor for human tumor necrosis factor. Cell 61: 361–370

Scheidt P, Waehneldt TV, Beuche W, Friede RL (1986) Changes of myelin proteins during Wallerian degeneration in situ and in millipore diffusion chambers preventing active phagocytosis. Brain Res 379: 380–384

Schleimer RP, Rutledge BK (1986) Cultured human vascular endothelial cells acquire adhesiveness for neutrophils after stimulation with interleukin 1, endotoxin, and tumor-promoting phorbol diesters. J Immunol 136: 649–654

Schmid J, Weissman C (1987) Induction of mRNA for serine protease and a β-thromboglobulin-like protein in mitogen-stimulated human leukocytes. J Immunol 139: 250–256

Schnyder J, Payne T, Dinarello CA (1987) Human monocyte or recombinant interleukin 1's are specific for the secretion of a metalloproteinase from chondrocytes. J Immunol 138:496–503

Schreiber AB, Winkler ME, Derynck R (1986) Transforming growth factor-α: a more potent angiogenic mediator than epidermal growth factor. Science 232: 1250–1253

Schröder, J-M (1989) The monocyte-derived neutrophil activating peptide (NAP/interleukin 8) stimulates human neutrophil arachidonate-5-lipoxygenase, but not the release of cellular arachidonate. J Exp Med 170: 847–863

Schultz GS, White M, Mitchell R, Brown G, Lynch J, Twardzik DR, Todaro GJ (1987) Epithelial wound healing enhanced by transforming growth factor-α and vaccinia growth factor. Science 235: 350–352

Schwarzbauer JE, Tamkun JW, Lemischka IR, Hynes RO (1983) Three different fibronectin mRNAs arise by alternative splicing within the coding region. Cell 35: 421–431

Scott J, Selby M, Urdea M, Quiroga M, Bell GI, Rutter WJ (1983) Isolation and nucleotide sequence of a cDNA encoding the precursor of mouse nerve growth factor. Nature 302: 538–540

Seelentag WK, Mermod J-J, Montesano R, Vassalli P (1987) Additive effects of interleukin 1 and tumour necrosis factor-α on the accumulation of the three granulocyte and macrophage colony-stimulating factor mRNAs in human endothelial cells. EMBO J 6: 2261–2265

Semon D, Kawashima E, Jongeneel CV, Shakhov AN, Nedospasov SA (1987) Nucleotide sequence of the murine TNF locus, including the TNF-α (tumor necrosis factor) and TNF-β (lymphotoxin) genes. Nucleic Acids Res 15: 9083–9084

Shaw G, Kamen R (1986) A conserved AU sequence from the 3′ untranslated region of GM-CSF mRNA mediates selective mRNA degradation. Cell 46: 659–667

Shaw GD, Boll W, Taira H, Mantei N, Lengyel P, Weissman C (1983) Structure and expression of cloned murine IFN-α genes. Nucleic Acids Res 11: 555–573

Sherry B, Tekamp-Olson P, Gallegos C et al. (1988) Resolution of the two components of macrophage inflammatory protein 1, and cloning and characterization of one of those components, macrophage inflammatory protein 1β. J Exp Med 168: 2251–2259

Shi DL, Savona C, Gagnon J, Cochet C, Chambaz EM, Feige JJ (1990) Transforming growth factor-β stimulates the expression of α_2-macroglobulin by cultured bovine adrenocortical cells. J Biol Chem 265: 2881–2887

Shimasaki S, Emoto N, Koba A et al. (1988) Complementary DNA cloning and sequencing of rat ovarian basic fibroblast growth factor and tissue distribution study of its mRNA. Biochem Biophys Res Commun 157: 256–263

Shimokado K, Raines EW, Madtes DK, Barrett EW, Benditt EP, Ross R (1985) A significant part of macrophage-derived growth factor consists of at least two forms of PDGF. Cell 43:277–286

Shoemaker CB, Mitsock LD (1986) Murine erythropoietin gene: cloning, expression, and human gene homology. Mol Cell Biol 6: 849–858

Shoyab M, Plowman GD, McDonald VL, Gradley JG, Todaro GJ (1989) Structure and function of human amphiregulin: a member of the epidermal growth factor family. Science 243: 1074–1076

Shoyab M, McDonald VL, Byles C, Todaro GJ, Plowman GD (1990) Epithelins 1 and 2: isolation and characterization of two cysteine-rich growth-modulating proteins. Proc Natl Acad Sci USA 87: 7912–7916

Sims JE, March CJ, Cosman D et al. (1988) cDNA expression cloning of the IL-1 receptor, a member of the immunoglobulin superfamily. Science 241: 585–589

Smith CA, Daris T, Anderson D et al. (1990) A receptor for tumor necrosis factor defines an unusual family of cellular and viral proteins. Science 248: 1019–1023

Smith EM, Morrill AC, Meyer WJ III, Blalock JE (1986) Corticotropin releasing factor induction of leukocyte-derived immunoreactive ACTH and endorphins. Nature 321: 881–882

Sonoda E, Matsumoto R, Hitoshi Y et al. (1989) Transforming growth factor β induces IgA production and acts additively with interleukin 5 for IgA production. J Exp Med 170: 1415–1420

Stempien MM, Fong NM, Rall LB, Bell GI (1986) Sequence of a placental cDNA encoding the mouse insulin-like growth factor II precursor. DNA 5: 357–361

Stiles CD (1983) The molecular biology of platelet-derived growth factor. Cell 33: 653–655

Stöckli KA, Lottspeich F, Sendtner M et al. (1989) Molecular cloning, expression and regional distribution of rat ciliary neurotrophic factor. Nature 342:920–923

Stoeckle MY, Barker KA (1990) Two burgeoning families of platelet factor 4-related proteins: mediators of the inflammatory responses. The New Biologist 2: 313–323

Stossel TP (1987) The molecular biology of phagocytes and the molecular basis of nonneoplastic phagocyte disorders. In: Stamatoyannopoulos G, Nienhuis A, Leder P, Maierus P (eds) The molecular basis of blood diseases. WB Saunders, Philadelphia, pp 499–533

Strieter RM, Kunkel SL, Showell HJ, Remick DG, Phan SH, Ward PA, Marks RM (1989) Endothelial cell gene expression of a neutrophil chemotactic factor by TNF-α, LPS, and IL-1β. Science 243: 1467–1469

Stuehr DJ, Nathan CF (1989) Nitric oxide. A macrophage product responsible for cytostasis and respiratory inhibition in tumor target cells. J Exp Med 169: 1543–1555

Subramanian N, Bray MA (1987) Interleukin 1 releases histamine from human basophils and mast cells in vitro. J Immunol 138: 271–275

Takahashi S, Yamashita T, Eto Y, Shibai H, Miyamoto K, Ogata E (1990) Inducible gene expression of activin A/erythroid differentiation factor in HL-60 cells. Biochem Biophys Res Commun 167: 654–658

Takemura R, Werb Z (1984a) Regulation of elastase and plasminogen activator secretion in resident and inflammatory macrophages by receptors for the Fc domain of immunoglobulin G. J Exp Med 159: 152–166

Takemura R, Werb Z (1984b) Secretory products of macrophages and their physiological functions. Am J Physiol 246: C1–C9

Tanabe O, Akira S, Kamiya T, Wong GG, Hirano T, Kishimoto T (1988) Genomic structure of the murine IL-6 gene. High degree conservation of potential regulatory sequences between mouse and human. J Immunol 141: 3875–3881

Tanabe T, Konishi M, Mizuta T, Noma T, Honjo T (1987) Molecular cloning and structure of the human interleukin-5 gene. J Biol Chem 262: 16580–16584

Taniguchi T, Matsui H, Fujita T, Takaoka C, Kashima N, Yoshimoto R, Hamuro J (1983) Structure and expression of a cloned cDNA for human interleukin-2. Nature 302: 305–310

ten Dijke P, Hansen P, Iwata KK, Pieler C, Foulkes JG (1988) Identification of another member of the transforming growth factor type β gene family. Proc Natl Acad Sci USA 85: 4715–4719

Thomas KA, Rios-Candelore M, Gimenez-Gallego G, DiSalvo J, Bennett C, Rodkey J, Fitzpatrick S (1985) Pure brain-derived acidic fibroblast growth factor is a potent angiogenic vascular endothelial cell mitogen with sequence homology to interleukin 1. Proc Natl Acad Sci USA 82: 6409–6413

Thorens B, Mermod JJ, Vassalli P (1987) Phagocytosis and inflammatory stimuli induce GM-CSF mRNA in macrophages through posttranscriptional regulation. Cell 48: 671–679

Tosato G, Seamon KB, Goldman ND et al. (1988) Monocyte-derived human B-cell growth factor identified as interferon-β (BSF-2, IL-6). Science 239: 502–504

Treiger BF, Leonard WJ, Svetlik P, Rubin LA, Nelson DL, Greene WC (1986) A secreted form of the human interleukin 2 receptor encoded by an "anchor minus" cDNA. J Immunol 136: 4099–4105

Tsuchiya M, Asano S, Kaziro Y, Nagata S (1986) Isolation and characterization of the cDNA for murine granulocyte colony-stimulating factor. Proc Natl Acad Sci USA 83: 7633–7637

Tsunawaki S, Sporn M, Ding A, Nathan C (1988) Deactivation of macrophages by transforming growth factor-β. Nature 334: 260–262

Ullrich A, Gray A, Berman C, Dull TJ (1983) Human β-nerve growth factor gene sequence highly homologous to that of mouse. Nature 303: 821–825

Unanue ER, Allen PM (1987) The basis for the immunoregulatory role of macrophages and other accessory cells. Science 236: 551–557

Underwood JL, Rappolee DA, Flannery ML, Werb Z (1990) A role for the tissue inhibitor of metalloproteinases (TIMP) in regeneration of peripheral nerve (abstract). J Cell Biol 111:15a

Urban JL, Shepard HM, Rothstein JL, Sugarman BJ, Schreiber H (1986) Tumor necrosis factor: a potent effector molecule for tumor cell killing by activated macrophages. Proc Natl Acad Sci USA 83: 5233–5237

Van Damme J, Opdenakker G, Simpson RJ et al. (1987) Identification of the human 26-kD protein, interferon β_2 (IFN-β_2), as a B cell hybridoma/plasmocytoma growth factor induced by interleukin 1 and tumor necrosis factor. J Exp Med 165: 914–919

Vanguri P, Farber J, Cloning of the mouse homolog of IP-10 from lymphokine-activated RAW 264.7 cells (abstract). Cytokine 1: 147

van Ooyen H, Nusse R (1984) Structure and nucleotide sequence of the putative mammary oncogene *int*-1; proviral insertions leave the protein-encoding domain intact. Cell 39: 233–240

Van Snick J, Cayphas S, Szikora J-P, Renauld J-C, Van Roost E, Boon T, Simpson RJ (1988) cDNA cloning of murine interleukin-HP1: homology with human interleukin 6. Eur J Immunol 18, 193–197 (1988)

Wahl SM, Hunt DA, Wakefield LM, McCartney-Francis N, Wahl LM, Roberts AB, Sporn MB (1987) Transforming growth factor type β induces monocyte chemotaxis and growth factor production. Proc Natl Acad Sci USA 84: 5788–5792

Walz A, Dewald B, von Tscharner V, Baggiolini M (1989) Effects of the neutrophil-activating peptide NAP-2, platelet basic protein, connective tissue-activating peptide III and platelet factor 4 on human neutrophils. J Exp Med 170: 1745–1750

Wang AM, Creasey AA, Ladner MB et al. (1985) Molecular cloning of the complementary DNA for human tumor necrosis factor. Science 228: 149–151

Warren MK, Ralph P (1986) Macrophage growth factor CSF-1 stimulates human monocyte production of interferon, tumor necrosis factor, and colony stimulating activity. J Immunol 137: 2281–2285

Weisbart RH, Kacena A, Schuh A, Golde DW (1988) GM-CSF induces human neutrophil IgA-mediated phagocytosis by an IgA Fc receptor activation mechanism. Nature 332: 647–648

Wentworth BM, Schaefer IM, Villa-Komaroff L, Chirgwin JM (1986) Characterization of the two nonallelic genes encoding mouse preproinsulin. J Mol Evol 23: 305–312

Werb Z (1987) Phagocytic cells: chemotaxis and effector functions of macrophages and granulocytes. I. Macrophages. In: Stites DP, Stobo JD, Wells JV (eds) Basic and clinical immunology, 6th edn. Appleton and Lange, Norwalk, Conn, pp 96–113

Werb Z, Chin JR (1983) Onset of apoprotein E secretion during differentiation of mouse bone marrow-derived mononuclear phagocytes. J Cell Biol 97: 1113–1118

Wiedermann CJ, Goldman ME, Plutchok JJ et al. (1986) Bombesin in human and guinea pig alveolar macrophages. J Immunol 137: 3928–3932

Wiktor-Jedrzejczak W, Bartocci A, Ferrante AW Jr, Ahmed-Ansari A, Sell KW, Pollard JW, Stanley ER (1990) Total absence of colony-stimulating factor 1 in the macrophage-deficient osteopetrotic (*op/op*) mouse. Proc Natl Acad Sci USA 87: 4828–4832

Williams DE, Morrissey PJ (1989) Alterations in megakaryocyte and platelet compartments following in vivo IL-1β administration to normal mice. J Immunol 142: 4361–4365

Willman CL, Stewart CC, Miller V, Yi T-L, Tomasi TB (1989) Regulation of MHC class II gene expression in macrophages by hematopoietic colony-stimulating factors (CSF). J Exp Med 170: 1559–1567

Wolpe SD, Sherry B, Juers D, Davatelis G, Yurt RW, Cerami A (1989) Identification and characterization of macrophage inflammatory protein 2. Proc Natl Acad Sci USA 86: 612–616

Wozney JM, Rosen V, Celeste AJ et al. (1988) Novel regulators of bone formation: molecular clones and activities. Science 242: 1528–1534

Yamamori T, Fukada K, Aebersold R, Korsching S, Fann MJ, Patterson PH (1989) The cholinergic neuronal differentiation factor from heart cells is identical to leukemia inhibitory factor. Science 246: 1412–1416

Yamasaki K, Taga T, Hirata Y et al. (1988) Cloning and expression of the human interleukin-6 (BSF-2/IFNβ 2) receptor. Science 241: 825–828

Yanagisawa M, Kurihara H, Kimura S et al. (1988) A novel potent vasoconstrictor peptide produced by vascular endothelial cells. Nature 332: 411–415

Yang Y-C, Ciarletta AB, Temple PA et al. (1986) Human IL-3 (multi-CSF): identification by expression cloning of a novel hematopoietic growth factor related to murine IL-3. Cell 47: 3–10

Ying SY (1988) Inhibins, activins, and follistatins: gonadal proteins modulating the secretion of follicle-stimulating hormone. Endocrine Rev 9: 267–293

Yokota T, Arai N, Lee F, Rennick D, Mosmann T, Arai K-I (1985) Use of a cDNA expression vector for isolation of mouse interleukin 2 cDNA clones: expression of T-cell growth-factor activity after transfection of monkey cells. Proc Natl Acad Sci USA 82: 68–72

Yoshida H, Hayashi S, Kunisada T et al. (1990) The murine mutation osteopetrosis is in the coding region of the macrophage colony stimulating factor gene. Nature 345: 442–444

Yoshimura T, Leonard EJ (1990) Secretion by human fibroblasts of monocyte chemoattractant protein-1, the product of gene JE. J Immunol 144: 2377–2383

Yoshimura T, Yuhki N, Moore SK, Appella E, Lerman MI, Leonard EJ (1989) Human monocyte chemoattractant protein-1 (MCP-1): full-length cDNA cloning, expression in mitogen-stimulated blood mononuclear leukocytes, and sequence similarity to mouse competence gene JE. FEBS Lett 244: 487–493

Zhan X, Bates B, Hu XG, Goldfarb M (1988) The human FGF-5 oncogene encodes a novel protein related to fibroblast growth factors. Mol Cell Biol 8:3487–3495

Zilberstein A, Ruggieri R, Korn JH, Revel M (1986) Structure and expression of cDNA and genes for human interferon-beta-2, a distinct species inducible by growth-stimulatory cytokines. EMBO J 5: 2529–2537

Zuckerman SH, Suprenant YM (1989) Induction of endothelial cell/macrophage procoagulant activity: synergistic stimulation by gamma interferon and granulocyte-macrophage colony stimulating factor. Thromb Haemost 61: 178–182

The Biology of CSF-1 and Its Receptor

P. ROTH[1] and E. R. STANLEY[2]

1 Introduction

The constant renewal of blood cells in vertebrate species depends on the proliferation and differentiation of hematopoietic stem cells in the bone marrow (HARRISON et al. 1988). These cells in turn give rise to progenitor cells which are committed to more restricted pathways of differentiation. The survival, proliferation, and differentiation of these progenitor cells are regulated by the colony stimulating factors (CSFs), so named because of their ability to promote the in vitro proliferation and differentiation of single progenitor cells into macroscopic colonies with discernible differentiated cell types (PLUZNIK and SACHS 1965; BRADLEY and METCALF 1966; ICHIKAWA et al. 1966). This group of hematopoietic growth factors includes interleukin-3 (IL-3), granulocyte-

Departments of [1]Pediatrics and [2]Developmental Biology and Cancer, Albert Einstein College of Medicine, 1300 Morris Park Avenue, Bronx, New York 10461, USA

Current Topics in Microbiology and Immunology, Vol. 181

macrophage colony stimulating factor (GM-CSF), granulocyte colony stimulating factor (G-CSF), interleukin-5 (IL-5), and colony stimulating factor-1 (CSF-1) (reviewed in STANLEY and JUBINSKY 1984; METCALF 1986).

Colony stimulating factor 1 is a lineage specific growth factor with the ability to control the survival, proliferation, and differentiation of the mononuclear phagocyte (reviewed in STANLEY et al. 1983). However, as will be discussed later in this chapter, it probably also plays an essential role in the regulation of the placenta during pregnancy (BARTOCCI et al. 1986; POLLARD et al. 1987; ARCECI et al. 1989; REGENSTREIF and ROSSANT 1989). The current review focuses initially on the biochemical, molecular, and cellular aspects of this growth factor and its receptor in order to provide a framework for a subsequent discussion of the in vitro and in vivo biologic activities of CSF-1.

2 CSF-1

2.1 Cellular Sources

Many methods, including radioimmunoassay, in situ hybridization, and Northern analysis, have been used to identify the numerous cell types that synthesize CSF-1 (reviewed in SHERR and STANLEY 1990). Fibroblasts, especially the conditioned medium from established cell lines of this type, e.g., mouse L cells, are a prominent source of this growth factor (STANLEY and HEARD 1977; YAN et al. 1990), as are bone marrow stromal cells, which compose the hematopoietic microenvironment (LANOTTE et al. 1982; FIBBE et al. 1988; YAN et al. 1990). While CSF-1 production is detectable in many tissues, including the uterus, there is a dramatic increase in this latter organ during pregnancy with production localized to the columnar epithelial cells of the glandular endometrium (POLLARD et al. 1987). Besides these uterine cells, there are additional reports of several other cell types which may exert local regulatory functions through the elaboration of this growth factor. These include thymic epithelium (LE et al. 1988), brain astrocytes (THERY et al. 1990), osteoblasts (ELFORD et al. 1987), and keratinocytes (CHODAKEWITZ et al. 1990). In addition, a variety of cells release CSF-1 following in vitro activation. These include endothelial cells (SEELENTAG et al. 1987), T lymphocytes (CERDAN et al. 1990), and B lymphocytes (REISBACH et al. 1989), as well as monocytes themselves (HORIGUCHI et al. 1986; RAMBALDI et al. 1987; GAFFNEY et al. 1988; HASKILL et al. 1988; LU et al. 1988; OSTER et al. 1989). The potential for CSF-1-responsive monocytes to produce CSF-1 raises the possibility of the involvement of autocrine or paracrine mechanisms in their activation. As is the case for activated normal human monocytes, a number of human tumors have been identified which coexpress CSF-1 and its receptor, including acute myeloblastic leukemia blasts (RAMBALDI et al. 1988; WANG et al. 1988) and adenocarcinoma cells of the breast, ovary, and endometrium (HORIGUCHI et al. 1988; KACINSKI et al. 1989a). Other tumors which have been shown to express CSF-1 both in vivo and in vitro include cells derived from patients with lymphoblastic leukemia (TAKAHASHI et al. 1988), diffuse large cell lymphoma (JANOWSKA-WIECZOREK et al. 1991), myeloma (NAKAMURA et al. 1989), Hodgkin's disease (PAIETTA et al. 1990), adenocarcinoma of the lung (HORIGUCHI et al. 1988), and pancreatic carcinoma (RALPH et al. 1986b).

2.2 Biochemistry

Colony stimulating factor-1, which has been extensively characterized in both man and the mouse, was originally purified from mouse L cell conditioned medium (STANLEY and HEARD 1977) and human urine (DAS et al. 1981), respectively. It is a disulfide-linked homodimeric molecule with a molecular weight of 45–90 kDa but also exists in higher molecular weight forms. Although the number of amino acids in the mature monomeric subunit varies due to alternative splicing of RNA transcripts (KAWASAKI et al. 1985; WONG et al. 1987; CERRETTI et al. 1988) and C-terminal proteolytic cleavage (RETTENMIER and ROUSSEL 1988) (Fig. 1), the N-terminal 158 amino acids are sufficient for biologic activity (KAWASAKI et al. 1985; HEARD et al. 1987a; WONG et al. 1987). The polypeptide is also modified by glycosylation (MANOS 1988; RETTENMIER and ROUSSEL 1988; PRICE and STANLEY, manuscript in preparation). The ultimate disposition of the molecule as either a rapidly secreted growth factor or as a biologically active membrane-bound (STEIN et al. 1990) growth factor, depends in large part on the presence or absence of an exon 6-encoded proteolytic cleavage site (HEARD et al. 1987a; RETTENMIER et al. 1987; MANOS 1988; RETTENMIER and ROUSSEL 1988) (see Sect. 2.3) (Fig. 1). The presence of these two distinct forms of CSF-1 may have important implications for its ability to function both as a circulating growth factor and as a local mediator of cell–cell interactions.

2.3 Molecular Biology

The organization of the CSF-1 gene is highly conserved between man and the mouse in those aspects that have been thus far investigated. The human gene is 21 kb in length and consists of ten exons and nine introns, with exons 1–8 coding for the CSF-1 polypeptide itself and exons 9 and 10 coding for alternative 3′-untranslated regions (LADNER et al. 1987; reviewed in KAWASAKI and LADNER 1990) (Fig. 1). Genetic analysis has revealed that the human gene is localized to the long arm of chromosome 5 at position 5q33.1 (PETTENATI et al. 1987), closely linked to the genes for GM-CSF (HUEBNER et al. 1985; LE BEAU et al. 1986; YANG et al. 1988; VAN LEEUWEN et al. 1989), IL-3 (YANG et al. 1988), IL-4 (LE BEAU et al. 1989; VAN LEEUWEN et al. 1989), IL-5 (SUTHERLAND et al. 1988; LE BEAU et al. 1989; VAN LEEUWEN et al. 1989), acidic fibroblast growth factor (JAYE et al. 1986), and the CSF-1 receptor (LE BEAU et al. 1986). However, the mouse CSF-1 gene maps

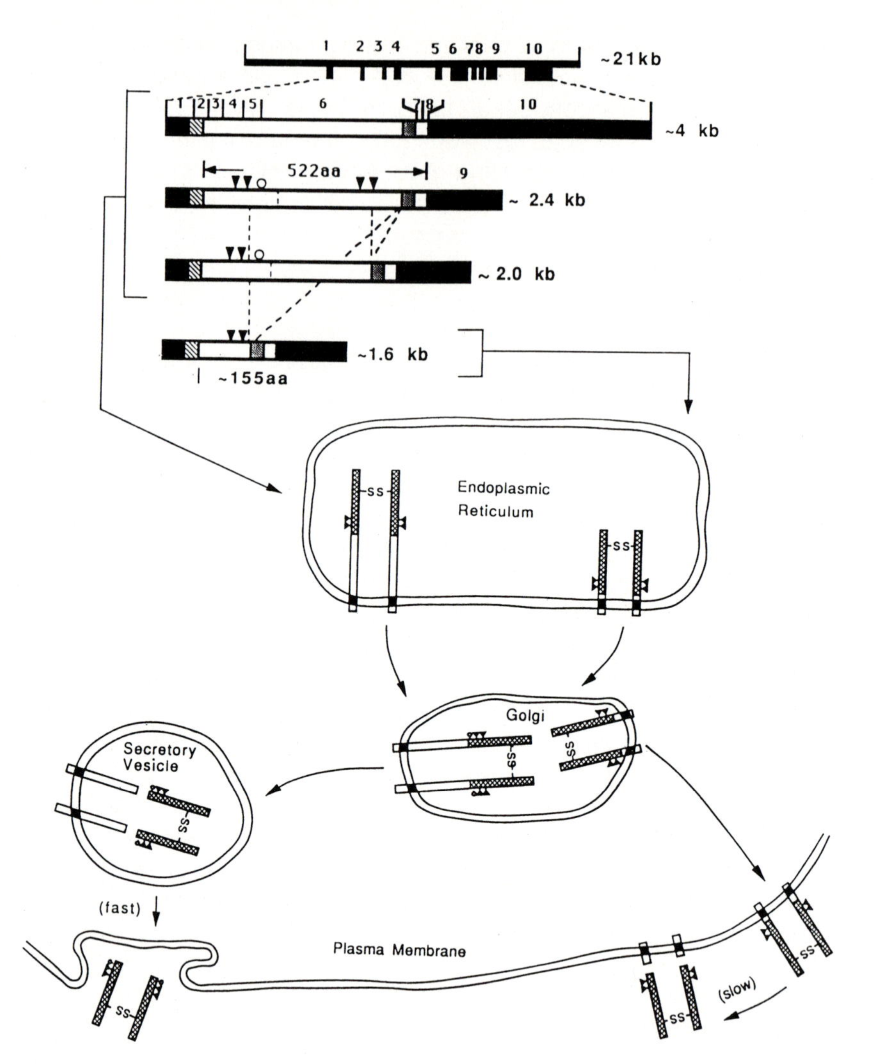

Fig. 1. Human CSF-1 genomic organization, transcripts, and expression. In the *upper part of the figure*, the intron–exon relationships of the human CSF-1 gene are shown, together with schematic representations of the four cDNA clones that have been sequenced. Exons (1–10), 5′ and 3′untranslated regions (*filled*), and coding region (*open*), including signal peptide (*hatched*) and transmembrane domain (*cross-hatched*), are indicated together with the N-linked (*arrowheads*) and O-linked (*open circles*) glycosylation sites and the intracellular proteolytic cleavage site (*dotted line*). The *lower part of the figure* depicts the later stages in the expression of homodimeric CSF-1. The short homodimers are derived from the 1.6-kb clone, the longer homodimers from the other clones. *Hatched regions*, mature CSF-1; *filled regions*, transmembrane domain

to chromosome 3 (GISSELBRECHT et al. 1989) on a different chromosome from the genes encoding its receptor (chromosome 18) (HOGGAN et al. 1988) and GM-CSF and IL-3 (chromosome 11) (BARLOW et al. 1987).

While several mRNA species (1.6–4.0 kb) have been observed in human cells and cDNAs corresponding to the 4.0-, 2.5-, 2.2-, and 1.6-kb species have been sequenced (KAWASAKI et al. 1985; LADNER et al. 1987; WONG et al. 1987; CERRETTI et al. 1988), two species of 4.0 and 2.3 kb, respectively, predominate in the mouse (LADNER et al. 1988). Although these mouse species differ from each other only in their alternatively spliced 3′- untranslated regions (LADNER et al. 1988), this seems to be the case only for the 4.0- and 2.5-kb human transcripts (LADNER et al. 1987; WONG et al. 1987). The other human mRNAs result from alternative splicing within exon 6 and are translated into different forms of CSF-1 (LADNER et al. 1987; CERRETTI et al. 1988) (Fig. 1). Lower molecular weight mRNAs have also been isolated from mouse cells (RAJAVASHISTH et al. 1987). In both the human and the mouse, the 3′-untranslated region of the "long" (4.0-kb) species is rich in AU sequences which may confer short half-lives on their respective mRNA transcripts (KAWASAKI and LADNER 1990) and may be important in regulation. Aside from these sequence characteristics, a number of additional transcriptional regulatory motifs have been identified within 400 base pairs of the mRNA initiation site in exon 1 of the CSF-1 gene (reviewed in KAWASAKI and LADNER 1990). Among these are (a) two TATA boxes with the sequences TTAAA and CATAAA at positions −26 to −22 and −54 to −49, respectively (KAWASAKI and LADNER 1990), (b) a series of alternating T and G residues at positions −126 to −80, which may favor the Z-DNA confirmation and affect transcription (RICH et al. 1984), (c) two GGCGGG sequences at positions −159 and −177, which may be involved in binding transcription factors (DYNAN et al. 1986), and (d) the sequences AGGAAAG and GGGAAAG at positions −317 and −377, respectively, which are both very similar to the consensus enhancer core sequence TGGAAAG that plays a role in the transcriptional level of genes with which it is associated (LAIMINS et al. 1983). While these various sequences offer the potential for regulation of CSF-1 expression at multiple levels, their definitive roles remain to be demonstrated.

3 CSF-1 Receptor

3.1 Occurrence

The receptor for CSF-1 (CSF-1R) is distributed in many tissues but is primarily expressed on cells of the mononuclear phagocytic lineage (GUILBERT and STANLEY 1980, 1986; BYRNE et al. 1981). With maturation, the surface expression of this receptor increases to maximal levels of approximately 5×10^4 molecules/ up-regulated tissue macrophage (GUILBERT and STANLEY 1986). In contrast, multipotent hematopoietic cells, which give rise to committed mononuclear phagocyte progenitors, express receptors at substantially lower densities of approximately 2×10^3 molecules/up-regulated cell (BARTELMEZ and STANLEY 1985). Additional cell types related to the mononuclear phagocyte that have also been shown to bear CSF-1R include subsets of acute myeloid leukemia cells (DUBREUIL et al. 1988; ASHMUN et al. 1989) and microglia of the central nervous system (SAWADA et al. 1990).

Although originally thought to be restricted to mononuclear phagocytes, the CSF-1R has more recently been demonstrated on maternal decidual cells and fetal trophoblastic cells during placental development (ARCECI et al. 1989; REGENSTREIF and ROSSANT 1989) and on human choriocarcinoma cell lines (RETTENMIER et al. 1986). The role of CSF-1 and its receptor in this biologic setting will be discussed further below (Sect. 4.4). Finally, one group has reported the presence of CSF-1R on both normal and neoplastic myoblasts (LEIBOVITCH et al. 1989).

3.2 Biochemistry

The CSF-1R, which is encoded by the c-*fms* proto-oncogene (SHERR et al. 1985), is a member of a family of growth factor receptors with intrinsic tyrosine kinase activity. In fact, the receptor is most closely related to the A- and B-type receptors for platelet-derived growth factor (PDGF) (YARDEN et al. 1986) and the c-*kit* proto-oncogene product (BESMER et al. 1986; YARDEN et al. 1987), whose ligand in the mouse has recently been shown to be the stem cell growth factor product of the Steel locus (HUANG et al. 1990; WILLIAMS et al. 1990; ZSEBO et al. 1990). The CSF-1R, like other members of this group of tyrosine kinase receptors, has a glycosylated extracellular ligand-binding domain, a hydrophobic membrane-spanning domain, and an intracellular tyrosine kinase domain (HAMPE et al. 1984; COUSSENS et al. 1986; ROTHWELL and ROHRSCHNEIDER 1987). Unlike other tyrosine kinase receptors, members of this family of receptors possess extracellular domains characteristic of members of the immunoglobulin gene superfamily and have unique hydrophilic spacer sequences of variable length within the tyrosine kinase domain that may play a role in substrate recognition (YARDEN et al. 1987). In the case of CSF-1R, the C-terminal tail also subserves a negative regulatory function as demonstrated by the increased transforming activity of both human and feline molecules mutated in this region (ROUSSEL et al. 1987) and the viral oncogene counterpart, v-*fms*, which lacks this sequence (BROWNING et al. 1986).

Studies of mononuclear phagocytes have demonstrated that these cells bear a single class of high affinity receptors for CSF-1 (GUILBERT and STANLEY 1980; STANLEY and GUILBERT 1981). Using a kinetic approach, this receptor has been shown to have dissociation constants of 4×10^{-10} M at 37 °C and of $\leq 10^{-13}$ M at 4 °C, respectively (GUILBERT and STANLEY 1986). While the molecular weight of the purified mouse CSF-1R (165 kDa, YEUNG et al. 1987) agrees well with its molecular weight determined by chemical cross-linking experiments in intact cells (165 kDa, MORGAN and STANLEY 1984), the human CSF-1R has a slightly lower apparent molecular weight (150 kDa, ROUSSEL et al. 1987). Although the density of receptors that can be expressed on individual cells varies between 2000 and 120 000 depending on cell type (BYRNE et al. 1981; STANLEY et al. 1983; BARTELMEZ and STANLEY 1985; Fig. 3), several other factors have been shown to regulate cell surface CSF-1R expression. The presence of CSF-1 results in receptor binding followed by internalization and degradation of receptor–ligand complexes, while removal of growth factor leads to increased cell surface receptor expression (STANLEY and GUILBERT 1981; GUILBERT and STANLEY 1986). Experiments involving other agents have shown that several bacterial products, including lipopolysaccharide (GUILBERT and STANLEY 1984), muramyl tripeptide-phosphatidylethanolamine, lipopeptide CGP 31362, and pertussis toxin (HUME and DENKINS 1989), result in decreased CSF-1R density. In addition, phorbol esters, presumably through activation of protein kinase C, down-regulate receptor expression through mechanisms distinct from those involved in ligand-induced changes (CHEN et al. 1983; GUILBERT et al. 1983; DOWNING et al. 1989). Finally, many cytokines such as IL-3 and GM-CSF, through a presumed "hierarchical down-modulation" (WALKER et al. 1985), as well as others such as IL-4, interferon-γ, and tumor necrosis factor-α (TNF-α), through as yet undefined mechanisms (HUME and DENKINS 1989; SHIEH et al. 1989), also result in decreased cell surface CSF-1R expression. Thus, many factors may influence the proliferation, differentiation, and activity of mononuclear phagocytic cells via their effects on CSF-1R expression.

3.3 Molecular Biology

As indicated above, the CSF-1R gene has been mapped to human chromosome 5 (5q33.3), in close proximity to the CSF-1 gene (GROFFEN et al. 1983; ROUSSEL et al. 1983; LEBEAU et al. 1986). In addition, this gene is the 3′ neighbor of the gene for the B-type PDGF receptor (YARDEN et al. 1986; ROBERTS et al. 1988). Both genes, which have a similar intron–exon organization, are juxtaposed in head-to-tail fashion with the CSF-1R promoter located within 0.5 kb of the B-type PDGF-R polyadenylation signal, raising the possibility that transcription of the former is affected by transcription of the latter (ROBERTS et al. 1988).

The CSF-1R gene (Fig. 2) which is 58 kb in length, consists of 21 introns and 22 exons (reviewed in SHERR 1990). Exon 1, which encodes nontranslated sequences, is 26 kb upstream of exon 2, which encodes the signal peptide. The transcription of the gene proceeds in a tissue-specific manner from two distinct promoters. While the precise locations of these promoters are unknown, it has been established that transcription is initiated upstream of exon 1 in placental trophoblasts, while it is initiated immediately upstream of exon 2 in mononuclear phagocytes (VISVADER and VERMA 1989). Increases in the steady state levels of CSF-1R mRNA observed with the maturation of precursors to more differentiated cells of the mononuclear phagocyte lineage (SARIBAN et al. 1985) appear to be

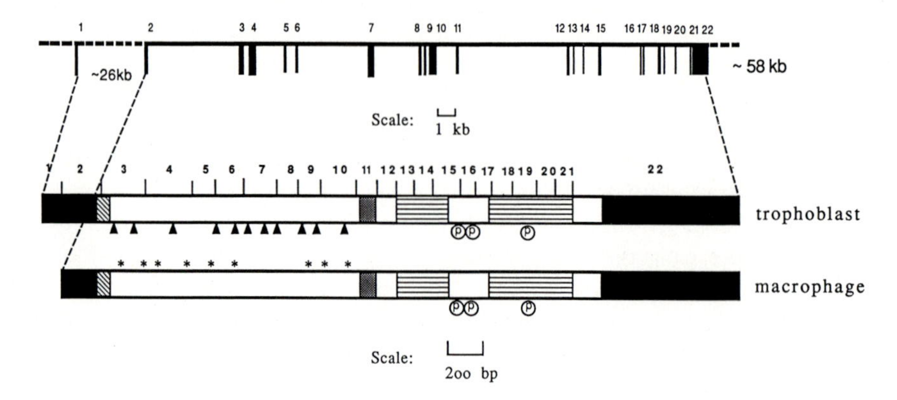

Fig. 2. CSF-1 receptor genomic organization and transcripts. The intron–exon (*arabic numerals*) relationships of the CSF-1 receptor gene are shown, together with a schematic representation of the mRNAs expressed in placental trophoblasts and macrophages. The untranslated regions (*filled*), signal sequence (*hatched*), transmembrane domain (*cross-hatched*), and interrupted tyrosine kinase catalytic domain (*striped*) are shown together with the potential N-linked glycosylation sites (*arrowheads*), positions of the extracellular cysteine residues (*asterisks*), and tyrosine phosphorylation sites (*P*). (Based on HAMPE et al. 1989)

induced by factors that increase receptor expression through alterations in mRNA half-lives (WEBER et al. 1989). Furthermore, hierarchical down-modulation of the CSF-1R (WALKER et al. 1985) also appears to be regulated via post-transcriptional mechanisms (GLINIAK and ROHRSCHNEIDER 1990).

4 Biologic Effects of CSF-1

4.1 Early Events in Response to CSF-1

Colony stimulating factor-1 has been shown to exert pleiotropic effects on cells of the mononuclear phagocyte series (reviewed in STANLEY et al. 1983.) In order to induce monocytic cells to divide, CSF-1 must be present during the entire G_1 phase to facilitate entry of cells into S phase and their subsequent progression through G_2 and M (TUSHINSKI and STANLEY 1985). While numerous cellular responses have been observed following ligand activation, it is difficult to determine how they contribute to the processes of survival, proliferation, and differentiation induced by the growth factor.

Among the earliest effects observed following CSF-1 binding are changes in the cell membrane itself, including ruffling and the formation of filopodia, vesicles, and vacuoles (TUSHINSKI et al. 1982; BOOCOCK et al. 1989). Soon after, several metabolic changes are observed, including increases in glucose uptake (HAMILTON et al. 1988) and protein synthesis with a concomitant decrease in protein degradation (TUSHINSKI and STANLEY 1985). In addition, there is an early influx of sodium by an amiloride-sensitive mechanism, presumably the Na^+/H^+ antiport, as well as an increase in Na^+/K^+-ATPase activity (IMAMURA and KUFE 1988; VAIRO and HAMILTON 1988). These latter changes are inhibited by pertussis toxin and are consequently believed to be mediated by G proteins (IMAMURA and KUFE 1988). The role of additional enzymatic systems in the cellular events leading to CSF-1- induced cell proliferation has been suggested by the increase in cyclooxygenase activity which results in increased prostaglandin E_2 and thromboxane A_2 production (KURLAND et al. 1979; ORLANDI et al. 1989), as well as by the association of the "activated" CSF-1R with phosphatidyl inositol-3 kinase activity, which results in the production of novel phosphoinositides, which may contribute to mitogenesis (VARTICOVSKI et al. 1989; SHURTLEFF et al. 1990). Other early changes noted have been at the transcriptional level with increased levels of expression of the c-*myc* and c-*fos* proto-oncogene products (BRAVO et al. 1987; ORLOFSKY and STANLEY 1987), which have generally been associated with subsequent increases in DNA synthesis (HAMILTON et al. 1989).

The very earliest cellular events following CSF-1 binding to the CSF-1R are the phosphorylation of the receptor and other cellular proteins on tyrosine residues (DOWNING et al. 1988; SENGUPTA et al. 1988). CSF-1-induced protein tyrosine phosphorylation is maximal within 30s of CSF-1 addition to cells at 37 °C (SENGUPTA et al. 1988). A particular sequence of events has been discerned by studying these reactions at 4 °C (SENGUPTA et al. 1988; LI and STANLEY 1991; M. BACCARINI, W. LI, P. DELLO SBARBA, E.R. STANLEY, submitted for publication), a temperature at which neither CSF-1 nor the CSF-1R is internalized but at which transmembrane-stimulated protein tyrosine phosphorylation occurs. It appears that CSF-1 binding stimulates or stabilizes the formation of noncovalent CSF-1R homodimers. This dimerization apparently leads to activation of the receptor kinase and receptor autophosphorylation on tyrosine, a process that can occur via intermolecular phosphorylation of adjacent kinase domains (OHTSUKA et al. 1990). These events are followed by the tyrosine phosphorylation of several, mostly cytoplasmic, proteins prior to the formation of disulfide bonds between monomeric units of the CSF-1R dimer, a modification of one of the monomeric units that increases its molecular weight from 165 kDa to approximately 250 kDa (LI and STANLEY 1991), and an increase in receptor tyrosine and serine phosphorylation (M. BACCARINI, W. LI, P. DELLO SBARBA, E.R. STANLEY, submitted for publication). These results and other data (LI and STANLEY 1991) suggest that the noncovalent homodimeric receptor–ligand complex is the active signaling species, whereas the covalent heterodimeric receptor–ligand species is inactive and selectively internalized. The identities of the proteins that are rapidly phosphorylated in tyrosine have not been established. It has been recently claimed that the proto-oncogene product RAF-1, a 72-kDa cytoplasmic serine/threonine kinase, is a substrate of the PDGF-R and therefore potentially involved in transducing the signal from membrane to nucleus (MORRISON et al. 1988, 1989). However, recent studies with the CSF-1R, PDGF-R (BACCARINI et al. 1990), and EGF-R (BACCARINI et al. 1991) indicate that while RFA-1 is serine phosphorylated and the RAF-1-associated kinase is activated following growth factor stimulation, it is not phosphorylated in tyrosine and therefore not a physiologic substrate of these receptor tyrosine kinases. Attempts to isolate cDNA clones encoding the genes for proteins rapidly phosphorylated in tyrosine in response to CSF-1 are currently underway. One of these proteins (57 kDa) is of particular interest as it is the only one that is phosphorylated in tyrosine in several cell types in response to different growth factors (Li et al. 1991). It should be noted that as a tyr-809 to phe-809 CSF-1R mutant which is unable to stimulate mitogenesis is able to stimulate protein tyrosine phosphorylation as effectively as the wild-type receptor (Roussel et al. 1990), it is likely that only one or a few of the tyrosine phosphorylated proteins are involved in the signal transduction pathway for proliferation.

4.2 Monocytopoiesis

Colony stimulating factor-1 is capable of exerting its biologic effects on the mononuclear phagocyte lineage beginning with the committed progenitor cell, which proliferates and differentiates to give rise in sequence to the following cells: monoblast → promonocyte → monocyte → macrophage (Fig. 3) (reviewed in Stanley 1990). These responses have been demonstrated in vitro using human CSF-1, which is capable of stimulating the formation of macrophage colonies from both mouse and human precursor cells (Das and Stanley 1982; Motoyoshi et al. 1982; Waheed and Shadduck 1982), as well as mouse CSF-1, which is species specific, stimulating macrophage production from mouse precursors only (Das and Stanley 1982). Interestingly, mouse cells show a greater proliferative response to human CSF-1 than do human precursor cells. However, it is not clear whether this phenomenon is due to poor definition of the culture conditions required for human cell proliferation (Das and Stanley 1982) or the requirement of additional growth factors (Ralph et al. 1986a; Chen et al. 1988).

The primary role of CSF-1 in regulating mononuclear phagocyte production in vivo has been demonstrated in mice (Hume et al. 1989) and nonhuman primates (Munn et al. 1990), in which administration of exogenous CSF-1 resulted in substantial increases in blood monocytes and tissue macrophages. Furthermore, work on the osteopetrotic *op/op* mouse has shown that its deficiency in cells of the mononuclear phagocyte lineage, including bone-resorbing osteoclasts, results from a total deficiency of circulating and tissue CSF-1 (Wiktor-Jedrzejczak et al. 1990). While data in humans are less extensive, we have shown that the elevated peripheral blood monocyte counts in newborns are associated with serum CSF-1 concentrations that are three times adult values at birth and go on to double over the first week of life (Roth 1990).

The steady state concentration of circulating growth factor in vivo is regulated by CSF-1R-mediated endocytosis and intracellular degradation by tissue macrophages, primarily in the liver (i.e., Kupffer cells) and spleen

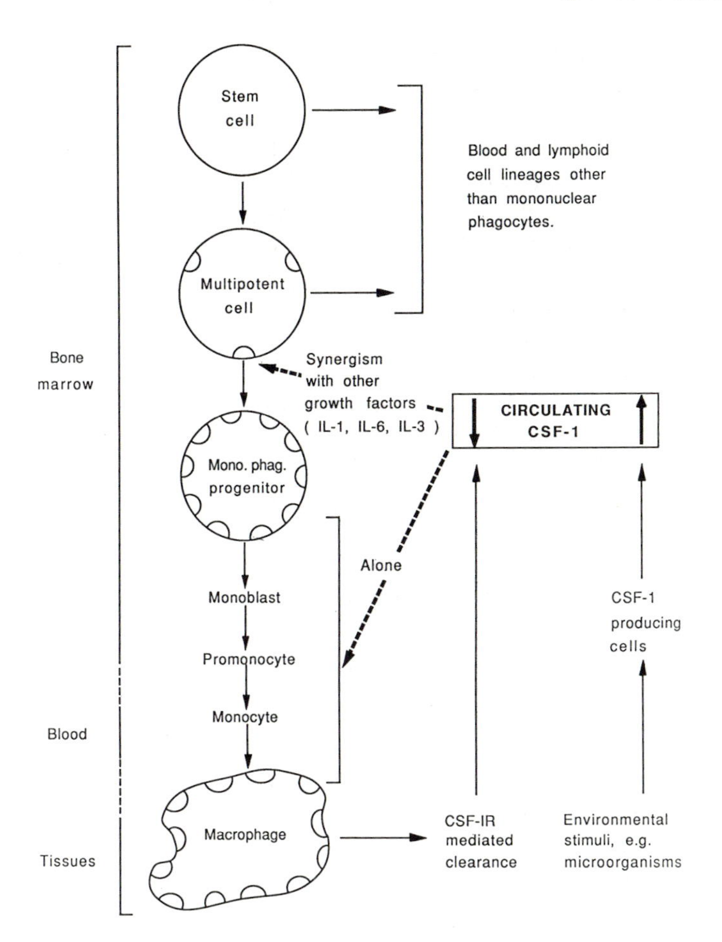

Fig. 3. Schematic representation of the role of circulating CSF-1 in the regulation of mononuclear phagocyte production. CSF-1R expression is represented by *small semi circles.* (STANLEY 1990)

(BARTOCCI et al. 1987). This mechanism provides a negative feedback loop regulating the production of mononuclear phagocytes. Rapid increases in circulating CSF-1 are brought about by the increased synthesis and release of the growth factor from the endothelial cells of several organs (P. ROTH, A. BARTOCCI, E.R. STANELY, unpublished).

Although CSF-1 alone is capable of regulating mononuclear phagocyte proliferation and differentiation, these events are probably regulated in vivo by more complex interactions involving several growth factors. In fact, CSF-1 has been shown to synergise with IL-3 (CHEN and CLARK 1986), GM-CSF (CARACCIOLO et al. 1987; CHEN et al. 1988), and TNF-α (BRANCH et al. 1989), as well as the neuropeptides substance P (MOORE et al. 1988) and neurotensin (MOORE et al. 1989), in its in vitro actions on monocytic cells. Aside from its effect on committed precursor cells, CSF-1 is also capable of acting in concert with other factors to direct primitive multipotent cells along the mononuclear phagocyte pathway of differentiation (BARTELMEZ and STANLEY 1985). This "channeling" of multipotent cells has been observed when CSF-1 is used in combination with IL-1 alone (MOCHIZUKI et al. 1987; WARREN and MOORE 1987; WILLIAMS et al. 1987), IL-1 + IL-3 (BARTELMEZ et al. 1989), GM-CSF (WILLIAMS et al. 1987; MCNIECE et al. 1988), and IL-6 (BOT et al. 1989) and results in an increase in the number of mononuclear phagocytes capable of being regulated by CSF-1 alone.

4.3 Mononuclear Phagocyte Differentiation

Colony stimulating factor-1 induces mature monocytes and macrophages to perform various differentiated cell functions, such as the production of other cytokines (reviewed in RALPH and WARREN 1989), including interferon, IL-1, TNF-α (RALPH et al. 1986a; WARREN and RALPH 1986), and G-CSF (MOTOYOSHI et al. 1982). In addition, other important cell products involved in monocyte/macrophage immune function that are produced following CSF-1 exposure are thromboplastin (LYBERT et al. 1987), plasminogen activator (HAMILTON et al. 1980), prostaglandins (ORLANDI et al. 1989), acidic isoferritin (BROXMEYER et al. 1987), and oxygen radicals (BECKER et al. 1987; KANADA et al. 1987).

Aside from its effects on the production of soluble factors, CSF-1 also induces changes in the expression of mononuclear phagocyte membrane glycoproteins, including Ia antigens (WILLMAN et al. 1989), MAC1, asialo-GM1 (AKAGAWA et al. 1988), type III Fc receptor, LFA-3 (MUNN et al. 1990), and MY-4 (GEISSLER et al. 1989), all of which are presumed to be involved in cell–cell interactions in the immune response. The impact of these and other as yet undescribed molecular changes has been manifested by the several functional changes seen in CSF-1-stimulated mononuclear phagocytes (reviewed in RALPH and WARREN 1989) (see Sect. 4.5).

Aside from these effects on macrophage differentiation, which affect immune function, CSF-1 appears to play an important role in bone resorption. However, the results are somewhat confusing in that CSF-1, which is produced by osteoblasts (ELFORD et al. 1987), is a potent inhibitor of bone resorption by osteoclasts, which are themselves part of the mononuclear phagocyte series (HATTERSLEY et al. 1988). On the other hand, production of this cell type is dependent on CSF-1, as evidenced by its absence in the CSF-1-deficient osteopetrotic *op/op* mouse (WIKTOR-JEDRZEJCZAK et al. 1990). Another novel effect of CSF-1 is its ability in vivo to lower plasma cholesterol, predominantly the low density lipoprotein fraction, presumably through augmented macrophage phagocytosis (reviewed in GARNICK and STOUDEMIRE 1990; ISHIBASHI et al. 1990; SHIMANO et al. 1990).

4.4 Placental Regulation

While the major target for CSF-1 is the mononuclear phagocyte, the placenta is another putative site for its actions (reviewed in STANLEY 1990). More than 15

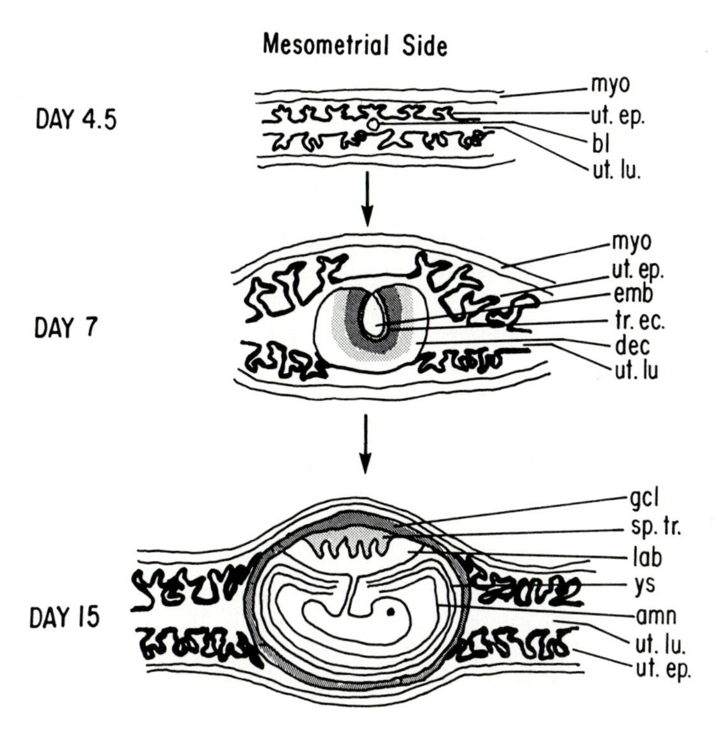

Fig. 4. Expression of CSF-1 and CSF-1R mRNA during placental development in the mouse. CSF-1 (*filled*) and CSF-1R (*stippled*, density of stippling proportional to intensity) mRNA expression is represented for days 4.5, 7, and 15 of pregnancy. Implantation takes place at day 5. *myo*, myometrium; *ut. ep.*, uterine epithelium; *bl*, blastocyst; *ut. lu.*, uterine lumen; *emb*, embryo; *tr. ec.*, trophoectoderm; *dec*, decidua; *gcl*, trophoblast giant cell layer; *sp. tr.*, spongiotrophoblast cells; *lab*, labyrinthine trophoblast cells; *ys*, yolk sac; *amn*, amnion. (STANLEY 1990)

years ago, it was shown that the pregnant uterus as well as the fetal and placental membranes were rich in colony stimulating activity (BRADLEY et al. 1971; ROSENDAAL 1975). Since then, CSF-1 has been identified as the growth factor responsible for this activity, with concentrations in the day 20 pregnant mouse uterus 1000 times higher than in the same nonpregnant organ (BARTOCCI et al. 1986). Further investigation using in situ hybridization has shown that expression of CSF-1 is confined to the luminal and glandular secretory epithelium (POLLARD et al. 1987) and is regulated by chorionic gonadotropin via synergistic action of the hormones, progesterone, and estrogen (BARTOCCI et al. 1986). Coincident with the increased expression of CSF-1, there is also substantial expression of CSF-1R mRNA in the murine placenta (MULLER et al. 1983; ARCECI et al. 1989) and human trophoblastic cells (HOSHINA et al. 1985). Studies employing in situ hybridization techniques have provided localization and temporal expression data that are compatible with CSF-1-regulated placental development (Fig. 4) (ARCECI et al. 1989; REGENSTREIF and ROSSANT 1989). CSF-1 mRNA expression in the uterine epithelium is increased above baseline by the third day of gestation, achieves substantially elevated levels by days 8.5–9.5, and peaks at day 15. CSF-1R mRNA, on the other hand, first appears in the maternal decidua on the sixth day of gestation and continues to be expressed at low levels in the decidua basalis. On the fetal side, there is limited expression in trophoectodermal cells by day 7.5 and in diploid trophoblasts during placentation. Once the placenta is mature by day 9.5 of gestation, CSF-1R mRNA expression is most prominent in fetally derived giant trophoblastic cells followed by moderate expression in the spongiotrophoblastic layer and lower expression in the labyrinthine layers (ARCECI et al. 1989; REGENSTREIF and ROSSANT 1989). Placental CSF-1R mRNA expression peaks at day 17 of gestation (ARCECI et al. 1989).

The expression of CSF-1 in the uterine epithelium of the pregnant mouse is very local. In contrast to the large increase in uterine CSF-1 concentration, the concentration of CSF-1 in other tissues and the serum of the pregnant mouse increases only twofold. The ability of CSF-1 to stimulate trophoblastic cell proliferation (ATHANASSAKIS et al. 1987) and the in vitro proliferation and differentiation of ectoplacental cones (REGENSTREIF and ROSSANT 1989) together with the close apposition of CSF-1 and CSF-1R producing cells in the placenta make it likely that this growth factor plays an important role in placental regulation. In fact, maximal expression of uterine CSF-1 and placental CSF-1R coincides with the period of maximal placental growth, which occurs on days 9–15 of gestation (ARCECI et al. 1989). The existence of a local placental function for CSF-1 in the regulation of nonmononuclear phagocytic cells as distinct from its humoral function in regulating mononuclear phagocytes is entirely compatible with data which demonstrate that different promoters regulate the expression of CSF-1R mRNA in trophoblastic cells and mononuclear phagocytes, respectively (VISVADER and VERMA 1989). Further evidence for a distinct role for CSF-1 in placental regulation is the observation that the predominant mRNA species for this growth factor in the gravid mouse uterus is 2.3 kb and not the 4.0-kb species seen in most other tissues (POLLARD et al. 1987). These transcripts differ only in their 3′-untranslated sequences, with the larger molecule containing three repeats of the AUUUA sequence, which in the right context has been shown to decrease mRNA stability (SHAW and KAMEN 1986). It is therefore possible that the larger species, which appears following immunologic stimuli, is rapidly eliminated once the demand for macrophages ceases, whereas the shorter molecule, which is required over a sustained period of time during gestation, is a more stable, long-lived species (POLLARD et al. 1987).

4.5 Macrophage Activation in Inflammation, Infection, and Cytotoxicity

Colony stimulating factor-1 is an important mediator of the inflammatory response, as evidenced by its increased presence in inflammatory exudates (STANLEY et al. 1976). In addition, as noted above (Sect. 4.3), it induces monocytic cells to release several cytokines, including IL-1, TNF-α, G-CSF, and interferon as well as proteolytic enzymes such as plasminogen activator. Furthermore, its ability to provide chemotactic signals for monocytic cells (WANG et al. 1988) makes it ideally suited for regulating the influx of cells to the site of an inflammatory stimulus.

The potential role for CSF-1 in host defenses against infection has been demonstrated in numerous systems. In a model system for the inflammatory response, we have shown that mice treated with lipopolysaccharide in vivo demonstrate rapid increases in serum and tissue CSF-1 concentrations, which are the result of increased growth factor synthesis rather than a decrease in the clearance of circulating growth factor (ROTH et al. 1990). Other investigators have focused on specific pathogens and have shown that CSF-1 enhances resistance to viral infection (LEE and WARREN 1987) as well as killing of the fungal species *Candida albicans* (KARBASSI et al. 1987; CHONG and LANGLOIS 1989; WANG et al. 1989; CENCI et al. 1991) and the intracellular pathogens *Mycobacterium avium–intracellulare* (BERMUDEZ and YOUNG 1990) and *Listeria monocytogenes* (CHEERS et al. 1989). Additional studies have shown that CSF-1 may play an important role in the response to parasitic infections, including leishmaniasis (HO et al. 1989) and schistosomiasis (CLARK et al. 1988). In contrast, CSF-1 facilitates replication of the human immunodeficiency virus (GENDLEMAN et al. 1988) and causes monocytic cells to suppress both antigen- and mitogen stimulated T cell proliferation (WING et al. 1986).

Colony stimulating factor-1-stimulated mononuclear phagocytes exert modest increases in cytotoxic activity against tumors in an antibody-independent (WARREN and RALPH 1986; RALPH and NAKOINZ 1987; SAMPSON-JOHANNES and GARLINO 1988) and antibody-dependent fashion (NAKOINZ and RALPH 1988; MUFSON et al. 1989; MUNN et al. 1990). However, in virtually all of these studies, cells are preincubated with CSF-1 for prolonged periods of time before assaying for cytotoxicity. Consequently, the effects observed may be the result of improved mononuclear phagocyte survival in vitro rather than a direct effect on tumor cytotoxicity.

4.6 Neoplasia

The v-*fms* oncogene derived from the Susan McDonough strain of the feline sarcoma virus is closely related to the c-*fms*-encoded CSF-IR (reviewed in SHERR 1990; SHERR and STANLEY 1990; STANLEY 1990). Despite its ability to bind ligand, v-*fms* is capable of stimulating cell growth and producing a transformed phenotype in the absence of CSF-1 (WHEELER et al. 1986). A number of differences between the amino acid sequences of v-*fms* and c-*fms* have been demonstrated and are thought to be responsible for differences in their respective abilities to transform cells which express them. First, there is a 3′-deletion in the v-*fms* gene resulting in replacement of the C-terminal 50 amino acids of c-*fms* with 11 unrelated residues encoded by the 3′-untranslated region of the feline c-*fms* (COUSSENS et al. 1986). This deletion is necessary for (WOOLFORD et al. 1988) or at least greatly enhances (ROUSSEL et al. 1988) the transforming ability of this oncogene, possibly by eliminating a peptide that serves to negatively regulate the tyrosine kinase activity of the native receptor (ROUSSEL et al. 1987). Further comparison of v-*fms* and c-*fms* has revealed nine additional amino acid substitutions, of which two at residues 301 and 374 are required for transformation (WOOLFORD et al. 1988). Substitution at leucine-301 is thought to alter the conformation of the extracellular domain in a manner that mimics ligand binding and results in constitutive phosphorylation of the v-*fms* molecule (WOOLFORD et al. 1988).

While the v-*fms* molecule is capable of transforming cells and inducing the formation of tumors derived from mononuclear phagocytes and other hematopoietic lineages (HEARD et al. 1987b), the c-*fms* gene results in transformation in situations where it and/or its ligand, CSF-1, are inappropriately expressed (reviewed in STANLEY 1990). BAUMBACH et al. (1987) have shown that mice injected with spleen cells infected with a c-*myc*-containing retrovirus developed monocytic tumors due to a rearrangement in the CSF-1 gene. This provided a "secondary event," which resulted in increased growth factor secretion and autocrine stimulation of receptor-bearing cells. A similar hypothesis has been invoked for the development of endometrial and ovarian tumors in humans, where CSF-1 and CSF-1R are frequently coexpressed in tumor cells (KACINSKI et al. 1989a, b). Similar coexpression of ligand and receptor has also been demonstrated in a subpopulation of patients with acute myeloblastic leukemia (WAKAMIYA et al. 1987; RAMBALDI et al. 1988). In another situation, it has been shown that many murine myeloblastic leukemias induced by the Friend murine leukemia virus, both in vivo and in vitro, actually result from proviral integration at a site that results in greatey enhanced c-*fms* expression, leading to the possibility of inappropriately increased responsiveness of primitive hematopoietic cells to CSF-1 (GISSELBRECHT et al. 1987).

4.7 Clinical Applications

On the basis of data derived from in vitro studies, in vivo experiments in animal models, and some preliminary observations in humans, it is clear that CSF-1 has tremendous potential for a wide array of clinical applications. The ability of CSF-1 to augment host responses to a variety of pathogens (see Sect. 4.5) makes it a prime candidate as an adjunct in the treatment of infection. While this growth factor has been shown to exert a protective effect in mice exposed to *E. coli* (CHONG and LANLOIS 1988), there are limited data as yet for its effects on human infection.

In the area of neoplastic diseases, CSF-1 and its receptor may be useful in several areas. First, the c-*fms*-encoded CSF-1R and serum CSF-1 concentrations may be useful as tumor markers to classify as well as to follow the progress of particular malignancies. In fact, several investigators have shown that a substantial proportion of patients with acute myeloblastic leukemia display the CSF-1R on their tumor cells, which in these instances generally show evidence of monocytic differentiation (DUBREUIL et al. 1988; ASHMUN et al. 1989). However, unlike the case of Friend virus-induced murine myeloblastic leukemias (GISSELBRECHT et al. 1987), there is no evidence of rearrangements in the region of the c-*fms* gene in these human leukemias (DUBREUIL et al. 1988; ASHMUN et al. 1989). Furthermore, recent studies have indicated that circulating CSF-1 concentrations are elevated in patients with myeloproliferative disease, especially those with peripheral bone marrow extension (GILBERT et al. 1989), and in a high proportion of patients with preleukemia, leukemia, and lymphoid malignancies (JANOWSKA-WIECZOREK et al. 1991).

Apart from lymphohemopoietic malignancies, this growth factor and its receptor may be important markers in the case of ovarian and endometrial cancers. Aside from their coexpression of growth factor and receptor in many cases, concentrations of circulating CSF-1 have been shown to reflect disease activity and can point to the presence of recurrent disease (KACINSKI et al. 1989a, b).

A second possible role for CSF-1 is as an adjunctive therapy in the treatment of malignancies. The ability of CSF-1 to induce terminal differentiation in vitro of blast cells from some patients with acute myeloblastic leukemia (MIYAUCHI et al. 1988) provided the basis for examining the role of this cytokine in vivo in several animal tumor models. CSF-1 has been shown to inhibit primary tumor growth in syngeneic mice (RALPH and WARREN 1989) as well as the development of metastases in a murine melanoma model (HUME et al. 1989). At the present time, data in humans are lacking.

A third role for CSF-1 in neoplastic disorders arises from its myelorestorative capabilities. In patients who have undergone ablative therapies followed by bone marrow transplantation for a variety of leukemias and neuroblastoma, CSF-1 accelerates the recovery of total leukocytes and granulocytes (MASAOKA et al. 1988). Similar results have also been seen in individuals undergoing intensive chemotherapy for other malignancies (MATSUMOTO et al. 1987). This therapeutic approach has also been employed in patients with primary deficiencies of granulocytes, e.g., chronic neutropenia of childhood, where CSF-1 treatment has resulted in increased neutrophil counts, which in some cases have been sustained (KOMIYAMA et al. 1988).

Aside from its hemopoietic activities, the role of CSF-1 in regulating the inflammatory response may be critical to clinical pathology. In both humans and experimental mice with systemic lupus erythematosus, there are increased concentrations of circulating CSF-1 and numbers of mononuclear phagocytes, which may contribute to the autoimmune injury seen in these individuals (A. BARTOCCI, V. PRALORAN, P. DELLO SBORBA, E.R. STANLEY, unpublished). Similarly, since circulating CSF-1 regulates osteoclast production (WIKTOR-JEDRZEJCZAK et al. 1990), it is conceivable that an increase in growth factor levels with age could in some way contribute to osteoporosis of the elderly. Thus, treatments for these disease entities may involve the use of antagonists capable of attenuating the CSF-1-mediated inflammatory response (reviewed in SHERR and STANLEY 1990). In contrast, the absence of CSF-1 in the *op/op* osteopetrotic mouse (WIKTOR-JEDRZEJCZAK et al. 1990) makes it conceivable that administration of exogenous CSF-1 may be therapeutic in at least some forms of the human disease. In fact, administration of this growth factor has been shown to correct many of the defects of the *op/op* mouse (WIKTOR-JEDRZEJCZAK et al. 1990; FELIX et al. 1990; W. WIKTOR-JEDRZEJCZAK, E. URBANOWSKA, S.L. AUKERMAN, J.W. POLLARD, E.R. STANLEY, P. RALPH, A.A. ANSARI, K.W. SELL, M. SZPERL, unpublished).

5 Conclusions

Through the use of the techniques of cell biology, biochemistry, and molecular biology, a great deal has been learned regarding the biology of CSF-1 and its receptor. While at the molecular and cellular level a great deal regarding the signal transduction pathways regulated by CSF-1 remains to be elucidated, methods have been developed to allow isolation and characterization of the various intermediates as well as the kinases and phosphatases that act upon them. From a biologic point of view, considerable information regarding the role of CSF-1 in development may come from analysis of, and attempts to reconstitute, the CSF-1-deficient *op/op* mouse. While providing insights into the physiology of growth factors and their receptors in general, these findings should also provide specific data regarding CSF-1 that will serve as a basis for the development of novel approaches for the management of inflammatory and neoplastic disorders.

Acknowledgements. The authors wish to thank their colleagues and collaborators who have been involved in the studies described in this review, as well as Ms. Marie Calabro for her help in preparing the manuscript. This work was supported by NIH grants CA 26504 and CA 32551, the Albert Einstein Core Cancer Grant NIH-NCI P30 CA 13330-16, and grants from the Lucille P. Markey Charitable Trust and the New York Lung Association.

References

Akagawa KS, Kamoshita K, Tokunaga T (1988) Effects of granulocyte-macrophage colony stimulating factor and colony stimulating factor-1 on the proliferation and differentiation of murine alveolar macrophages. J Immunol 141 : 3383–3390

Arceci RJ, Shanahan F, Stanley ER, Pollard JW (1989) Temporal expression and locations of colony stimulating factor-1 (CSF-1) and its receptor in the female reproductive tract are consistent with CSF-1 regulated placental development. Proc Natl Acad Sci USA 86 : 8818–8822

Ashmun RA, Look AT, Roberts WM, Roussel MF, Seremetis S, Ohtsuka M, Sherr CJ (1989) Monoclonal antibodies to the human CSF-1 receptor (c-*fms* proto-oncogene product) detect epitopes on normal mononuclear phagocytes and on human myeloid leukemic blast cells. Blood 73 : 827–837

Athanassakis I, Bleackley RC, Poetkau V, Guilbert L, Barr PJ, Wegman TG (1987) The immunostimulatory effect of T cells and T cell lymphokines on murine fetally derived placental cells. J Immunol 138: 37–44

Baccarini M, Sabatini DM, App H, Rapp UR, Stanley ER (1990) Colony stimulating factor-1 (CSF-1) stimulates temperature dependent phosphorylation and activation of the RAF-1 proto-oncogene product. EMBO J 9: 3649–3657

Baccarini M, Gill GN, Stanley ER (1991) Epidermal growth factor stimulates phosphorylation of RAF-1 independently of receptor autophosphorylation and internalization. J Biol Chem (in press)

Barlow DP, Bucan M, Lebract H, Hogan BLM, Gough NM (1987) Close genetic and physical linkage between the murine haematopoietic growth factor genes GM-CSF and multi-CSF (IL-3). EMBO J 6: 617–623

Bartelmez SH, Stanley ER (1985) Synergism between hemopoietic growth factors (HGFs) detected by their effects on cells bearing receptors for a lineage-specific HGF: assay of hemopoietin-1. J Cell Physiol 122: 370–378

Bartelmez SH, Bradley TR, Bertoncello I et al. (1989) Interleukin-1 plus interleukin-3 plus colony stimulating factor-1 are essential for clonal proliferation of primitive myeloid bone marrow cells. Exp Hematol 17: 240–245

Bartocci A, Pollard JW, Stanley ER (1986) Regulation of colony-stimulating factor-1 during pregnancy. J Exp Med 164: 956–961

Bartocci A, Mastrogiannis DS, Migliorati G, Stockert RJ, Wolkoff AW, Stanley ER (1987) Macrophages specifically regulate the concentration of their own growth factor in the circulation. Proc Natl Acad Sci USA 84: 6173–6183

Baumbach WR, Colston EM, Cole MD (1987) Integration of the BALB/c ecotropic provirus into the colony stimulating factor-1 growth factor locus in a *myc* retrovirus-induced murine monocyte tumor. J Virol 62: 3151–3155

Becker SJ, Warren MK, Haskill S (1987) Colony stimulating factor-induced monocyte survival and differentiation into macrophages in serum-free cultures. Immunology 139: 3703–3709

Bermudez L, Young LS (1990) Macrophage colony stimulating factor (CSF-1) stimulates murine and human macrophages to kill mycobacterium avium complex (MAC). (submitted for publication)

Besmer P, Murphy JE, George PC et al. (1986) A new acute transforming feline retrovirus and relationship of its oncogene v-*kit* with the protein kinase gene family. Nature 320: 415–421

Boocock CA, Jones GE, Stanley ER, Pollard JW (1989) Colony stimulating factor-1 induces rapid behavioral responses in the mouse macrophage cell line, BAC 1.2F5. J Cell Sci 93: 447–456

Bot FJ, Van Eijk L, Broeders L, Aarden LA, Lowenberg B (1989) Interleukin-6 synergizes with M-CSF in the formation of macrophage colonies from purified human marrow progenitor cells. Blood 73: 435–437

Bradley TR, Metcalf D (1966) The growth of mouse bone marrow cells in vitro. Aust J Exp Biol Med Sci 44: 287–299

Bradley TR, Stanley ER, Sumner MT (1971) Factors from mouse tissues stimulating colony growth of mouse bone marrow cells in vitro. Aust J Exp Biol Med Sci 49: 595–603

Branch DR, Turner AR, Guilbert LJ (1989) Synergistic stimulation of proliferation by the monokines tumor necrosis factor-alpha and colony stimulating factor-1. Blood 73: 307–311

Bravo R, Neuberg M, Burckhardt J, Almendral J, Wallich R, Muller R (1987) Involvement of common and cell-type specific pathways in c-*fos* gene control: stable induction by cAMP in macrophages. Cell 48: 251–260

Browning PJ, Bunn HF, Cline A, Shuman M, Nienhuis AW (1986) "Replacement" of COOH-terminal truncation of v-*fms* with c-*fms* sequences markedly reduces transformation potential. Proc Natl Acad Sci USA 83: 7800–7804

Broxmeyer HE, Williams DE, Cooper S et al. (1987) Comparative effects in vivo of recombinant murine interleukin-3, natural murine colony stimulating factor-1 and recombinant murine granulocyte-macrophage colony stimulating factor on myelopoiesis in mice. J Clin Invest 79: 721–730

Byrne PV, Guilbert LJ, Stanley ER (1981) Distribution of cells bearing receptors for a colony-stimulating factor (CSF-1) in murine tissues. J Cell Biol 91: 848–853

Caracciolo D, Shirsat N, Wong GG, Lange B, Clark S, Rovera G (1987) Recombinant human macrophage colony stimulating factor (M-CSF) requires subliminal concentrations of granulocyte-macrophage (GM)-CSF for optimal stimulation of human macrophage colony formation in vitro. J Exp Med 166: 1851–1860

Cenci E, Bartocci A, Puccetti P, Mocci S, Stanley ER, Bistoni F (1991) Macrophage colony stimulating factor in murine candidiasis: serum and tissue levels during infection and protective effect of exogenous administration. Infect Immun (in press)

Cerdan C, Courcoul M, Rozanajaona D et al. (1990) Activated but not resting T cells or thymocytes express colony-stimulating factor 1 mRNA without co-expressing c-*fms* mRNA. Eur J Immunol 20: 331–335

Cerretti DP, Wignall J, Anderson D et al. (1988) Human macrophage colony stimulating factor: alternative RNA and protein processing from a single gene. Mol Immunol 25: 761–770

Cheers C, Hill M, Haight AM, Stanley ER (1989) Stimulation of macrophage phagocytic but not bactericidal activity by colony stimulating factor-1. Infect Immun 57: 1512: 1516

Chen BD, Clark CR (1986) Interleukin-3 (IL-3) regulates the in vitro proliferation of both blood monocytes and peritoneal exudate macrophages. Synergism between a macrophage lineage-specific colony stimulating factor (CSF-1) and IL-3. J Immunol 137: 563–570

Chen BDM, Clark CH, Chou T (1988) Granulocyte-macrophage colony stimulating factor stimulates monocyte and tissue macrophage proliferation and enhances their responsiveness to macrophage colony stimulating factor. Blood 71: 997–1002

Chodakewitz JA, Lacy J, Edwards SE, Birchall N, Coleman DL (1990) Macrophage colony stimulating factor production by murine and human keratinocytes: enhancement by bacterial lipopolysaccharide. J Immunol 144: 2190–2196

Chong KT, Langlois L (1988) Enhancing effect of M-CSF on leukocytes and host defense in normal and immunosuppressed mice. FASEB J 2: A1474

Chong KT, Langlois L (1989) Recombinant human macrophage colony stimulating factor (M-CSF) enhanced murine host defense against lethal *Candida albicans* infections. Annual Meeting of the American Society of Microbiology 89: 134A

Chen BDM, Lin HS, Hsu S (1983) Tumor-promoting phorbol esters inhibit the binding of colony-stimulating factor (CSF-1) to murine peritoneal exudate macrophages. J Cell Physiol 116: 207–212

Clark CR, Chen BDM, Boros DL (1988) Macrophage progenitor cell and colony stimulating factor production during granulomatous schistosomiasis mansoni in mice. Intect Immun 56: 2680–2685

Coussens L, Van Beveren C, Smith D et al. (1986) Structural alteration of viral homologue of receptor protooncogene *fms* at carboxyl terminus. Nature 320: 277–280

Das SK, Stanley ER (1982) Structure-function studies of a colony stimulating factor (CSF-1). J Biol Chem 257: 679–684

Das SK, Stanley ER, Guilbert LJ, Forman LW (1981) Human colony stimulating factor (CSF-1) radioimmunoassay: resolution of three subclasses of human colony stimulating factors. Blood 58: 630–641

Downing JR, Rettenmier CW, Sherr CJ (1988) Ligand-induced tyrosine kinase activity of the colony stimulating factor-1 receptor in a murine macrophage cell line. Mol Cell Biol 8: 1795–1799

Downing JR, Roussel MF, Sherr CJ (1989) Ligand and protein kinase C downmodulate the colony stimulating factor-1 receptor by independent mechanisms. Mol Cell Biol 9: 2890–2896

Dubreuil P, Torres H, Courcoul MA, Birg F, Mannoni P (1988) c-*fms* expression is a molecular marker of human acute myeloid leukemias. Blood 72: 1081–1085

Dynan WS, Sazer S, Tjian R, Schimke RT (1986) Transcription factor Spl recognizes a DNA sequence in the mouse dihydrofolate reductase promoter. Nature 349: 246–248

Elford PR, Felix R, Cecchini M, Trechsel U, Fleisch H (1987) Murine osteoblast-like cells and the osteogenic cell MC 3T3-E1 release a macrophage colony-stimulating activity in culture. Calcif Tissue Int 41: 151–156

Felix R, Cecchiri MG, Fleisch H (1990) Macrophage colony stimulating factor restores in vivo bone resorption in the *op/op* osteopetrotic mouse. Endocrinology 127: 2592–2594

Fibbe WE, Van Damme J, Billian A et al. (1988) Interleukin-1 induces human marrow stromal cells in long-term culture to produce granulocyte colony-stimulating factor and macrophage colony-stimulating factor. Blood 71: 430–435

Gaffney EV, Lingenfelter SE, Koch GA, Lisi PJ, Chu CW, Tsai SC (1988) Regulation by interferon-γ of function in the acute monocytic leukemia cell line, THP-1. J Leukocyte Biol 43: 248–255

Garnick MB, Stoudemire JB (1990) Preclinical and clinical evaluation of recombinant human macrophage colony stimulating factor. Int J Cell Cloning 8 [Suppl 1]: 356–373

Geissler K, Harrington M, Srivastaua C, Leemhais T, Tricot G, Broxmeyer HE (1989) Effects of recombinant human colony stimulating factors (CSF) (GM-CSF, G-CSF and CSF-1) on human monocyte/macrophage differentiation. J Immunol 143: 140–146

Gendleman H, Orenstein JM, Martin MA et al. (1988) Efficient isolation and propagation of human immunodeficiency virus on recombinant colony stimulating factor-1-treated monocytes. J Exp Med 167: 1428–1441

Gilbert HS, Praloran V, Stanley ER (1989) Increased circulating CSF-1 (M-CSF) in myeloproliferative disease: association with myeloid metaplasia and peripheral bone marrow extension. Blood 74: 1231–1234

Gisselbrecht S, Fichelson S, Sola B et al. (1987) Frequent c-*fms* activation by proviral insertion in mouse myeloblastic leukemias. Nature 329: 259–261

Gisselbrecht S, Sola B, Fichelson S et al. (1989) The murine M-CSF gene is localized on chromosome 3. Blood 73: 1742–1746

Gliniak BC, Rohrschneider LR (1990) Expression of the M-CSF receptor is controlled posttranscriptionally by the dominant actions of GM-CSF or multi-CSF. Cell 63: 1073–1083

Groffen J, Heisterkamp N, Spurr N, Dana S, Wasmuth JJ, Stephenson JR (1983) Chromosomal localization of the human c-*fms* oncogene. Nucleic Acids Res 11: 6331–6339

Guilbert LJ, Stanley ER (1980) Specific interaction of murine colony- stimulating factor with mononuclear phagocytic cells. J Cell Biol 85: 153–159

Guilbert LJ, Stanley ER (1984) Modulation of receptors for the colony stimulating factor CSF-1 by bacterial lipopolysaccharide and CSF-1. J Immunol Methods 73: 17–28

Guilbert LJ, Stanley ER (1986) The interaction of ^{125}I-colony-stimulating factor-1 with bone marrow derived macrophages. J Biol Chem 261: 4024–4032

Hamilton JA, Stanley ER, Burgess AW, Shadduck R (1980) Stimulation of macrophage plasminogen activator by colony stimulating factors. J Cell Physiol 103: 435–445

Hamilton JA, Vairo G, Lingelbach SR (1988) Activation and proliferation signals in murine macrophages. Stimulation of glucose uptake by hemopoietic growth factors and other agents. J Cell Physiol 134: 405-412

Guilbert LJ, Nelson DJ, Hamilton JA, Williams N (1983) The nature of 12-0-tetradecanoylphorbol-13-acetate (TPA) stimulated hemopoiesis. Colony stimulating factor (CSF) requirement for colony formation and the effect of ^{125}I-CSF-1 binding. J Cell Physiol 115: 226–282

Hamilton JA, Veis N, Bordun AM, Vairo G, Gonda TJ, Phillips WA (1989) Activation and proliferation signals in murine macrophages: relationships among c-*fos* and c-*myc* expression, phosphoinositide hydrolysis, superoxide formation and DNA synthesis. J Cell Physiol 141: 618–626

Hampe A, Gobet M, Sherr CJ, Galibert F (1984) The nucleotide sequence of the feline retroviral oncogene v-*fms* shows unexpected homology with oncogenes encoding tyrosine-specific protein kinases. Proc Natl Acad Sci USA 81: 85–89

Hampe A, Shamoon BM, Gobet M, Sherr CJ, Galibert F (1989) Nucleotide sequence and structural organization of the human *fms* proto-oncogene. Oncogene Res 4: 9–17

Harrison D, Astle C, Lerner C (1988) Number and continuous proliferative pattern of transplanted premature immunohematopoietic stem cells. Proc Natl Acad Sci USA 85: 822–826

Haskill S, Johnson C, Eierman D, Becker S, Warren K (1988) Adherence induces selective mRNA expression of monocyte mediators and proto-oncogenes. J Immunol 140: 1690–1694

Hattersley G, Dorey E, Horton MA, Chambers TJ (1988) Human macrophage colony stimulating factor inhibits bone resorption by osteoclasts disaggregated from rat bone. J Cell Physiol 137: 199–203

Heard JM, Roussel MF, Rettenmier CW, Sherr CJ (1987a) Synthesis, post-translational processing and autocrine transforming activity of a carboxylterminal truncated form of colony stimulating factor-1. Oncogene Res 1: 423–440

Heard JM, Roussel MF, Rettenmier CW, Sherr CJ (1987b) Multilineage hematopoietic disorders induced by transplantation of bone marrow cells expressing the v-*fms* oncogene. Cell 51: 663–673

Ho JL, Reed S, Wick EA (1989) Effect of macrophage-colony stimulating factor (M-CSF) on intramacrophage (MO) survival of *Leishmania mexicana amazonensis* (LMA). Clin Res 37: 431A

Hoggan MD, Halden NF, Buckler CE, Kozak CA (1988) Genetic mapping of the mouse c-*fms* proto-oncogene to chromosome 18. J Virol 62: 1055–1056

Horiguchi J, Warren MK, Ralph P, Kufe D (1986) Expression of the macrophage specific colony-stimulating factor (CSF-1) during human monocytic differentiation. Biochem Biophys Res Commun 141: 924–930

Horiguchi J, Sherman ML, Sampson-Johannes A, Weber BL, Kufe DW (1988) CSF-1 and c-*fms* gene expression in human carcinoma cell lines. Biochem Biophys Res Commun 157: 395–401

Hoshina M, Hishio A, Bo M, Boime I, Mochizuki M (1985) The expression of oncogene *fms* in human chorionic tissue. Acta Obstet Gynecol Jpn 37: 2791–2798

Huang E, Nocka K, Beier DR et al. (1990) The hematopoietic growth factor KL is encoded by the Sl locus and is the ligand of c-kit receptor, the gene product of the W locus. Cell 63: 225–233

Huebner K, Isobe M, Croce CM, Golde DW, Kaufman SE, Gasson JC (1985) The human gene encoding GM-CSF is at 5q21-q32, the chromosome region deleted in the 5q- anomaly. Science 230: 1282–1285

Hume DA, Denkins YM (1989) Activation of macrophages to express cytocidal activity correlates with inhibition of their responsiveness to macrophage colony stimulating factor (CSF-1): involvement of a pertussis-toxin sensitive reaction. Immunol Cell Biol 67: 243–249

Hume DA, Pawli P, Donahue RE, Fidler IJ (1988) The effect of human recombinant macrophage colony stimulating factor (CSF-1) on the murine mononuclear phagocyte system in vivo. J Immunol 141: 3405–3409

Hume DA, Donohue RE, Fidler IJ (1989) The therapeutic effect of human recombinant macrophage colony stimulating factor (CSF-1) in experimental murine metastatic melanoma. Lymphokine Res 8: 69–77

Ichikawa Y, Pluznik DH, Sachs L (1966) In vitro control of the development of macrophage and granulocyte colonies. Proc Natl Acad Sci USA 56: 488–495

Imamura K, Kufe D (1988) Colony stimulating factor-1-induced Na^+ influx into human monocytes involves activation of a pertussis toxin-sensitive GTP-binding protein. J Biol Chem 263: 14093–14098

Ishibashi S, Inaba T, Shimano H et al. (1990) Monocyte colony stimulating factor enhances uptake and degradation of acetylated low density lipoproteins and cholesterol esterification in human monocyte derived macrophages. J Biol Chem 265: 14109–14117

Janowska-Wieczorek A, Belch AR, Jacobs A, Bowen D, Paietta E, Stanley ER (1991) Increased circulating colony stimulating factor-1 in patients with preleukemia, leukemia and lymphoid malignancies. Blood (in press)

Jaye M, Howk R, Burgess W et al. (1986) Human endothelial cell growth factor: cloning, nucleotide sequence and chromosome localization. Science 233: 541–545

Jubinsky PT, Yeung YG, Sacca R, Li W, Stanley ER (1988) Colony stimulating factor-1 stimulated macrophage membrane protein phosphorylation. In: Kudlow JE, Maclennan DH, Bernstein A, Gottlieb AI (eds) Biology of Growth factors: molecular biology, oncogenes, signal transduction and clinical applications. Plenum, New York, pp 75–90

Kacinski BM, Bloodgood RS, Schwartz PE, Carter D, Stanley ER (1989a) The macrophage colony stimulating factor CSF-1 is produced by human ovarianand endometrial adenocarcinoma-derived cell lines and is present at abnormally high levels in the plasma of ovarian carcinoma patients with active disease. Cold Spring Harbor Symp Quant Biol, Cancer Cells 7: 333–337

Kacinski BM, Stanley ER, Carter D, Chambers JT, Chambers SK, Kohorn EI, Schwartz PE (1989b) Circulating levels of CSF-1 (M-CSF), a lymphohematopoietic cytokine, may be a useful marker of disease status in patients with malignant ovarian neoplasms. Int J Radiat Oncol Biol Phys 17: 159–164

Kanada T, Shikita M, Tsuneoka K, Sakai N, Tomita T, Kanegashi S (1987) J Pharmacobio-Dyn 10: 215

Karbassi A, Becker JM, Foster JJ, Moore RN (1987) Enhanced killing of *Candida albicans* by murine macrophages treated with macrophage-colony stimulating factor: evidence for augmented expression of mannose receptors. J Immunol 139: 417–421

Kawasaki ES, Ladner MB (1990) Molecular biology of macrophage colony stimulating factor. In: Dexter TM, Garland JM, Testa NG (eds) Colony stimulating factors. Marcel Dekker, New York, pp 155–176

Kawasaki ES, Ladner MB, Wang AM et al. (1985) Molecular cloning of a complementary DNA encoding human macrophage-specific colony stimulating factor (CSF-1). Science 230: 291–296

Komiyama A, Ishiguro A, Kubo T et al. (1988) Increases in neutrophil counts by purified human urinary colony-stimulating factor in chronic neutropenia of childhood. Blood 71: 41–45

Kurland JI, Pelus LM, Ralph P, Bockman RS, Moore MAS (1979) Induction of PGE synthesis in normal and neoplastic macrophages: role of CSF(s) distinct from effects on myeloid progenitor cell proliferation. Proc Natl Acad Sci USA 76: 2326–2330

Ladner MB, Martin GA, Noble JA, Nikoloff DM, Tal R, Kawasaki ES, White TJ (1987) Human CSF-1: gene structure and alternative splicing of mRNA precursors. EMBO J 6: 2693–2698

Ladner MB, Martin GA, Noble JA, Wittman VP, Warren MK, McGrogan M, Stanley ER (1988) cDNA cloning and expression of murine macrophage colony-stimulating factor from 1929 cells. Proc Natl Acad Sci USA 85: 6706–6710

Laimins LA, Kessel M, Rosenthal N, Khoury G (1983) Viral and cellular enhancer elements. In: Gluzman Y, Shenk T (eds) Communications in molecular biology, enhancers and eukaryotic gene expression. Cold Spring Harbor Laboratory, New York, pp 28–87

Lanotte M, Metcalf D, Dexter TM (1982) Production of monocyte/macrophage colony stimulating factor by preadipocyte cell lines derived from murine marrow stroma. J Cell Physiol 112: 123–127
Le PT, Kurtzberg J, Brandt SJ, Niedel JE, Haynes BF, Singer KH (1988) Human thymic epithelial cells produce granulocyte and macrophage colony-stimulating factors. J Immunol 141: 1211–1217
Le Beau MM, Westbrook CA, Diaz MO et al. (1986) Evidence for the involvement of GM-CSF and *fms* in the deletion (5q) in myeloid disorders. Science 231: 984–987
Le Beau MM, Epstein ND, O'Brien SJ, Nienhuis AW, Yang YC, Clark SC, Rowley, JD (1987) The interleukin-3 gene is located on human chromosome 5 and is deleted in myeloid leukemias with a deletion of 5q. Proc Natl Acad Sci USA 84: 5913–5917
Le Beau MM, Lemons RS, Espinosa R, Larson RA, Arai N, Rowley JD (1989) Interleukin-4 and interleukin-5 map to human chromosome 5 in a region encoding growth factors and receptors and are deleted in myeloid leukemias with a del (5q). Blood 73: 647–650
Lee MT, Warren MK (1987) CSF-1-induced resistance to viral infection in murine macrophages. J Immunol 138: 3019–3022
Leibovitch SA, Leibovitch MP, Borycki AG, Harel J (1989) Expression of CSF-1- receptor-related proteins in muscular stem cells. Oncogene Res 4: 157–162
Li W, Stanley ER (1991) Role of dimerization and modification of the CSF-1 receptor in its activation and internalization during the CSF-1 response. EMBO J (in press)
Li W, Yeung YG, Stanley ER (1991) Tyrosine phosphorylation of a common 57-kDa protein in growth factor stimulated and transformed cells. J Biol Chem (in press)
Lu L, Walker D, Graham CD, Waheed A, Shadduck RK, Broxmeyer HE (1988) Enhancement of release from MHC class II antigen-positive monocytes of hematopoietic colony-stimulating factors CSF-1 and G-CSF by recombinant human tumor necrosis factor-α: synergism with recombinant human interferon-gamma. Blood 72: 34–41
Lyberg T, Stanley ER, Prydz H (1987) Colony stimulating factor-1 induces thromboplastin activity in murine macrophages and human monocytes. J Cell Physiol 132: 367–370
Monos MM (1988) Expression and processing of a recombinant human macrophage colony-stimulating factor in mouse cells. Mol Cell Biol 8: 5035–5039
Masaoka T, Motoyoshi K, Takaku F (1988) Administration of human urinary colony stimulating factor after bone marrow transplantation. Bone Marrow Transplantation 3: 121–127
Matsumoto K, Kakizoe T, Nakagami Y et al. (1987) Clinical trials of CSF-HU (colony stimulating factor derived from human urine: p-100) on granulocytopenia induced by anticancer chemotherapy in urogenital cancer patients. Hinyokika Kiyo 33: 972–982
McNiece IK, Kriegler AB, Bradley TR, Hodgson GS (1986) Subpopulations of mouse bone marrow high-proliferative-potential colony forming cells (HPP-CFC). Exp Hematol 14: 856–860
McNiece IK, Robinson BE, Quesenberry PJ (1988) Stimulation of murine colony forming cells with high proliferative potential by the combination of GM-CSF and CFS-1. Blood 72: 191–195
Metcalf D (1986) The molecular biology and function of the granulocyte-macrophage colony-stimulating factors. Blood 67: 257–267
Miyauchi J, Wang C, Kelleher CA, Wong GG, Clark SC, Minden MD, McCulloch EA (1988) The effects of recombinant CSF-1 on the blast cells of acute myeloblastic leukemia in suspension culture. J Cell Physiol 135: 55–62
Mochizuki DY, Eisenman JR, Conlon PJ, Larsen AD, Tushinski RJ (1987) Interleukin-1 regulates hematopoietic activity, a role previously ascribed to hemopoietin-1. Proc Natl Acad Sci USA 84: 5267–5271
Moore RN, Osmand AP, Dunn JA, Joshi JG, Ronse BT (1988) Substance P augmentation of CSF-1 stimulated in vitro myelopoiesis: a two-signal progenitor restricted, tuftsin-like effect. J Immunol 141: 2699–2703
Moore RN, Osmand AP, Dunn JA, Joshi JG, Koontz JW, Ronse BT (1989) Neurotensin regulation of macrophage colony stimulating factor-stimulated in vitro myelopoiesis. J Immunol 142: 2689–2694
Morgan CJ, Stanley ER (1984) Chemical crosslinking of the mononuclear phagocyte specific growth factor CSF-1 to its receptor at the cell surface. Biochem Biophys Res Commun 119: 35–41
Morrison DK, Kaplan DR, Rapp U, Roberts TM (1988) Signal transduction from membrane to cytoplasm: growth factors and membrane-bound oncogene products increase raf-1 phosphorylation and associated protein kinase activity. Proc Natl Acad Sci USA 85: 8855–8859
Morrison DK, Kaplan DR, Escobedo JA, Rapp U, Roberts TM, Williams LT (1989) Direct activation of the serine/threonine kinase activity of RAF-1 through tyrosine phosphorylation by the PDGF-b receptor. Cell 58: 649–657

Motoyoshi K, Suda T, Kusumoto K, Takaku F, Miura Y (1982) Granulocyte-macrophage colony stimulating and binding activities of purified human urinary colony stimulating factor to murine and human bone marrow cells. Blood 60: 1378–1386

Mufson RA, Aghajanian T, Wong G, Woodhouse C, Morgan AL (1989) Macrophage colony stimulating factor enhances monocyte and macrophage antibody-dependent cell-mediated cytotoxicity. Cell Immunol 119: 182–192

Muller R, Slamon DJ, Adamson ED, Tremblay JM, Muller D, Cline MJ, Verma IM (1983) Transcription of c-*onc* genes c-*ras*^{ki} and c-*fms* during mouse development. Mol Cell Biol 3: 1062–1069

Munn DH, Garnick MB, Cheung NKV (1988) Effects of parenteral macrophage colony stimulating factor (M-CSF) on circulating monocyte number, immunophenotype and anti-tumor activity in cynomolgus monkeys. Blood 72: 127a

Munn DH, Garnick MB, Cheung NV (1990) Effects of parenteral recombinant human macrophage colony stimulating factor on monocyte number, phenotype and antitumor cytotoxicity in non-human primates. Blood 75: 2042–2048

Nakamura M, Merchau S, Carter A, Ernst TJ, Demetri GD, Furukawa Y, Anderson W, Freedman AS, Griffin JD (1989) Expression of a novel 3.5kb macrophage colony stimulating factor transcript in human myeloma cells. J Immunol 143: 3543–3547

Nakoinz I, Ralph P (1988) Stimulation of macrophage antibody-dependent killing of tumor targets by recombinant lymphokine factors and M-CSF. Cell Immunol 116: 331–340

Ohtsuka M, Roussel MF, Sherr CJ, Downing JR (1990) Ligand-induced phosphorylation of the colony stimulating factor-1 receptor can occur through an intermolecular reaction that triggers receptor downmodulation. Mol Cell Biol 10: 1664–1671

Orlandi M, Barbolini G, Minghetti L, Luchetti S, Giuliucci B, Chiricolo M, Tomasi V (1989) Prostaglandin and thromboxane biosynthesis in isolated platelet-free human monocytes. III. The induction of cyclo-oxygenase by colony stimulating factor-1. Prostaglandins, Leukotrienes and Essential Fatty Acids 36: 101–106

Orlofsky A, Stanley ER (1987) CSF-1-induced gene expression in macrophages; dissociation from the mitogenic response. EMBO J 6: 2947–2952

Oster W, Lindemann A, Mertelsmann R, Herrmann F (1989) Production of macrophage-, granulocyte-, granulocyte-macrophage- and multi-colony-stimulating factor by peripheral blood cells. Eur J Immunol 19: 543–547

Paietta E, Racevsbis J, Stanley ER, Andreef M, Papenhausen P, Wiernik PH (1990) Expression of the macrophage growth factor CSF-1 and its receptor c-*fms* by a Hodgkin's disease-derived cell line and its variants. Cancer Res 50: 2049–2055

Pettenati MJ, LeBeau MM, Lemons RS et al. (1987) Assignment of CSF-1 to 5q 33.1: evidence for clustering of genes regulating hematopoiesis and for their involvement in the deletion of the long arm of chromosome 5 in myeloid disorders. Proc Natl Acad Sci USA 84: 2970–2974

Pluznik DH, Sachs L (1965) The cloning of normal "mast" cells in tissue culture. J Cell Comp Physiol 66: 319–324

Pollard JW, Bartocci A, Arceci R, Orlofsky A, Ladner MB, Stanley ER (1987) Apparent role of the macrophage growth factor, CSF-1, in placental development. Nature 330: 484–486

Rajavashisth TB, Eng R, Shadduck RK, Waheed A, Ben-Avram CM, Shively JE, Lusis AJ (1987) Cloning and tissue specific of mouse macrophage colony stimulating factor mRNA. Proc Natl Acad Sci USA 84: 1157–1161

Ralph P, Nakoinz I (1987) Stimulation of macrophage tumoricidal activity by CSF-1. Cell Immunol 105: 270–279

Ralph P, Warren MK (1989) Molecular biology, cell biology and clinical future in myeloid growth factors. In: Cruse JM, Lewis RE Jr (eds). Immunoregulatory cytokines and cell growth. Year Immunol, Karger, Basel, vol 5, pp 103–125

Ralph P, Warren MK, Nakoinz I et al. (1986a) Biological properties and molecular biology of the human macrophage growth factor, CSF-1. Immunobiology 172: 194–204

Ralph P, Warren MK, Lee MT et al. (1986b) Inducible production of human macrophage growth factor, CSF-1. Blood 68: 633–639

Rambaldi A, Young DC, Griffin JD (1987) Expression of the M-CSF (CSF-1) gene by human monocytes. Blood 69: 1409–1413

Rambaldi A, Wakiyama N, Vellenger E, Horiguchi J, Warren MK, Kufe D, Griffin JD (1988) Expression of the macrophage colony-stimulating factor and c-*fms* genes in human acute myeloblastic leukemia cells. J Clin Invest 81: 1030–1035

Regenstreif LJ, Rossant J (1989) Expression of the *c-fms* proto-oncogene and of the cytokine, CSF-1, during mouse embryogenesis. Dev Biol 133: 284–294

Reisbach G, Sindermann J, Kremer JP, Hultner L, Wolf H, Dormer P (1989) Macrophage colony stimulating factor is expressed by spontaneously outgrown EBV-B cell lines and activated normal B lymphocytes. Blood 74: 959–964

Rettenmier CW, Roussel MF (1988) Differential processing of colony stimulating factor-1 precursors encoded by two human cDNAs. Mol Cell Biol 8: 5026–5034

Rettenmier CW, Sacca R, Furman WL et al. (1986) Expression of the human c-*fms* protooncogene product (colony stimulating factor-1 receptor) on peripheral blood mononuclear cells and choriocarcinoma cell lines. J Clin Invest 77: 1740–1746

Rettenmier CW, Roussel MF, Ashmun RA, Ralph P, Price K, Sherr CJ (1987) Synthesis of membrane bound colony-stimulating factor-1 (CSF-1) and down-modulation of CSF-1 receptors in NIH 3T3 cells transformed by cotransfection of the human CSF-1 and c-*fms* (CSF-1 receptor) genes. Mol Cell Biol 7: 2378–2387

Rich A, Nordheim A, Wang A (1984) The chemistry and biology of left-handed Z-DNA. Annu Rev Biochem 53: 792–846

Roberts WM, Look AT, Roussel MF, Sherr CJ (1988) Tandem linkage of human CSF-1 receptor (c-*fms*) and PDGF receptor genes. Cell 55: 655–661

Rosendaal M (1975) Colony stimulating factor (CSF) in the uterus of the pregnant mouse. J Cell Sci 19: 411–423

Roth P (1990) Human colony stimulating factor-1 is elevated in the fetus and newborn. Clin Res 38: 772A

Roth P, Bartocci A, Stanley ER (1990) Lipopolysaccharide induces synthesis of murine colony stimulating factor-1 (CSF-1) in vivo. Ped Res 27: 49A

Rothwell VM, Rohrschneider LR (1987) Murine c-*fms* cDNA: cloning, sequence analysis, and retroviral expression. Oncogene Res 1: 311–324

Roussel MF, Sherr CJ, Barker PE, Ruddle FH (1983) Molecular cloning of the c-*fms* locus and its assignment to human chromosome 5. J Virol 48: 770–773

Roussel MF, Dull TJ, Rettenmier CW, Ralph P, Ullrich A, Sherr CJ (1987) Transforming potential of the c-*fms* protooncogene (CSF-1 receptor). Nature 325: 549–552

Roussel MF, Downing JR, Rettenmier CW, Sherr CJ (1988) A point mutation in the extracellular domain of the human CSF-1 receptor (c-*fms*) proto-oncogene product) activates its transforming potential. Cell 55: 979–988

Roussel MF, Shurtleff S, Downing JR, Sherr CJ (1990) A point mutation at tyrosine-809 in the human colony stimulating factor-1 receptor impairs mitogenesis without abrogating tyrosine kinase activity, association with phosphatidylinositol-3-kinase, or induction of c-*fos* and *jun* B genes. Proc Natl Acad Sci USA 87: 6738–6742

Sampson-Johannes A, Carlino JA (1988) Enhancement of human monocyte tumoricidal activity by recombinant M-CSF. J Immunol 141: 3680–3686

Sariban E, Mitchell T, Kufe D (1985) Expression of the c-*fms* protooncogene during human monocytic differentiation. Nature 316: 64–66

Sawada M, Suzumura A, Yamamoto H, Marunouchi T (1990) Activation and proliferation of the isolated microglia by CSF-1 and possible involvement of protein kinase C. Brain Res 509: 119–124

Seelentag WK, Mermod JJ, Montesano R, Vassalli P (1987) Additive effects of interleukin-1 and tumor necrosis factor-alpha on the accumulation of the three granulocyte and macrophage colony-stimulating factor mRNAs in human endothelial cells. EMBO J 6: 2261–2265

Sengupta A, Liu WK, Yeung YG, Yeung DCY, Frackelton AR, Stanley ER (1988) Identification and subcellular localization of proteins that are rapidly phosphorylated in tyrosine in response to colony stimulating factor-1. Proc Natl Acad Sci USA 85: 8062–8066

Shaw G, Kamen R (1986) A conserved AU sequence from the 3′-untranslated region of GM-CSF mRNA mediates selective degradation. Cell 46: 659–667

Sherr CJ (1990) Colony stimulating factor-1 receptor. Blood 75: 1–12

Sherr CJ, Stanley ER (1990) Colony stimulating factor-1. In: Sporn MB, Roberts AB (eds) Peptide growth factors and their receptors. Springer, Berlin Heidelberg New York, pp 667–698

Sherr CJ, Rettenmier CW, Sacca R, Roussel MF, Look AT, Stanley ER (1985) The c-*fms* proto-oncogene product is related to the receptor for the mononuclear phagocyte growth factor CSF-1. Cell 41: 665–676

Shieh JH, Peterson RHF, Warren DJ, Moore MAS (1989) Modulation of colony stimulating factor-1 receptors on macrophages by tumor necrosis factor. J Immunol 143: 2534–2539

Shimano H, Yamada N, Motoyoshi K, Matsumoto A, Ishibashi S, Mori N, Takaku F (1990) Plasma cholesterol-lowering activity of monocyte colony stimulating factor (M-CSF). Ann NY Acad Sci 587: 362–370

Shurtleff SA, Downing JR, Rock CO, Hawkins SA, Roussel MF, Sherr CJ (1990) Structural features of the colony stimulating factor-1 receptor that affect its association with phosphatidylinositol-3-kinase. EMBO J 9: 2415–2421

Stanley ER (1990) Role of colony stimulating factor-1 in monocytopoiesis and placental development. In: Mahowald A (ed) Genetics of pattern formation and growth control. Wiley-Liss, New York, pp 165–180

Stanley ER, Guilbert LJ (1981) Methods for the purification, assay, characterization and target cell binding of a colony stimulating factor (CSF-1). J Immunol Methods 42: 253–284

Stanley ER, Heard PM (1977) Factors regulating macrophage production and growth. Purification and some properties of the colony stimulating factor from medium conditioned by mouse L-cells. J Biol Chem 252: 4305–4312

Stanley ER, Jubinsky PT (1984) Factors affecting growth and differentiation of hematopoietic cells in culture. Clin Haematol 13: 329–336

Stanley ER, Cifone M, Heard PM, Defendi V (1976) Factors regulating macrophage production and growth: identity of colony stimulating factor and macrophage growth factor. J Exp Med 143: 631–646

Stanley ER, Chen DM, Lin HS (1978) Induction of macrophage production and proliferation by a purified colony stimulating factor. Nature 274: 168–170

Stanley ER, Guilbert LJ, Tushinski RJ, Bartelmez SH (1983) CSF-1: a mononuclear phagocyte lineage-specific growth factor. J Cell Biochem 21: 151–159

Stein J, Borzillo GV, Rettenmier CW (1990) Direct stimulation of cells expressing receptors for macrophage colony stimulating factor (CSF-1) by a plasma membrane-bound precursor of human CSF-1. Blood 76: 1308–1314

Sutherland GR, Baker E, Callen DJ et al. (1988) Interleukin-5 is at 5q 31 and is deleted in the 5q-syndrome. Blood 71: 1150–1152

Takahashi M, Hong YM, Yasuda S, Takano M, Kawai K, Nakai S, Hirai Y (1988) Macrophage colony stimulating factor is produced by human T lymphoblastoid cell line, CEM-ON: identification by aminoterminal amino acid sequence analysis. Biochem Biophys Res Commun 152: 1401–1409

Thery C, Hetier E, Evrard C, Mallat M (1990) Expression of macrophage colony-stimulating factor gene in mouse brain during development. J Neurosci Res 26: 129–133

Tushinski RJ, Stanley ER (1985) The regulation of mononuclear phagocyte entry into S phase by the colony stimulating factor CSF-1. J Cell Physiol 122: 221–228

Tushinski RJ, Oliver IT, Guilbert LJ, Tynan PW, Warner JR, Stanley ER (1982) Survival of mononuclear phagocytes depends on a lineage-specific growth factor that the differentiated cells selectively destroy. Cell 28: 71–81

Vairo G, Hamilton JA (1988) Activation and proliferation signals in murine macrophages: stimulation of Na^+/k^+ ATPase activity by hematopoietic growth factors and other agents. J Cell Physiol 134: 13–24

Van Leeuwen BH, Martinson ME, Webb GC, Young IG (1989) Molecular organization of the cytokine gene cluster, involving human IL-3, IL-4, IL-5 and GM-CSF genes, on human chromosome 5. Blood 73: 1142–1148

Varticovski L, Druker B, Morrison D, Cantley L, Roberts T (1989) The colony stimulating factor-1 receptor associates with and activates phosphatidylinositol-3 kinase. Nature 342: 699–702

Visvader J, Verma IM (1989) Differential transcription of exon 1 of the human c-*fms* gene in placental trophoblasts and monocytes. Mol Cell Biol 9: 1336–1341

Waheed A, Shadduck RK (1982) Purification of colony stimulating factor by affinity chromotography. Blood 60: 238–244

Wakamiya N, Horiguchi J, Kufe D (1987) Detection of c-*fms* and CSF-1 RNA by in situ hybridization. Leukemia 1: 518–520

Walker F, Nicola NA, Metcalf D, Burgess AW (1985) Hierarchical down-modulation of hemopoietic growth factor receptors. Cell 43: 269–276

Wang C, Kelleher CA, Cheng GYM et al. (1988) Expression of the CSF-1 gene in the blast cells of acute myeloblastic leukemia: association with reduced growth capacity. J Cell Physiol 135: 133–138

Wang FM, Friedman H, Djen J (1989) Enhancement of human monocyte action against *Candida albicans* by colony stimulating factors (CSF): IL-3, granulocyte-macrophage CSF and macrophage CSF. J Immunol 143: 671–677

Wang JM, Griffin JD, Rambaldi A, Chen ZG, Mantovani A (1988) Induction of monocyte migration by recombinant macrophage colony stimulating factor. J Immunol 141: 575–579

Warren D, Moore MAS (1987) Synergy of interleukin-1 and granulocyte stimulating factor: in vivo stimulation of stem cell recovery and hematopoietic regeneration following 5-fluorouracil treatment of mice. Proc Natl Acad Sci USA 84: 7134–7138

Warren MK, Ralph P (1986) Macrophage growth factor CSF-1 stimulates human monocyte production of interferon, tumor necrosis factor and colony stimulating activity. J Immunol 137: 2281–2285

Weber B, Horiguchi J, Luebbers R, Sherman M, Kufe D (1989) Post-translational stabilization of c-*fms* mRNA by a labile protein during human monocyte differentiation. Mol Cell Biol 9: 769–775

Wheeler EF, Rettenmier CW, Look AT, Sherr CJ (1986) The v-*fms* oncogene induces factor independence and tumorigenicity in CSF-1 dependent macrophage cell line. Nature 324: 377–380

Wiktor-Jedrzejczak W, Bartocci A, Ferrante AW, Ahmed-Ansari A, Sell KW, Pollard JW, Stanley ER (1990) Total absence of colony stimulating factor-1 in the macrophage deficient osteopetrotic (*op/op*) mouse. Proc Natl Acad Sci USA 87: 4828–4832

Williams DE, Stranera JE, Cooper S et al. (1987) Interactions between purified murine colony stimulating factors (natural CSF-1, purified recombinant GM-CSF and purified recombinant IL-3) on the in vitro proliferation of purified murine granulocyte-macrophage progenitor cells. Exp Hematol 15: 1007–1012

Williams DE, Eisenman J, Baird A et al. (1990) Identification of a ligand for the c-*kit* proto-oncogene. Cell 63: 167–174

Willman CL, Stewart CC, Miller V, Yi TL, Tomasi TB (1989) Regulation of MHC class II gene expression in macrophages by hematopoietic colony stimulating factors (CSF). Induction by GM-CSF and inhibition by CSF-1. J Exp Med 170: 1559–1567

Wing EJ, Magee DM, Pearson AC, Waheed A, Shadduck RK (1986) Peritoneal macrophages exposed to purified macrophage-colony stimulating factor (M-CSF) suppress mitogen- and antigen-stimulated lymphocyte proliferation. J Immunol 137: 2768–2773

Wong GG, Temple PA, Leary AC et al. (1987) Human CSF-1: molecular cloning and expression of 4-kb cDNA encoding the human urinary protein. Science 235: 1504–1508

Woolford J, McAuliffe A, Rohrschneider LR (1988) Activation of the feline c-*fms* protooncogene: multiple alterations are required to generate a fully transformed phenotype. Cell 55: 965–977

Yan ZJ, Wang QR, McNiece IK, Wolf NS (1990) Dissecting the hematopoietic microenvironment. VII. The production of an autostimulatory factor as well as a CSF by unstimulated murine marrow fibroblasts. Exp Hematol 18: 348–354

Yang YC, Kovacic S, Kriz R et al. (1988) The human genes for GM-CSF and IL-3 are closely linked in tandem on chromosome 5. Blood 71: 958–961

Yarden Y, Escobedo JA, Kuang WJ et al. (1986) Structure of the receptor for platelet-derived growth factor helps define a family of closely related growth factor receptors. Nature 323: 226–232

Yarden Y, Kuang WJ, Yuang-Feng T et al. (1987) Human protooncogene c-*kit*: a new cell surface receptor tyrosine kinase for an unidentified ligand. EMBO J 6: 3341–3351

Yeung YG, Jubinsky PT, Sengupta A, Yeung DCY, Stanley ER (1987) Purification of the colony stimulating factor-1 receptor and demonstration of its tyrosine kinase activity. Proc Natl Acad Sci USA 84: 1268–1271

Zsebo KM, Wypych J, Yuschenkoff VN, Lu H, Hunt P, Dukes PP, Langley KE (1988) Effects of hemopoietin-1 and interleukin-1 activities on early hematopoietic cells of the bone marrow. Blood 71: 962–968

Zsebo KM, Williams DA, Geissler EN et al. (1990) Stem cell factor is encoded at the S1 locus of the mouse and is the ligand for the c-*kit* tyrosine kinase receptor. Cell 63: 213–224

Lipopolysaccharide Receptors and Signal Transduction Pathways in Mononuclear Phagocytes

T.-Y. CHEN, M.-G. LEI. T. SUZUKI, and D. C. MORRISON

1 Introduction

It is generally accepted that bacterial products are among the most potent stimuli leading to the activation of monocytes and macrophages. Of all the bacterial products which have been investigated during the past 50 years, the endotoxic lipopolysaccharides (LPS), derived from gram-negative microorganisms, have become recognized as the microbial activator of choice for many studies. This has, in part, been due to the fact that relatively low concentrations of LPS are required to effect macrophage stimulation. In addition, however, LPS may be obtained in highly purified form and active principle of LPS responsible for biologic activity has been identified and chemically characterized.

There is good experimental evidence from both in vivo and in vitro studies that LPS will elicit in macrophages most of the cellular responses usually attributable to these multifunctional inflammatory cells. In recent years, particular attention has focused upon the capacity of LPS to stimulate macrophages to synthesize and secrete immunologically important cytokines, including interferon-α/β (IFN-α/β) (HAVELL and SPITALNY 1983), interleukin-1 (IL-1) (GERY et al. 1972), and cachectin/tumor necrosis factor-α (OLD 1985; AGGARWAL et al. 1985). Evidence has also accumulated which indicates that LPS-stimulated macrophages will release metabolic end products of both the cyclooxygenase and lipoxygenase pathways of arachidonic acid (SCHADE et al. 1987). In addition, LPS is now recognized to modulate the surface expression of macrophage receptors and other markers (HOTER et al. 1987; DING et al. 1989).

Department of Microbiology, Molecular Genetics, and Immunology, University of Kansas Medical Center, Kansas City, KS 66103, USA

Current Topics in Microbiology and Immunology, Vol. 181

Finally, LPS, either alone or in conjunction with IFN-γ, will stimulate macrophages to a fully activated state in which these cells achieve the capacity to kill tumor cells (ALEXANDER and EVANS 1971; PACE and RUSSELL 1981).

Investigations into the structural components of LPS which contribute to their capacity to stimulate macrophages have, for the most part, centered upon the structurally conserved lipid A region. That isolated, highly purified lipid A can, under the appropriate circumstances, activate macrophages has been well documented (DOE et al. 1978). Further, it has been shown by many investigators that agents such as polymyxin B, which bind the lipid A region of LPS, will inhibit virtually all of the LPS-dependent macrophage responses including synthesis and secretion of cytokines (CHIA et al. 1989), generation of procoagulant activity (NIEMETZ and MORRISON 1977), and macrophage-dependent tumor cell cytotoxicity (DOE et al. 1978). Final proof that lipid A alone can serve as a stimulus for macrophages has come from studies using synthetic lipid A and related structures (KOTANI et al. 1985; LOPPNOW et al. 1989). These latter studies have provided important new clues about structure–function relationship of lipid A-dependent macrophage activation.

The importance of lipid A to LPS-dependent activation of macrophages and monocytes is, therefore, undisputed. What is perhaps less clear is whether mechanisms involved in lipid A activation of these cells using isolated highly purified lipid A can always be extrapolated to results obtained with more intact polysaccharide-containing LPS, and several recently published studies in variety of experimental systems would indicate that this issue is far from resolved. For example, LUDERITZ et al. (1989) have reported major differences in relative amounts of lipoxygenase vs. cyclooxygenase products of arachidonate metabolism following stimulation of murine macrophages with various chemotypes of LPS and/or lipid A. One of the more significant findings of these studies was that stimulation of macrophages with a mixture of purified lipid A and biologically inactive deacylated Re-LPS reproduced results obtained with intact LPS. These studies suggest the provocative concept that, although lipid A may be necessary for LPS stimulation of macrophages, oligosaccharide components of LPS, and more specifically 2-keto-2-deoxyoctulosonic acid (KDO) determinants, may facilitate either the initial interaction or some biochemical signal transductive mechanism. These interesting results might, therefore, suggest the existence of carbohydrate-specific receptors for LPS in addition to lipid A-specific receptors.

Although space limitations preclude a more extensive discussion of this issue, several relevant studies merit comment. OHNO and MORRISON have reported that lysozyme, which binds with high affinity to the lipid A region of LPS (OHNO and MORRISON 1989a), will abrogate the capacity of Re-LPS and lipid A to stimulate murine macrophages to secrete IL-1 (OHNO and MORRISON 1989b); however, no inhibition of Ra-LPS, Rc-LPS was detectable. Similarly, FLEBBE et al. (1990) have demonstrated significant activation of endotoxin-hyporesponsive C3H/HeJ macrophages by R-chemotype LPS even though such cells remain refractory to stimulation with S-LPS or purified lipid A. Curiously, as will be discussed below, significant differences have also been reported between LPS and lipid A in their capacity to bind to proteins suggested to function as potential receptors for lipid A. Therefore, while the use of purified lipid A as a tool by which to investigate LPS stimulation of macrophages and monocytes has certain advantages, including the investigation of lipid A-specific receptors, it should be recognized that the information obtained may not always be used to extrapolate to LPS-dependent mechanisms of macrophage activation.

In this respect, there is persistent experimental evidence in the scientific literature suggesting that totally lipid A-independent mechanisms also contribute to LPS activation of macrophages. The majority of such studies have documented polysaccharide-dependent activation of macrophages, and different investigators have reported a possible specific role for O antigen-specific carbohydrates as well as more compelling evidence for chemically conserved core oligosaccharide determinants (reviewed in HAEFNER-CAVAILLON 1985a). More recent studies by WRIGHT and colleagues (1989) have, in addition, defined a novel acute phase protein-dependent pathway for LPS activation of macrophages (to be discussed below). Thus, it would appear that multiple mechanisms exist by which LPS may stimulate macrophages, a concept consistent with the complex and diverse chemical structure of these unique microbial products. The initial interactions between LPS and putative macrophage membrane receptors which trigger the activation signal (s), as well as the biochemical nature of the signal (s), may therefore be diverse, and depend on the source of the LPS. We shall, in the following discussion, focus upon the more general lipid A-dependent mechanisms of macrophage activation, while recognizing that there are some important, and as yet not fully understood, differences between the activities manifested by lipid A and LPS, as well as between different LPS preparations. It should be appreciated, therefore, that multiple structural components of LPS, as well as physical-chemical factors which contribute to the formation of different aggregate structures, all may contribute via specific interactions with different membrane receptor components. We shall consequently also briefly discuss the evidence for potential carbohydrate-binding receptors and physical-chemical properties of LPS which may regulate nonspecific membrane interactions.

2 Lipopolysaccharide Receptors on Monocytes and Macrophages

It is clear that macrophages are only one of a variety of mammalian cells which can bind and/or be stimulated by LPS and lipid A. There are, as a consequence, a number of published studies which describe evidence for specific LPS receptors on B and T lymphocytes, hepatocytes, polymorphonuclear leukocytes, and even erythrocytes. While it may be compelling, from a conceptual point of view, to postulate a biochemical conservation of LPS receptor molecules expressed on different mammalian cells, and there is at least some experimental evidence to support this concept, this fact has not been established unequivocally. As a consequence, we shall limit our discussion to the current evidence for LPS receptors on macrophages and monocytes. For a discussion of LPS receptors expressed on other mammalian cell types, the reader is referred to several recent review articles (HAEFFNER-CAVAILLON et al. 1985a; MORRISON and RUDBACH 1981; MORRISON 1989; LEI et al. 1990).

Studies of LPS binding to macrophages/monocytes in addition are complicated by the fact that these cells may have the capacity to take up LPS from the surrounding medium by both phagocytosis and pinocytosis, and such contributions to the total LPS–macrophage interactions may complicate efforts to define specific LPS binding. Further, although specific, presumably lipid A-dependent binding of LPS to human monocytes has been reported using radiolabeled preparations of LPS, many investigators have reported that binding of LPS to many cell types is nonsaturable (reviewed in MORRISON 1985). In the light of these high levels of nonspecific binding—presumably to the membrane phospholipid bilayer—studies to identify specific LPS receptors have been technically more difficult to perform.

While highly specific and structurally unique receptors for LPS/lipid A may well exist on the macrophage membrane, it is, in addition, clear that several already well-defined macrophage receptors may also function as LPS receptors. For example, both the D-mannose-specific receptor characterized by STEPHENSON and SHEPHERD (1987) and EZEKOWITZ et al. (1988) and the D-galactose-specific receptor defined by ROOS et al. (1985) may function to bind LPS preparations in which D-mannose or D-galactose represents a major component of the O antigen-specific carbohydrate. Specific interactions of D-mannose-containing LPS with murine macrophages have, in fact, been demonstrated. Since many LPS preparations do not contain either galactose or mannose as a terminal residue in either the O antigen or core oligosaccharide (LIANG-TAKASAKI et al. 1982), it is unlikely that such carbohydrate-specific binding proteins are the major functional class of LPS receptors responsible for LPS activation of macrophages.

Perhaps the best experimental evidence to support the existence of carbohydrate-specific LPS receptors on macrophages is the work of HAEFFNER-CAVAILLON and his collaborators (HAEFFNER-CAVAILLON et al. 1985b, 1989; LEBBAR et al. 1986). These investigators have explored the binding of *Bordetella pertussis* LPS to rabbit macrophages, and about 8 years ago reported saturable and specific binding to peritoneal macrophages (HAEFFNER-CAVAILLON et al. 1982). It is of interest that similar studies with alveolar macrophages did not yield any detectable specific binding. Of particular importance was the authors' demonstration that binding could be inhibited by a lipid A-free polysaccharide fraction of *B. pertussis* LPS. Binding was not inhibited by *B. pertussis* lipid A, although possible degradation of critical lipid A structures during the chemical isolation procedures precludes definitive conclusions regarding such studies. Additional studies by HAEFFNER-CAVAILLON et al. (1985b) using *B. pertussis* LPS showed apparently specific binding of this LPS to murine peritoneal macrophages and human monocytes. Again, binding specificity was suggested to reside in the polysaccharide component of the LPS. Of some interest was the fact that LPS binding could be enhanced by serum complement but not by immunoglobulin, albumin, or fibronectin. These results are consistent with the facts that LPS is well recognized as an activator of serum complement (MORRISON and KLINE 1977), and that macrophages and monocytes express a variety of complement receptors (FEARON and WONG 1983).

In more recent studies, the potential polysaccharide-dependent interaction of LPS with human monocytes has been examined. The collective results of these studies have implicated the KDO residues, common carbohydrate components of virtually all gram-negative bacterial LPS, as a potentially relevant structural feature for stimulation of monocytes for IL-1 production (LEBBAR et al. 1986). Chemical dissection of the KDO molecule has revealed that the aldehyde group in the 2-position of the KDO may be an important residue involved in monocyte activation (HAEFFNER-CAVAILLON et al. 1989). These latter studies suggest the potential existence of inner core oligosaccharide, KDO-specific binding sites on the monocyte and are consistent with earlier studies showing enhanced phagocytosis of latex particles covalently conjugated with Re-chemotype LPS relative to untreated particles or particles conjugated with S-LPS (LUBINSKY et al. 1983). There is, however, as yet no direct evidence to support the existence of specific KDO (or other inner core oligosaccharide) receptors on mononuclear phagocytes. Recent studies have suggested that the contribution of KDO-dependent interactions relative to those of lipid A/LPS-dependent mononuclear phagocyte activation is not impressive (LOPPNOW et al. 1989); however, the multivalency of structures in LPS aggregates and potential conformational accessibility of various determinants in LPS macromolecular aggregates versus partial structures complicate the interpretation of these results.

Several of the findings of HAEFFNER-CAVAILLON and his co-workers have been confirmed independently by other investigators using different experimental systems. In this respect, perhaps one of the more noteworthy findings of HAEFFNER-CAVAILLON discussed above was the striking differences between rabbit peritoneal macrophages and alveolar macrophages as determined by specific binding of LPS. Examination of equivalent subpopulations of murine macrophages by AKAGAWA and TOKUNAGA (1985) indicated that alveolar macrophages were unresponsive to stimulation by LPS, as assessed by induction of tumor cell cytotoxicity. In contrast, peritoneal macrophages were readily activated by LPS to become tumoricidal. Importantly, when these cells were probed with fluoresceinated LPS to assess LPS binding, more than 90% of either resident or elicited peritoneal macrophages were positive for LPS binding, in comparison to less than 5% of alveolar macrophages. The addition of polymyxin B to cultures, an agent known to bind to the lipid A region of LPS (MORRISON and JACOBS 1976), inhibited LPS binding, suggesting, but not proving, a lipid A dependence of the interaction. This finding contrasts with the earlier studies of HAEFFNER-CAVAILLON et al. (1982). Nevertheless, the fact that significant differences exist in LPS binding between different subpopulations of macrophages and/or macrophages from different species is of potential importance and should be considered in comparisons of LPS receptor studies. Of potential significance are the observations of AKAGAWA and TOKUNAGA that after treatment of alveolar macrophages with recombinant IFN-γ for 20 h, up to 60% of these cells would bind fluoresceinated LPS by a polymyxin B-inhibitable mechanism. These results may suggest that IFN-γ can up-regulate the expression of LPS receptors. However, in view of the well-recognized capacity of LPS to interact nonspecifically with cells, it is possible that IFN-γ treatment simply alters macrophage membrane fluidity such that nonspecific binding of the LPS probe is markedly increased.

In more recent studies AKAGAWA and her colleagues reported that macrophage colonies generated by alveolar macrophages in response to colony stimulating factor 1 also contained more than 90% FITC-LPS binding cells (AKAGAWA et al. 1988). Using a somewhat similar approach, ERROI and his colleagues showed that certain tumor-associated macrophages (TAMs), isolated from a murine tumor which was poorly immunogenic in vivo, were unresponsive to LPS as assessed by activation for the production of procoagulant activity. The authors also demonstrated that these TAMs contained only 6% LPS positive cells as determined by fluorescence, whereas more than 80% of resident peritoneal macrophages were positive for the FITC-LPS (ERROI et al. 1988). Although these experimental results provide some evidence for specific binding of LPS to macrophages/monocytes, the membrane surface macromolecule(s) responsible for LPS binding, which would presumably be an LPS receptor(s), has not yet been established.

A second point which arises from the studies of HAEFFNER-CAVAILLON et al. relates to the interrelationship between complement and LPS in the activation of mononuclear phagocytes. Since LPS is well recognized as an activator of serum complement (MORRISON and KLINE 1977) and since these cells both secrete complement components (NATHAN 1987) and express complement receptors (FEARON and WONG 1983), an activation pathway which depends upon these factors was proposed many years ago by DIERICH et al. (1973). Indeed, there is good experimental evidence that LPS will initiate the enhanced secretion of complement components, particularly C3, from both human monocytes (STRUNK et al. 1985) and murine macrophages (GOODRUM 1987). However, an obligate role for complement components in LPS-dependent activation has not yet been established.

Studies by WRIGHT and his colleagues have, on the other hand, provided rather convincing evidence for a role for complement receptors as potential targets in LPS stimulation of human monocytes. Using LPS bound to erythrocytes, these authors showed that a monoclonal antibody with specificity for the CD11/18 surface complement receptor antigen (CR3) on human monocytes inhibited binding of LPS-coated erythrocytes (WRIGHT and JONG 1986). Further, depletion of surface expression of CD11/18 by monoclonal antibody-dependent capping abrogated LPS–erythrocyte binding, whereas depletion of a variety of other surface antigens defined by a panel of monoclonal antibodies had no detectable effect. Additional studies to explore the specificity of this interaction suggested that the LPS binding site on CD11/18 was distinct from the C3bi binding site characterized by the Arg-Gly-Asp peptide (WRIGHT and LEVIN 1989).

While these studies clearly establish a solid relationship between the expression of CD11/18 surface receptors and binding of LPS-coated particles, more recent studies by WRIGHT and his colleagues have cast strong doubts on the obligate expression of these receptors for LPS-dependent human monocyte activation (WRIGHT et al. 1990). In these more recent studies, mononuclear cells from a number of patients genetically deficient in CD11/18 were shown to manifest in vitro cellular responses to isolated LPS indistinguishable from those of monocytes obtained from normal human volunteers; however, responses of these cells to LPS-erythrocytes was not investigated. Thus, whereas CD11/18 may contribute to the mononuclear phagocyte response to particulate LPS, it would appear not to be essential in order to achieve stimulation. It would be of interest to compare the relative efficacy of erythrocyte-bound LPS versus soluble LPS in initiating LPS-dependent monocyte responses, since the possibility of different LPS molecular forms mediating distinct responses via interactions with different surface structures remains an interesting possibility. Moreover, the question of whether monoclonal antibody to CD11/18 would either block or mimic LPS activation of these cells by either soluble or particulate preparations of LPS would be an interesting question to pursue.

Recent experiments published by TOBIAS et al. (1986) have shown that LPS will complex with a specific LPS-binding acute phase protein. Further studies by these authors have established that the induction of inflammatory responses in vivo results in the rapid synthesis of this acute phase protein, which binds with high affinity to all chemotypes of LPS (TOBIAS et al. 1989) Interestingly, this acute phase protein, termed LPS-binding protein (LBP), shares remarkable sequence homology (TOBIAS et al. 1988) with a neutrophil granule protein characterized previously by WEISS, ELSBACH, and their colleagues (1978) for its capacity to increase bacterial permeability (BPI) via interactions with LPS on the microbial surface. Of particular importance with respect to LPS-macrophage interactions, WRIGHT et al. have shown that complexes of LPS and LBP, but not LPS and BPI, will interact with human monocytes via specific binding to the CD14 molecule. These authors have further shown that CD14 binding requires the LPS–LBP complex and that high affinity binding of either component alone could not be demonstrated. This novel interaction of LPS with LBP and CD14 may be of considerable importance in the overall host response to endotoxin; however, the question as to whether CD14 serves an obligate receptor function for LPS-dependent monocyte activation remains to be established. In this respect, it is clear that LPS can activate mononuclear phagocytes in the absence of LBP and thus, it can be concluded that the CD14-dependent pathway may not be essential for LPS interaction with and stimulation of macrophages and monocytes. It is possible, in this respect, that LBP, by complexing with LPS, passively focuses these complexes on the monocyte surface via interactions with CD14, where it then interacts with specific LPS/lipid A receptors. Again, the question of whether antibody to CD14 will either mimic or inhibit the effects of LPS on monocytes would appear relevant.

Probably the dominant issue within the framework of LPS interactions with mononuclear phagocytes as it relates to fundamental mechanisms of LPS-initiated host responses, is the identity of LPS/lipid A-specific receptors expressed on these cells. Although this has proven a difficult experimental problem, recent evidence from several laboratories has suggested that lipid A-specific binding proteins can, in fact, be identified on mononuclear phagocytes and lymphocytes and that at least one of these may serve as a true receptor in the activation of these cells. While from a historical perspective, the earliest studies to define LPS receptors focused upon the latter cells, we will, as indicated earlier, restrict our comments to a review of macrophage receptors for LPS/lipid A. It is, of course, conceivable that more than one specific receptor may exist for LPS/lipid A and there is at least one suggestion in the scientific literature that lymphocyte receptors for LPS may be different than those on macrophages (JAKWAY and DEFRANCO 1985). Finally, the potential contribution of hydrophobic interactions of LPS with the membrane lipid bilayer as a necessary prerequisite to effective lipid A–receptor interactions remains as a potentially significant factor in LPS activation of macrophages and will be briefly discussed.

Early experiments yielding evidence that specific LPS binding sites might exist on human peripheral blood monocytes were carried out by LARSEN and SULLIVAN (1984a). These investigators used tritiated LPS to show that the binding of LPS to human monocytes was specific and saturable. The binding of radiolabeled LPS to human monocytes firstly involved a rapid, reversible, temperature-independent surface adsorption. Then there was a slower, irreversible, temperature-dependent uptake of radiolabeled LPS by the monocytes. Interestingly, the LPS-binding sites appeared to be decreased when monocytes were pretreated with an appropriate concentration of unlabeled LPS, suggesting competitive binding. In an accompanying report (LARSEN and SULLIVAN 1984b), the authors tested the binding of the same radiolabeled LPS to Percoll density gradient-isolated monocyte membranes. They reported that the binding of LPS to isolated membranes was considerably lower than binding to viable monocytes. The authors concluded that, besides the presence of specific surface receptor(s), "precise spatial arrangements and structural interrelationships of phospholipids and proteins as well as the surface charge distribution, the surface area, and the resting membrane potential of the cell are factors which may dictate membrane interactions with these macromolecules by directly affecting hydrophobic and ionic associations between endotoxin molecules and the cell membrane which are necessary for effective binding." This interesting hypothesis, although yet to be established unequivocally, receives at least indirect support through the well-established capacity of LPS to interact with membrane bilayers (reviewed in MORRISON 1985).

A different experimental approach taken by HAMPTON and his colleagues to investigate the existence of specific receptors for LPS was predicated upon fact that a lipid A precursor (lipid IVa) can be obtained in radiolabeled form with high specific radioactivity and can be used to probe solubilized extracts of a macrophage cell line, RAW264.7. HAMPTON et al. (1988) fractionated solubilized cell extracts by gel electrophoresis and transferred the fractionated extracts to nitrocellulose by electroblotting. When blots were subsequently probed with ^{32}p-lipid IVa, two major lipid IVa-binding proteins of approximately 95 kDa and 31 kDa were identified by autoradiography. A weaker lipid IVa-binding protein of approximately 80 kDa was also apparent in the published autoradiographs of these experiments but was not commented on by the authors. HAMPTON et al. stated that the binding of ^{32}p-lipid IVa to these proteins could be inhibited both by intact lipid A and by Re-chemotype LPS, suggesting that the binding specificity was for the lipid A. However, the binding of lipid IVa could not be inhibited by intact polysaccharide-containing LPS, although a number of factors, including the potential role for membrane phospholipids suggested above, could contribute to this observation. The authors suggested that the 30-kDa LPS-binding protein is most likely a nuclear histone, but that the 96-kDa protein may serve as a candidate LPS receptor. Experimental evidence that this 96-kDa protein may function as an LPS receptor has derived from studies showing saturable binding of lipid IVa to the macrophage cell line which is inhibited by Re-LPS and lipid A but not by unrelated phospholipids (RAETZ et al. 1988).

An alternative experimental approach, taken by MORRISON and co-workers, was to synthesize photoactivatable, radioiodinated LPS derivatives with which to probe LPS-binding sites on target cells. WOLLENWEBER and MORRISON (1985) established that the derivatized LPS behaved indistinguishably from native LPS in respect of a number of LPS biologic activities in vitro. Using this methodology, fractionation of LPS-photocrosslinked, reduced and solubilized mouse macrophage extracts on two-dimensional polyacrylamide gels allowed LEI and MORRISON (1988a) to identify an LPS-binding protein with an approximate pI of 6.5 and a molecular mass of approximately 80 kDa.

It appeared that the LPS-binding protein fractionated from mouse macrophages was indistinguishable from that fractionated from mouse lymphocytes as defined by mobility of this protein on two-dimensional polyacrylamide gels. The extension of these experimental protocols to peripheral blood mononuclear cells of a variety of mammalian species established that, as assessed by electrophoretic mobility on two-dimensional polyacrylamide gels, the 80-kDa LPS-binding protein was relatively highly conserved (ROEDER et al. 1989). Evidence in support of this 80-kDa protein serving as a potential receptor for LPS was provided by the facts that the protein was membrane localized and that binding of LPS was saturable and inhibited by underivatized homologous and heterologous LPS and by lipid A (LEI and MORRISON 1988b). Binding was not inhibited by a variety of peptidoglycans even at 100-fold excess by weight (unpublished observations). Binding was dependent upon both the time and the temperature of incubation, with maximal binding observed after 15 min at 37 °C.

More recently, CHEN et al. (1989) have described a purification strategy to obtain the 80-kDa LPS-binding protein in partially purified form from C3Heb/FeJ splenocytes using photoaffinity labeling, butanol extraction, preparative SDS polyacrylamide gel electrophoresis, and electroelution. Using this partially purified 80-kDa protein as antigen, BRIGHT and his colleagues (1990) developed a panel of Armenian hamster monoclonal antibodies which react with the partially purified 80-kDa protein. Extensive characterization of one such monoclonal antibody (MoAb5D3) showed binding specificity for highly purified 80-kDa protein, as assessed by ELISA. Competitive binding assays have established that LPS and MoAb5D3 will reciprocally compete for binding to 80-kDa protein on intact cells, although MoAb5D3 more readily inhibits LPS binding than vice versa.

An evaluation of the functional properties of MoAb5D3 has recently provided evidence that this 80-kDa protein may function as an LPS receptor on the mouse macrophage surface. Using MoAb5D3 as the ligand, CHEN and his colleagues (1990) demonstrated that the addition of ion exchange column-purified MoAb5D3 to cultures of bone marrow-derived C3H/HeN macrophages led to activation of these cells for tumor cell killing. In contrast, this antibody failed to activate macrophages from the C3H/HeJ mouse even though quantitative ELISA established equivalent binding of MoAb to these two cell populations. The fact that the activity of MoAb5D3 was heat labile and was not inhibited by polymyxin B suggested that the observed agonist activity was not the result of endotoxin contamination. In addition, significantly lower concentrations of MoAb5D3 were required to effect equivalent macrophage activation in the presence of IFN-γ. This is consistent with the previously defined role of IFN-γ as a "priming" signal and LPS as a "triggering" signal (PACE and RUSSELL 1981). These experimental results suggest that the 80-kDa membrane protein recognized by MoAb5D3 may be at least one of the entities through which activation for tumor cell killing is regulated by LPS. These studies also suggest that the LPS-like effect on macrophage activation may occur in the absence of potential secondary nonspecific hydrophobic interactions with the membrane bilayer. Although nonspecific interactions of this particular monoclonal antibody with the macrophage membrane cannot be formally excluded, this possibility is unlikely.

A very recent report by HARA-KUGE and collaborators (1990) has described an interesting and potentially important experimental approach for the identification of LPS receptors. These investigators have identified and characterized a mutant cell line from the murine J774 macrophage parental cell line which is phenotypically unresponsive to LPS stimulation. Examination of LPS-binding proteins by photoaffinity LPS crosslinking probes similar to those used by LEI and MORRISON (1988a) identified two proteins with molecular masses of approximately 65 kDa and 55 kDa on the parent cell line which were not detectable on the LPS-unresponsive mutant. These data provide provocative evidence for a role for one or both of these proteins in LPS responses. It is possible that the 65-kDa protein identified by HARA-KUGE and colleagues and the 80-kDa protein characterized by LEI and MORRISON are identical, given that the LPS and photocrosslinking probe used in the two studies were identical; the differences in apparent molecular weight might have been due to differences in gel electrophoresis conditions. Experiments to determine whether this is the case are currently underway between the two laboratories. Very preliminary results (unpublished) suggest that MoAb5D3 will inhibit binding of photolabeled LPS to both the 65-kDa and the 55-kDa macrophage LPS-binding proteins.

Collectively, these results indicate that one or more specific glycoproteins may be present on the membrane surface of mononuclear phagocytes which can function as specific binding sites and/or receptors for LPS or lipid A. There is also evidence to suggest that specific membrane-localized gangliosides may also function either directly as target binding sites for LPS or as accessory targets in modulating LPS responsiveness. As reported by MORRISON et al. (1985), LPS will form molecular complexes with a variety of purified gangliosides via interactions primarily with the lipid A region. Several investigators have reported that specific gangliosides will modulate LPS responsiveness in cultures of mouse macrophages, and recent studies by Ryan and co-workers have reported that only selected gangliosides on the macrophage membrane will bind LPS (J. RYAN, personal communication, 1990). It is also noteworthy that marked changes in macrophage ganglioside profiles are obtained upon stimulation with LPS (BERENSON et al. 1989). The precise relationship between ganglioside–LPS interactions and macrophage activation, and the potential role of these membrane-localized gangliosides as receptors, remain to be defined.

3 Biochemical Mechanisms of Macrophage Signal Transduction by LPS

Despite recent intensive studies by a number of laboratories, the LPS-triggered biochemical sequence of events leading to tumoricidal activation of macrophages remains to be fully defined. The problem arises mainly from the ambiguity as to whether any or all of the LPS-induced events studied in the macrophage actually serve as the primary inducer of macrophage activation or represent secondary processes. We will therefore briefly review and discuss the recent progress in studies of LPS-induced signal transduction pathways, which suggest probable involvement of guanine nucleotide-binding (G) protein, myristoyl transferase, phospholipase C (PLC), and protein kinase C (PKC) in macrophage activation.

The probable involvement of pertussis toxin-sensitive Gi protein, which inhibits adenylate cyclase, in mediating the effects of LPS on a mouse macrophage cell line was first demonstrated by JAKWAY and DEFRANCO (1986). Several lines of experimental evidence supporting this conclusion include: (a) inhibition of macrophage membrane adenylate cyclase by LPS; (b) abrogation of this inhibition by prior treatment of the cells with pertussis toxin, which inactivates Gi function by ADP-ribosylation of the Giα subunit; (c) inhibition of ADP-ribosylation of Giα by prior treatment of macrophage cell membranes with LPS; and (d) inhibition of LPS-induced membrane-associated IL-1 production. Collectively, these data suggest that the ability of LPS to activate Gi protein may be crucial for its ability to regulate these cells. More recently, DANIEL-ISSAKANI et al. (1989) suggested that the LPS response of phorbol myristate acetate (PMA)-pretreated U937 cells is linked to a specific pertussis toxin-sensitive G protein which was tentatively identified as Gi_2. However, it is recognized that pertussis toxin will inhibit not only adenylate cyclase-coupled Gi protein, but also other members of the G protein family which may be involved in the regulation of ion channels of neural cells (PFAFFINGER et al. 1985), histamine release by mast cells (NAKAMURA and UI 1985), and migration and lysosomal enzyme release from neutrophils (SMITH et al. 1985) and macrophages (BACKLUND et al. 1985). Since mast cells, neutrophils, and macrophages have been shown to respond to these stimuli by activation of PLC rather than adenylate cyclase, pertussis toxin could have modulated the function of a PLC-coupled G protein (BERRIDGE 1987).

In addition to these potential concerns, if adenylate cyclase-coupled Gi protein were to be the initial target of LPS action, then it would be predicted that LPS should regulate intracellular cAMP levels. However, there is no unequivocal evidence that the critical action of LPS involves regulation of cAMP levels. Furthermore, the specific mechanism by which LPS might modulate the function of G proteins is also unclear. In this regard, the suggestion by DANIEL-ISSAKANI et al. (1989) that LPS may activate a macrophage protein kinase which, in turn, might then phosphorylate Gi_2, is interesting, particularly in the light of several recent reports suggesting activation of macrophage PKC in response to LPS (described below). An interesting possibility that one of the LPS receptors discussed in the preceding paragraphs may be directly coupled to a Gi_2 protein warrants further investigation. Such studies would clarify the interpretation of the observed inhibitory action of pertussis toxin, since it is possible that phosphorylation of Gi_2 may result in the dissociation of Gi_2 from the putative LPS receptor and thus render the latter sensitive to pertussis toxin.

Another early biochemical event that follows LPS stimulation of macrophages is probable activation of myristoyl transferase, which catalyzes myristoylation of proteins of 68 and 36–42-kDa (ADEREM 1988). This myristoylation appears to play an important role in subsequent arachidonate metabolic cascade triggered by other stimuli. The LPS-stimulated 68-kDa protein translocates from cytosol to membrane, where it may be phosphorylated by PKC (ADEREM et al. 1988). However, the exact role of this 68-kDa protein in signal transduction is unknown at present.

Several studies have recently suggested that LPS-triggered activation of phosphoinositide (PI)-specific PLC may be an early, critical biochemical event during LPS-triggered macrophage activation. Evidence that the treatment of mouse peritoneal macrophages with LPS or lipid A promptly leads to activation of PI-specific PLC, which is accompanied by Ca^{2+} fluxes and activation of PKC, has been presented by PRPIC et al. (1987). In these studies, the activation of PLC, as measured by the formation of inositol triphosphate (IP_3), was rapid but rather modest in magnitude. The IP_3 formed in response to LPS was found to plateau within 1 min of stimulation, with a maximal change of about 130% from the time zero value. Significant PLC activation was shown to require relatively large doses ($>1\,\mu g/ml$) of LPS or lipid A.

Using two macrophage cell lines, J774 and P388D$_1$, CHANG et al. (1990) partly confirmed the findings of PRPIC et al. (1987) by showing LPS-induced formation of IP_3 in both cell lines. The time required for maximal response of J774 cells was about 5 min, whereas that for P388D$_1$ cells was about 20 min. The degree of the maximal response was again modest (maximal response being about 130% of the control) and the dose of LPS required for significant PLC activation was more than 1 µg/ml. It is of interest that only the J774 cell line responded to LPS by becoming activated for tumor cell killing; however as discussed below, the relevance of this information to the above-cited differences in the kinetics of IP_3 formation is unclear.

Thus, the data from several laboratories have shown that LPS treatment of macrophages can result in modest activation of PI-specific PLC within 1–20 min. However, a critical question remains as to whether or not LPS-triggered activation of PLC is a primary inducer for tumoricidal activation of macrophages. Addressing this question, CHANG et al. (1990) reported that the treatment of macrophages with neomycin (SCHNACHT 1976) inhibited LPS-triggered PLC activity of LPS-activated J774 cells, but failed to block LPS-induced tumoricidal activation of the cells. In addition, GRABAREK et al. (1988) recently showed that lipid A may activate PKC in platelets, without concomitant activation of PLC. These results suggest that LPS-triggered activation of PLC may not be an obligatory cellular event in response to LPS or lipid A.

Lipopolysaccharide-stimulated activation of PLC should result in production of two different second messengers, namely 1,2-diacylglycerol (DAG) and IP_3 (PRPIC et al. 1987; CHANG et al. 1990). DAG and Ca^{2+} released from intracellular stores by the action of IP_3 would activate PKC. WEIEL et al. (1986) earlier suggested probable activation of PKC following stimulation of mouse macrophages by LPS, based upon the similarity of the LPS-induced phosphorylation pattern of cellular proteins to that induced by the well- characterized PKC activator, PMA. PKC-catalyzed phosphorylation of cellular proteins could lead to the modulation of gene expression of macrophages, which is probably essential for LPS-induced development of tumoricidal activity. In this regard, it is interesting to note that some investigators have reported that tumoricidal activation of macrophages induced by either IFN-γ or IFN-β could be blocked by inhibitors of PKC, such as H-7 (CELADA and SCHREIBER 1986; RADIOCH and VARESIO 19; HAMILTON et al. 1985) (although it should also be noted that there were some discrepancies in the reported results). However, the basic question of the mechanism by which LPS interaction with the macrophage leads to activation of

PKC remains unanswered. In considering this question, we will first briefly review the known properties of PKC and calpain.

Protein Kinase C has been shown to be involved in a wide variety of cellular functions, such as secretion and exocytosis, modulation of ion conductance, regulation of receptor interaction with components of the signal transduction apparatus, smooth muscle contraction, gene expression, and cell proliferation (NISHIZUKA 1986). The PKC enzymes in various tissues are recognized to be polymorphic, and at least seven isoenzymes, denoted as α, βI and βIII, γ, δ, ε, and ξ have so far been identified (PARKER et al. 1986; NISHIZUKA 1988). The genes encoding these isoenzymes manifest a great degree of sequence homology, not only among themselves, but also with many other protein kinases. The N-terminal half of the single chain 80-kDa PKC is considered to be the regulatory domain. It contains two highly conserved regions, termed C_1 and C_2, and two highly variable regions, v_1 and v_2. The C_1 region contains a tandem repeat of a cysteine-rich sequencem, C-X_2-C-X_{13}-C-X_2-C, which is reminiscent of the cysteine-zinc-DNA-binding finger found in many metalloproteins and DNA-binding proteins related to transcriptional regulation, although there is no evidence that PKC binds DNA. The sites involved in binding of PKC regulatory molecules, such as Ca^{2+}, 1,2-DAG, and phosphatidylserine, are thought to be contained in C_1 and C_2, but have not yet been identified with certainty. The C-terminal half of the enzyme contains two conserved regions, C_3 and C_4, and is considered to contain the catalytic domain. The conserved region C_3 has an ATP-binding sequence, Gly-X-Gly-X_2-Gly, which marks the beginning of the kinase domain. A second ATP-binding site may be present in the conserved regions C_4 with the consensus sequence. The site which is susceptible to the Ca^{2+}-dependent neutral protease, calpain (SUZUKI et al. 1987; KISHIMOTO et al. 1989), has been shown to be within the V_3 region which separates the regulatory and the catalytic domains.

The properties of PKC present in mouse macrophages have not yet been fully characterized. The LPS-mediated activation of PLC discussed above, which results in the generation of DAG and IP_3, should lead to the activation of PKC. Since the pretreatment of J774 cells with either H-7 or PMA clearly blocked LPS-induced cytotoxicity in a dose- dependent manner (NOVOTNEY et al. 1990), LPS-induced activation of PKC may well be a critical step in tumoricidal activation, as already suggested for IFN-γ- or -β-induced macrophage activation (CELADA and SCHREIBER 1986; RADIOCH and VARIESO 1988). The *critical questions* are whether LPS activates macrophage PKC only through the generation of DAG and IP_3, or also by some alternative pathway which involves the calpain/calpastatin system. NOVOTNEY et al. (1991) have recently presented evidence that certain types of PKC in LPS-activatable J774 cells undergo proteolytic cleavage which results in production of a 40-kDa regulatory domain fragment. This fragment could be recognized in Western blots of two-dimensional gels by specific rabbit anti-PKC regulatory domain serum (JAKWAY and DE FRANCO 1986). No proteolytic cleavage in response to LPS was observed in P388D_1 cells (see above). The cytosol of J774 cells was found clearly to contain two types of protein kinase (one basic and the other acidic) which were readily separable by HPLC. In contrast, the cytosol of P388D$_1$ cells was found to contain only one protein kinase, corresponding to the basic fraction of J774 cells. The acidic protein kinase, which is present in only J774 cells, may represent the catalytic domain fragment, because it bound the radiolabeled PKC-specific inhibitor [^{3}H]staurosporine, and because its activity did not appear to be augmented by phosphatidyserine, diolein, and Ca^{2+}. These data thus suggested that LPS treatment of macrophages may activate PKC by two distinct pathways: (a) via a classic DAG- IP_3-mediated pathway which occurs in both LPS-activatable and LPS-nonactivatable cell lines; and (b) via the activation of Ca^{2+}-dependent neutral protease (calpain) in LPS-activatable cells, which cleaves PKC to generate a catalytic domain fragment (denoted as PKM), the activity of which does not require regulatory molecules such as DAG, Ca^{2+}, and phospholipid. The lack of generation of a 40-kDa protein recognizable by antiregulatory domain antibody in LPS-nonactivatable P388D$_1$ cells might be due to structural alteration of the V_3 region of PKC, which renders PKC resistant to the action of calpain. Although there is no direct evidence at present to indicate that LPS activates calpain, it is interesting to consider this as a possible pathway.

Calpain consists of two subunits, an 80-kDa and a 30-kDa protein (EMORI et al. 1986a, b). Upon autolysis, the 80-kDa calpain subunit is converted to the protease-active 76- to 78-kDa protein. In addition, calpain has been shown to be able to associate with plasma membranes through an N-terminal, glycine-rich hydrophobic domain of the 30-kDa subunit, which has been shown to interact with phospholipids and biologic membranes (IMAJOH et al. 1986) as well as galactosyl residue-containing polysaccharide (ZIMMERMAN and SCHLAEPFER 1988). LPS is expected to interact with membranes through its hydrophobic lipid A moiety and there is now good evidence for LPS binding to at least one membrane-localized 80-kDa receptor protein perhaps either related to, or identical with, the 80-kDa subunit of calpain. This interaction may facilitate the binding of LPS, through its polysaccharide component, with the N-terminal regulatory domain of the 30-kDa subunit, which, together with the 80-kDa subunit, is associated with the inner side of the plasma membrane. The interaction between the 30-kDa calpain subunit and LPS may then promote autolysis of the 80-kDa subunit to produce the proteolytically active 76- to 78-kDa calpain fragment. This type of calpain activation was observed by ZIMMERMAN and SCHLAEPFER (1988) during the binding of calpain to agarose matrix.

Evidence that calpain may regulate the activity of PKC has been presented by many laboratories (TAPLEY and MURRAY 1985; MURRAY et al. 1987). Incubation of platelets or neutrophils with PMA was shown to result in the formation of a 50-kDa protein kinase (PKM), active in the absence of added Ca^{2+} and phospholipid, which eluted from DE52 cellulose at a higher salt concentration than PKC. Formation of this kinase was blocked by preincubation of permeabilized platelets with leupeptin, a calpain inhibitor (CELADA and SCHREIBER 1986). In addition, the patterns of phosphorylation of myosin light chain catalyzed by either PKC or PKM were basically identical. As described above, purified PKC could be cleaved at its V_3 region by the purified calpain (NISHIZUKA 1988; SUZUKI et al. 1987). The cleavage of PKC to PKM by calpain frees it from its regulatory constraint and releases active kinase (PKM) into the cytosol. Thus, PKM should have access to a different range of substrates than membrane-bound PKC. However, the questions remain as to whether or not: (a) LPS actually activates calpain by interacting with either type of calpain subunit and leads to the generation of PKM; and (b) the photoaffinity-labeled 80-kDa LPS-binding protein represents the 80-kDa calpain subunit.

4 Summary and Conclusions

There is little question but that bacterial lipopolysaccharides (LPS) remain one of the most potent stimuli which can affect macrophage activation. Although the precise biochemical mechanisms responsible for this remain to be fully defined, there is now evidence accumulating from a number of laboratories that functional receptors for these bacterial products do exist and may contribute to the initial triggering event. Unfortunately, there is currently no consensus as to which of the candidate receptors identified to date serves as the primary binding target for LPS, and it is possible that the difference in macrophage cell types, LPS probes, and detection systems will all influence the nature of the binding. At the present time, therefore, macromolecules of 96-kDa, 95-kDa (adhesion β chain), 80-kDa, 65-kDa, and 55-kDa may be considered as possible LPS targets. With the exception of the 96-kDa protein identified by HAMPTON and his co-workers, there exists some experimental evidence for a functional role for each of the molecules so far identified. It is apparent that the molecular cloning and sequencing and subsequent biochemical characterization of these LPS receptors will be required to determine unequivocally their role in LPS-mediated triggering events. Such information will be invaluable in sorting out the relevant biochemical second signals involved in macrophage activation. Although much new information has recently been accumulated on potential signaling pathways for LPS, the definitive events remain far from unequivocally established. In view of the obvious importance of LPS–macrophage interactions in the overall capacity of the mammalian host to respond appropriately to the potentially hostile prokaryotic environment, a precise delineation of LPS-mediated macrophage activation is critical to our understanding of this important inflammatory mediator cell.

Acknowledgements. This research was supported by NIH grants AI23447-05 (to D.C.M.), AI22948-06 (to D.C.M.), AI22742-05 (to T.S.), and CA35977-06 (to T.S.). Dr. Tai-Ying Chen is a Scholar of the Wesley Foundation Cancer Research and Training Grant to the Kansas Universities. We also thank Ms. Janet Hollands for expert secretarial assistance.

References

Aderem AA (1988) Protein myristoylation as an intermediate step during signal transduction in macrophages: its role in arachidonic acid metabolism and in responses to interferon γ. J Cell Sci [Suppl] 9: 151–167

Aderem AA, Albert KA, Keum MM, Wang JKT, Greengard P, Cohn ZA (1988) Stimulus-dependent myristoylation of a major substrate for protein kinase C. Nature 332: 362–364

Aggarwal BB, Kohr WJ, Hass PE et al. (1985) Human tumor necrosis factor. Production, purification and characterization. J Biol Chem 260: 2345–2354

Akagawa KS, Tokunaga T (1985) Lack of binding of bacterial lipopolysaccharide to mouse lung macrophages and restoration of binding by γ-IFN. J Exp Med 162: 1444–1459

Akagawa KS, Kamoshita K, Tokunaga T (1988) Effects of granulocyte-macrophage colony-stimulating factor and colony-stimulating factor-1 on the proliferation and differentiation of murine alveolar macrophages. J Immunol 141: 3383–3390

Alexander P, Evans R (1971) Endotoxin and double stranded RNA render macrophages cytotoxic. Nature New Biol 232: 76–78

Backlund PS, Meade BD, Manclark CR, Cantoni GL, Akasmit RR (1985) Pertussis toxin inhibition of chemotaxis and the ADP-ribosylation of a membrane protein in a human-mouse hybrid cell line. Proc Natl Acad Sci USA 82: 2637–26421

Berenson CS, Yohe HC, Ryan JL (1989) Factors mediating lipopolysaccharide- induced ganglioside expression in murine peritoneal macrophages. J Leukocyte Biol 45: 221–230

Berridge MJ (1987) Inositol triphosphate and diacylglycerol: Two interacting second messengers. Annu Rev Biochem 56: 159–193

Bright SW, Chen TY, Flebbe LM, Lei MG, Morrison DC (1990) Generation and characterization of hamster-mouse hybridomas secreting monoclonal antibodies with specificity for lipopolysaccharide receptor. J Immunol 145: 1–7

Celada A, Schreiber RD (1986) Role of protein kinase C and intracellular calcium mobilization in the induction of macrophage tumoricidal activity by interferon-γ. J Immunol 137: 2373–2379

Chang ZL, Novotney M, Suzuki T (1990) Phospholipase C and A_2 in tumoricidal activation of murine macrophage-like cell lines. FASEB J 4: A1753

Chen TY, Lei MG, Morrison DC (1989) Partial purification of specific LPS binding proteins from murine splenocytes. FASEB, 3: A1082

Chen TY, Bright SW, Pace JL, Russell SW, Morrison DC (1990) Induction of macrophage mediated tumor cytotoxicity by a hamster monoclonal antibody with specificity for LPS receptor. J Immunol 145: 8–12

Chia JK, Pollack M, Guelde G, Koles NL, Miller M, Evans ME (1989) Lipopolysaccharide (LPS)-reactive monoclonal antibodies fail to inhibit LPS-induced tumor necrosis factor secretion by mouse-derived macrophage. J Infect Dis 159: 872–880

Daniel-Issakani S, Spiegel AM, Strulovici B (1989) Lipopolysaccharide response is linked to the GTP binding protein, Gi_2 in the promonocytic cell line U937. J Biol Chem 264: 20240–20247

Dierich MP, Bitter-Suermann D, Konig W (1973) Analysis of bypass activation of C3 by endotoxic LPS and loss of this potency. Immunology 24: 721–733

Ding AH, Sanchez E, Srimal S, Nathan CF (1989) Macrophages rapidly internalize their tumor necrosis factor receptors in response to bacterial lipopolysaccharide. J Biol Chem 264: 3924–3929

Doe MF, Yang D, Morrison DC, Beta SJ, Henson PM (1978) Macrophage stimulation by bacterial lipopolysaccharides. II. Evidence for independent differentiation signals delivered by lipid A and by a protein rich fraction of LPS. J Exp Med 148: 557–568

Emori Y, Kawasaki H, Imajoh S, Kawashima S, Suzuki K (1986a) Isolation and sequence analysis of cDNA clones for the small subunit of rabbit calcium- dependent protease. J Biol Chem 261: 9472–9474

Emori Y, Kawasaki H, Sugihara H, Imajoh S, Kawashima S, Suzuki K (1986b) Isolation and sequence analyses of cDNA clones for the large subunits of two isozymes of rabbit calcium-dependent protease. J Biol Chem 261: 9465–9471

Erroi A, Casali B, Donati MB, Mantovani A, Semeraro N (1988) Mouse tumor-associated macrophages do not generate procoagulant activity in response to different stimuli. Int J Cancer 41: 65–68

Ezekowitz RAB, Day LE, Herman GA (1988) A human mannose-binding protein is an acute-phase reactant that shares sequence homology with other vertebrate lectins. J Exp Med 167:1034–1046

Fearon DT, Wong WW (1983) Complement ligand-receptor interactions that mediate biological responses. Annu Rev Immunol 1: 243–271

Flebbe LM, Chapes SK, Morrison DC (1990) Activation of C3H/HeJ macrophage tumoricidal activity and cytokine release by R-chemotype LPS preparations: differential effects of interferon gamma. J Immunol 145: 1505–1511

Gery I, Gersho RK, Waksman GH (1972) Potentiation of the T-lymphocyte response to mitogenes. I. The responding cell. J Exp Med 136: 128–142

Goodrum KJ (1987) Complement component C3 secretion by mouse macrophage-like cell lines. J Leukocyte Biol 41: 295–301

Grabarek J, Timons S, Hawiger J (1988) Modulation of human platelet protein kinase C by endotoxic lipid A. J Clin Invest 82: 954–971

Haeffner-Cavaillon N, Chaby R, Cavaillon JM, Szabo L (1982) Lipopolysaccharide receptor on rabbit peritoneal macrophages. I. Binding characteristics. J Immunol 128: 1950–1954

Haeffner-Cavaillon N, Cavaillon JM, Szabo L (1985a) Cellular receptors for endotoxin. In: Berry LJ (ed). Handbook of endotoxin, vol 3, Elsevier Science, New York, pp 1–25

Haeffner-Cavaillon N, Cavaillon JM, Elievant M, Lebbar S, Szabo L (1985b) Specific binding of endotoxin to human monocytes and mouse macrophages: serum requirement. Cell Immunol 91: 119–131

Haeffner-Cavallion N, Caroff M, Cavaillon JM (1989) Interleukin-1 induction by lipopolysaccharides: structural requirements of the -deoxy-D- manno-2-octulosonic acid (KDO). Mol Immunol 26: 485–494

Hamilton TA, Becton DL, Somers SD (1985) Interferon-γ modulates protein kinase C activity in murine peritoneal macrophages. J Biol Chem 260: 1378–1381

Hampton RY, Golenbock DT, Raetz RHC (1988) Lipid A binding sites in membranes of macrophage tumor cells. J Biol Chem 263: 14802–14807

Hara-Kuge S, Amano F, Nishijima M, Akamatsu Y (1990) Isolation of an LPS-resistant mutant, with defective LPS binding of cultured macrophage-like cells. J Biol Chem. 265:6606–6610

Havell EA, Spitalny GL (1983) Endotoxin-induced interferon synthesis in macrophage cultures. J Reticuloendothel Soc 33: 369–380

Hoter W, Goldman CK, Casabo L, Nelson DL, Greene WC, Waldmann TA (1987) Expression of functional IL-2 receptors by lipopolysaccharide and interferon- γ stimulated human monocytes. J Immunol 138: 2917–2922

Imajoh S, Kawasaki H, Suzuki K (1986) The amino-terminal hydrophobic region of the small subunit of calcium-activated neutral protease (CANP) is essential for its activation by phosphatidylinositol. J Biochem (Tokyo) 99: 1281–1284

Jakway JP, DeFranco AL (1985) Development of monoclonal antibodies which may recognize the receptor for lipopolysaccharide on mouse B cells. Fed Proc 44: 1297

Jakway JP, DeFranco AL (1986) Pertussis toxin inhibition of B cell and macrophage responses to bacterial lipopolysaccharide. Science 234: 734–746

Kishimoto A, Mikawa K, Hashimoto K et al. (1989) Limited proteolysis of protein kinase C subspecies by calcium-dependent neutral protease (Calpain). J Biol Chem 264: 4088–4092

Kotani S, Takada H, Tsujimoto M et al. (1985) Synthetic lipid A with endotoxic and related biological activities comparable to those of a natural lipid A from an *Escherichia coli* re-mutant. Infect Immun 49: 225–237

Larsen NE, Sullivan R (1984a) Interaction between endotoxin and human monocytes: characteristics of the binding of ^{3}H-labelled lipopolysaccharides and ^{51}Cr-labelled lipid A before and after the induction of endotoxin tolerance. Proc Natl Acad Sci USA 81: 3491–3495

Larsen NE, Sullivan R (1984b) Interaction of radiolabelled endotoxin molecules with human monocyte membranes. Biochim Biophys Acta 774: 261–268

Lebbar S, Cavaillon JM, Caroff M, Ledar A, Brade H, Sarfati R, Haeffner- Cavaillon N (1986) Molecular requirement for interleukin 1 induction by lipopolysaccharide-stimulated human monocytes: involvement of the heptosyl-2-keto-3-deoxy-octulosonate region. Eur J Immunol 16: 87–91

Lei MG, Morrison DC (1988a) Specific endotoxic lipopolysaccharide binding proteins on murine splenocytes. I. Detection of LPS binding sites on splenocytes and splenocyte subpopulations. J Immunol 141: 996–1005

Lei MG, Morrison DC (1988b) Specific endotoxic lipopolysaccharide binding proteins on murine splenocytes II. Membrane localization and binding characteristics. J Immunol 141: 1006–1011

Lei MG, Chen TY, Morrison DC (1990) Lipopolysaccharide receptors on mammalian lymphoreticular cells. In: Jirillo E (ed). International reviews of immunology, Vol. 6. Harwood Academic, New York pp 223–235

Liang-Takasaki CJ, Makela H, Leive L (1982) Phagocytosis of bacteria by macrophages changing the carbohydrate of lipopolysaccharide alters interaction with complement and macrophages. J Immunol 128: 1229–1235

Loppnow H, Brade H, Durrbaum I, Dinarello CA, Kusumoto S, Rietschel ET, Flad HD (1989) IL-1 induction-capacity of defined lipopolysaccharide partial structures. J Immunol 142: 3229–3238

Lubinsky S, Munkenbeck P, Morrison DC (1983) Interaction of latex insolubilized endotoxins with murine macrophages. I. Phagocytic responses of endotoxin responsive (C3Heb/FeJ) and unresponsive (C3H/HeJ) macrophages in vitro. J Reticuloendothelial Soc 33: 353–368

Luderitz T, Brandenburg K, Seydel U, Roth A, Galanos C, Rietschel ET (1989) Structural and physiochemical requirements of endotoxins for the activation of arachidonic acid metabolism in mouse peritoneal macrophage in vitro. Eur J Biochem 179: 11–16

Morrison DC (1985) Non-specific interactions of bacterial lipopolysaccharides with membranes and membrane components. In: Proctor RA (ed). Handbook of endotoxin, vol 3. Elsevier Science, New York, pp 25–55

Morrison DC (1989) Minireview–the case for specific lipopolysaccharide receptors expressed on mammalian cells. Microbial Pathogenesis 7: 389–398

Morrison DC, Jacobs DM (1976) Binding of polymyxin B to the lipid A portion of bacterial lipopolysaccharides. Immunochemistry 13: 813–818

Morrison DC, Kline LF (1977) Activation of the classical and properdin pathways of complement by bacterial lipopolysaccharides. J Immunol 118: 362–368

Morrison DC, Rudbach JA (1981) Endotoxin-cell membrane interactions leading to transmembrane signals. In: Mandy WJ, Imman FP (eds) Contemporary topics in molecular immunology, vol 8. Plenum, New York, pp 187–218

Morrison DC, Brown DE, Vukajlovich SW, Ryan JL (1985) Ganglioside modulation of lipopolysaccharide initiated complement activation. Mol Immunol 22: 1169–1176

Murray AW, Fournier A, Hardy SJ (1987) Proteolytic activation of protein kinase C: a physiological reaction? Trends Biochem Sci 12: 53–54

Nakamura T, Ui M (1985) Simultaneous inhibition of inositol phospholipid break down, arachidonic acid release, and histamine secretion in mast cells by islet-activating protein, pertussis toxin: A possible involvement of the toxin-specific substrate in the Ca^{2+}- mobilizing receptor-mediated biosignaling system. J Biol Chem 260: 3584–3593

Nathan CF (1987) Secretory products of macrophages. J Clin Invest 79: 319–326

Niemetz J, Morrison DC (1977) Lipid A as the biologically active moiety in bacterial endotoxin (LPS) initiated generation of procoagulant activity by peripheral blood leukocytes. Blood 49: 947–956

Nishizuka Y (1986) Studies and perspective of protein kinase C. Science 233: 305–312

Nishizuka Y (1988) The molecular heterogeneity of protein kinase C and its implications for cellular regulation. Nature 334: 661–665

Novotney M, Chang ZL, Uchiyama H, Suzuki T (1991) Protein kinase C in tumoricidal activation of murine macrophage-like cell lines. Biochemistry 30: 5597–5604

Ohno N, Morrison DC (1989a) Lipopolysaccharide interaction with lysozyme. I. Binding of LPS to lysozyme and inhibition of lysozyme enzymatic activity. J Biol Chem 264: 4434–4441

Ohno N, Morrison DC (1989b) Lipopolysaccharide interactions with lysozyme differentially affect LPS immunostimulatory activity. Eur J Biochem 186: 629–636

Old LJ (1985) Tumor necrosis factor (TNF). Science 120: 630–632

Pace JL, Russell SW (1981) Activation of mouse macrophages for tumor cell killing: I. Quantitative analysis of interactions between lymphokine and lipopolysaccharide. J Immunol 126: 1863–1867

Parker P, Coussens L, Totty N et al. (1986) The complete primary structure of protein kinase C—the major phorbol ester receptor. Science 233: 853–859

Pfaffinger PJ, Martin JM, Hunter DD, Nathanson NM, Hille B (1985) GTP-binding protein couple cardiac muscarinic receptors to a K channel. Nature 317: 536–538

Prpic V, Weiel JE, Somers SD et al. (1987) Effects of bacterial lipopolysaccharide on the hydrolysis of phosphatidylinositol-4, 5-bisphosphate in murine peritoneal macrophages. J Immunol 139: 526–533

Radioch D, Varesio L (1988) Protein kinase C inhibitors block the activation of macrophages by INF-β but not by INF-γ. J Immunol 140: 1259–1263

Raetz CRH, Brozek KA, Clementz T, Coleman JD, Galloway SM, Golenbock DT, Hampton RY (1988) Gram-negative endotoxin: a biologically active lipid. Cold Spring Harbor Symposia on Quantitative Biology, vol LIII, pp 973–982

Roeder D, Lei MG, Morrison DC (1989) Specific endotoxic lipopolysaccharide (LPS) binding proteins on lymphoid cells of various animal species—correlation with endotoxin susceptibility. Infect Immun 57: 1054–1058

Roos PH, Hartman HG, Schlepper-Schafer J, Kolb H, Kolb-Bachofen V (1985) Galactose-specific receptors on liver cells. II. Characterization of the purified receptor from macrophages reveals no structural relationship to the hepatocyte's receptor. Biochem Biophys Res Commun 847: 115–121

Schade UF, Burmeister I, Engl R (1987) Increased 13-hydroxyoctadecadienoic acid content in lipopolysaccharide stimulated macrophages. Biochem Biophys Res Commun 147: 695–700

Schnacht J (1976) Inhibition by neomycin of polyphosphoinositide turnover in subcellular fractions of guinea pig cerebral cortex in vitro. J Neurochem 27: 1119–1124

Smith CD, Lane BJ, Kusaka I, Vergehese MW, Snyderman R (1985) Chemoattractant receptor-induced hydrolysis of phosphatidylinositol 4, 5-bisphosphate in human polymorphonuclear leukocyte membranes. Requirement for a guanine nucleotide regulatory protein. J Biol Chem 260: 5875–5878

Stephenson JD, Shepherd VL (1987) Purification of the human alveolar macrophage mannose receptor. Biochem Biophys Res Commun 148: 883–889

Strunk RC, Whitehead AS, Cobe FS (1985) Pretranslation and regulation of the synthesis of the third component of complement in human mononuclear phagocytes by the lipid A portion of lipopolysaccharide. J Clin Invest 76: 985–990

Suzuki K, Imajo S, Emori Y, Kawasaki H, Minami Y, Ohno S (1987) Calcium-activated neutral protease and its endogenous inhibitor. FEBS Lett 220: 271–277

Tapley PM, Murray AW (1985) Evidence that treatment of platelets with phorbol ester causes proteolytic activation of ca^{2+} activated, phospholipid-dependent protein kinase. Eur J Biochem 151: 419–423

Tobias PS, Soldau K, Ulevitch RJ (1986) Isolation of a lipopolysaccharide-binding acute phase reactant from rabbit serum. J Exp Med 164: 777–793

Tobias PS, Mathison JC, Ulevitch RJ (1988) A family of lipopolysaccharide binding proteins involved in responses to gram-negative sepsis. J Biol Chem 263: 13479–13481

Tobias PS, Soldau K, Ulevitch RJ (1989) Identification of a lipid A binding site in the acute phase reactant lipopolysaccharide binding protein. J Biol Chem 264: 10867–10871

Weiel JE, Hamilton TA, Adams AO (1986) LPS induces altered phosphate labeling of proteins in murine peritoneal macrophages. I Immunol 136: 3012–3018

Weiss J, Elsbach P, Olsson I, Odeberg H (1978) Purification and characterization of a potent bactericidal and membrane-active protein from the granules of human polymorphonuclear leukocytes. J Biol Chem 253: 2664–2672

Wollenweber HW, Morrison DC (1985) Synthesis and biochemical characterization of a photoactivatable, iodinatable, cleavable bacterial lipopolysaccharide derivative. J Biol Chem 260: 15068–15074

Wright SD, Jong MTC (1986) Adhesion-promoting receptor on human macrophages recognize *Escherichia coli* by binding to lipopolysaccharide. J Exp Med 164: 1876–1888

Wright SD, Levin SD (1989) CR3 (CD11b/CD18) expresses one binding site for Arg-Gly-Asp-containing peptides and a second site for bacterial lipopolysaccharide. J Exp Med 169: 175–182

Wright SD, Tobias PS, Ulevitch RJ, Ramos RA (1989) Lipopolysaccharide (LPS) binding protein opsonizes LPS-bearing particles for recognition by a novel receptor on macrophages. J Exp Med 170: 1231–1241

Wright SD, Detmers PA, Aida Y, Adamowski R, Anderson DC, Chad Z, Kabbash LG, Pabst MJ (1990) CD18-deficient cells respond to lipopolysaccharide in vitro. J Immunol 144: 2566–2571

Zimmerman UJ, Schlaepfer WW (1988) Calcium-activated Neutral proteases (calpains) are carbohydrate binding proteins. J Biol Chem 263: 11609–11612

The Role of Myristoylated Protein Kinase C Substrates in Intracellular Signaling Pathways in Macrophages

A. ADEREM

1 Introduction

The protein kinases C (PKC) are a family of diacylglycerol-activated, calcium-dependent protein kinases that regulate diverse cellular pathways in macrophages, including those leading to phagocytosis and the secretion of arachidonic acid metabolites and reactive oxygen intermediates (KIKKAWA et al. 1989; ADEREM 1988; CLARK 1990). Very little is known about the molecular mechanism by which PKC mediates such diverse responses. It is known,

The Rockefeller University, 1230 York Avenue, New York, NY 10021, USA however, that PKC-mediated events in macrophages are profoundly influenced by inflammatory mediators such as bacterial lipopolysaccharide (LPS) and by cytokines such as gamma-interferon (IFN-γ) and tumor necrosis factor (TNF-α) (HAMILTON and ADAMS 1987). Thus, while LPS alone is incapable of activating PKC, it can prime macrophages for vastly increased PKC-dependent responses such as the release of eicosanoids and prostanoids (ADEREM et al. 1986a, ADEREM and COHN 1988). Concomitant with priming, LPS also induces the transcription, translation, and myristoylation of three macrophage proteins with apparent molecular masses of 40, 42, and 68 kDa (ADEREM et al. 1986b, 1988a). All three proteins are substrates for PKC and are therefore excellent candidate effectors of PKC-induced responses. In addition, we have characterized a 48-kDa myristoylated PKC substrate which is induced in macrophages by IFN-γ and which appears likely to mediate some of the responses of this macrophage-activating cytokine (ADEREM et al. 1988b).

This chapter will focus primarily on the properties of the 68-kDa and 48-kDa myristoylated PKC substrates, which are likely to act as effectors of PKC-dependent responses in macrophages. I will also include recent information on protein myristoylation which will situate the discussion in a broader context.

2 Protein N-Myristoylation

In all cases described, N-myristoylation of proteins occurs via an amide linkage to the α-amino group of an N-terminal glycine (SCHMIDT 1989; SCHULTZ et al. 1988; TOWLER et al. 1988). N-myristoylation occurs cotranslationally or very soon after the completion of polypeptide synthesis, as demonstrated by the incorporation of myristic acid into nascent peptides, and blockage of this process by protein synthesis inhibitors (WILCOX et al. 1987; OLSON and SPIZZ 1986). The precise function of protein-bound myristic acid is not fully understood. In some cases it is required to target proteins to the plasma membrane, but myristoylated proteins are also found in the cytosol, endoplasmic reticulum, and nucleus (SCHULTZ et al. 1985; BUSS and SEFTON 1985; TOWLER et al. 1988; SCHMIDT 1989). The diversity of proteins known to be myristoylated suggests a number of possibilities regarding the functional role of this modification. The list of myristoylated proteins includes several known to be involved in intracellular signaling pathways, such as the tyrosine kinases $p60^{src}$ (SCHULTZ et al. 1985; BUSS and SEFTON 1985) and $p56^{lck}$ (MARCHILDON et al. 1984), the catalytic subunit of cAMP-dependent protein kinase (CARR et al. 1982), the regulatory subunit of protein phosphatase 2B (AITKEN et al. 1982), a prominent protein kinase C substrate (ADEREM et al. 1988a) (which will be the major focus of this review), and the G-proteins (BUSS et al. 1987; SCHULTZ et al. 1987). A number of viral proteins are also myristoylated and in most cases, this modification is required for membrane attachment and for an intact replication cycle (TOWLER et al. 1988; SCHULTZ et al. 1988; SCHMIDT 1989).

2.1 Enzymology

The acylation reaction is catalyzed by a myristoyl CoA: protein *N*-myristoyl transferase (NMT) which utilizes myristoyl CoA exclusively (TOWLER et al. 1988). An NMT has been purified and cloned from *Saccharomyces cerevisiae* (TOWLER et al. 1987; DURONIO et al. 1989). It exhibits a high degree of selectivity for the sequence of its substrate peptide, and a loose consensus sequence required for effective myristoylation has been elucidated (TOWLER et al. 1988).

2.2 Regulation

There are a number of potential points of regulation of N-myristoylation. First, since acylation occurs cotranslationally or very soon after the completion of polypeptide synthesis (WILCOX et al. 1987), and since all cells examined contain active NMT, the transcription and translation of the candidate protein may directly regulate its N-myristoylation. Second, since the initiator methionine must first be removed by an *N*-methionine aminopeptidase to expose the acceptor glycine, this peptidase could potentially regulate N-myristoylation. Third, the activity or subcellular location of the NMT might be regulated. Thus, certain proteins bearing a myristoylation consensus sequence might be translated on ribosomes that are not in the proximity of the NMT. In this context it is interesting to note that a subset of G-proteins contains the N-terminal glycine but is not myristoylated (MUMBY et al. 1990; JONES et al. 1990). Finally, although all known N-myristoylated proteins that have been sequenced bear a glycine after the initiator methionine, it is possible that proteins could be acylated at a cryptic site exposed by proteolytic cleavage.

Lipopolysaccharide and IFN-γ treatment of macrophages does not greatly influence NMT activity, and the increased levels of myristoylated PKC substrates observed in response to LPS and IFN-γ are due to the cotranslational myristoylation of LPS- and IFN-γ-induced gene products (J. SEYKORA and A. ADEREM, unpublished observations).

3 The Myristoylated, Alanine-Rich C Kinase Substrate (MARCKS)

The 68-kDa protein whose synthesis and myristoylation is induced by LPS proved to be an acidic PKC substrate which is the murine macrophage homologue of the 80- to 87-kDa PKC substrate. The protein was first described as an 87-kDa protein in rat brain synaptosomes, where it was found to be phosphorylated in response to phorbol esters and potassium depolarization (WU et al. 1982; ALBERT et al. 1987; PATEL and KLIGMAN 1987). In fibroblasts, where it is called the 80-kDa protein, it is the major protein phosphorylated when the cells are treated with growth factors or with phorbol esters (ROZENGURT et al. 1983; BLACKSHEAR et al. 1986). Indeed, the 68- to 87-kDa protein is so ubiquitously distributed that its phosphorylation is synonymous with PKC activation, and its phosphorylation has been used as an assay for the intracellular activation of the kinase. The protein was first shown to be myristoylated in macrophages (ADEREM et al. 1988a) and this observation has been confirmed in a variety of cell types (JAMES and OLSON 1989; THELEN et al. 1990). This modification, together with its high proportion of alanine, led to the acronym MARCKS (for *m*yristoylated, *a*lanine- *r*ich *C* *k*inase *s*ubstrate) (STUMPO et al. 1989).

3.1 Primary Structure of MARCKS

The murine macrophage MARCKS gene encodes a 309 amino acid protein with a calculated molecular mass of 29.6 kDa and a theoretical pl of 4.1 (SEYKORA et al. 1991). The calculated pl is identical to that of the purified murine brain protein and to that of MARCKS immunoprecipitated from murine macrophages (ROSEN et al. 1989). On the other hand, the calculated molecular mass of 29.6 kDa is at variance with the apparent molecular mass of 68 kDa obtained from SDS-PAGE analyis (ADEREM et al. 1988a) Transfection of the complete coding region into TK-L cells produces a protein which migrates with an apparent molecular mass of 68 kDa on SDS-PAGE (SEYKORA et al. 1991). The anomalous migration of MARCKS in SDS-PAGE, which results in the discrepancy between the actual and apparent molecular mass of the protein, is attributable to its large Stoke's radius and to its rod-shaped dimensions (ALBERT et al. 1987; J. HARTWIG et al. 1992). The cDNAs for bovine brain and chicken brain MARCKS encode proteins of 31.9 and 28.7 kDa, respectively, which migrate on SDS-PAGE with apparent molecular masses of 87 and 67 kDa (STUMPO et al. 1989; GRAFF et al. 1989b). The coding region of the murine MARCKS gene exhibits a highly repetitive structure that is manifested in the protein as short amino acid motifs which are repeated numerous times throughout the length of the molecule. For example, the element Pro-Ala-Ala-(Ala) is repeated seven times in the molecule and the element (Ala)-Ala-Ala-Pro is repeated three times. These repetitive motifs are reminiscent of highly structured linear molecules such as collagen (PROCKOP 1990) and analysis of the protein's secondary structure by the method of Garnier suggests that it is approximately 76% helical, consistent with our rotary shadowing data, which define MARCKS as a rod-shaped molecule with the dimension of 33 nm × 2.5 nm (J. HARTWIG et al. 1992). The amino acid composition of murine MARCKS is unusual: alanine accounts for 28.8% of the residues while glutamate and proline constititute 16.5% and 11.0% of the residues. Ten amino acids (Ala, Glu, Pro, Gly, Ser, Lys, Gln, Thr, Asp, and Phe) represent more than 95% of the residues.

3.2 Domain Structure of MARCKS

Comparison of the primary sequences of the murine, bovine, and chicken MARCKS reveals that the N-terminal and the phosphorylation domains are highly conserved, whereas the remainder of the protein is divergent (STUMPO et al. 1989; GRAFF et al. 1989b; SEYKORA et al. 1991). It would be reasonable to predict that these two conserved domains are important for the basic function of the MARCKS molecule. This appears to be the case. The N-terminal domain contains a myristoylation consensus sequence consisting of Gly-Ala-Gln-Phe-Ser-Lys-Thr-Ala which is necessary, but not sufficient, for membrane attachment of the protein (GRAFF et al. 1989; M. THELEN et al. 1991) (Fig. 1). The conserved domain spanning amino acids 128–180 of murine MARCKS contains all the known, phosphorylation sites of the protein (serines 152, 156, 163), as well as the calmodulin- (GRAFF et al. 1989c) and actin-binding sites (see below) (Fig. 1). The predicted secondary structure of the non-conserved regions of MARCKS is primarily helical, and this structure is further

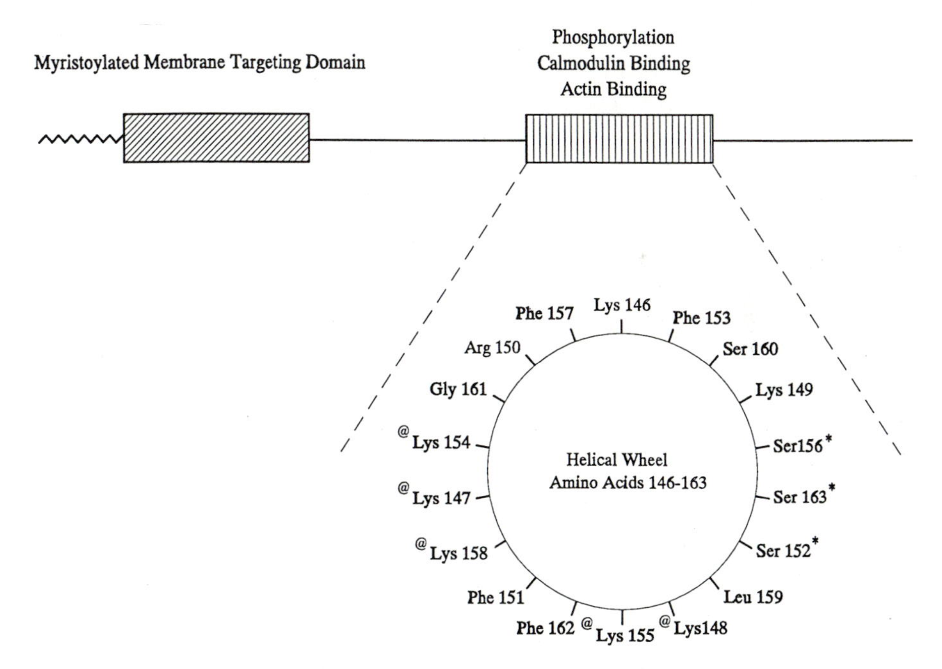

Fig. 1. The domain structure of MARCKS. MARCKS has two highly conserved domains. The myristolated N-terminal membrane-targeting domain is indicated by *diagonal lines*, and the highly charged phosphorylation domain, which also bears the calmodulin- and actin-binding sites, is denoted by *vertical lines*. The phosphorylation domain is predicted to be α-helical and a helical wheel representation of amino acids 146–163 of the murine protein is depicted (SEYKORA et al. 1991). The five lysine residues (indicated with @ positioned on one side of the helix form the calmodulin- and actin-binding sites. The phosphorylatable serines (indicated with *) are positioned at the opposite side of the helix. The depicted peptide binds calmodulin and actin in vitro and this binding is regulated by PKC-dependent phosphorylation

reinforced by rod-shaped dimensions of the protein. All the data point to a functional domain organization of MARCKS consisting of a membrane-binding, N-terminal, region which is separated by a helical region from the calmodulin- and actin-binding, phosphorylation domain (Fig. 1). Mutational analysis is currently underway, which will define precisely these structure–function relationships.

3.3 The MARCKS Gene

Southern analysis of murine genomic DNA reveals that the MARCKS gene is most likely present at a single copy per haploid genome and has a simple gene structure (STUMPO et al. 1989; GRAFF et al. 1989b; SEYKORA et al. 1991). The appearance of weakly hybridizing bands on Southern blot analysis shows that some genes bear a moderate degree of homology to the MARCKS gene and suggest the possibility of a family of related proteins.

3.4 Regulation of MARCKS Gene

Highest levels of MARCKS mRNA are found in brain and spleen, intermediate levels are found in kidney and heart, and very low levels are expressed in the liver. All mouse tissues examined express multiple MARCKS transcripts, which are due to differential polyadenylation and incomplete processing (SEYKORA et al. 1991).

The steady state levels of MARCKS mRNA in murine macrophages are increased 20- to 50-fold by prior exposure of the cells to LPS (SEYKORA et al. 1991). This is consistent with our previous demonstration that the MARCKS protein is strongly induced by LPS in murine macrophages and in human neutrophils (ADEREM et al. 1988a; THELEN et al. 1990). In addition, MARCKS is also induced by TNF-α in human neutrophils, where it constitutes approximately 90% of all protein synthesized in response to this cytokine (THELEN et al. 1990). The fact that MARCKS constitutes the majority of all protein synthesized in response to TNF-α suggests that it has a role to play in TNF-α-dependent signal transduction. This is supported by the observation that chemotactic peptide-dependent activation of PKC results in much higher levels of phosphorylated MARCKS in TNF-α-primed neutrophils, compared with unprimed cells (THELEN et al. 1990). Interestingly, while TNF-α stimulates the transcription and translation of MARCKS, it does not promote its phosphorylation in the absence of a second stimulus (THELEN et al. 1990). Since the phosphorylation of MARCKS is synonymous with the activation of PKC, it follows that TNF-α does not directly activate PKC, but rather, that it modifies PKC-dependent pathways by inducing the synthesis of an effector substrate of the kinase.

3.5 LPS Regulates the Phosphorylation of MARCKS

Lipopolysaccharide promotes the synthesis and myristoylation of MARCKS and concomitantly alters its profile of phosphorylation (ADEREM et al. 1988a). Activators of PKC induce the much more rapid phosphorylation of MARCKS in LPS-treated macrophages than in control cells. In addition, thermolytic mapping of the phosphorylated protein reveals that treatment of macrophages with LPS induces the phosphorylation of the protein on a novel site (ROSEN et al. 1989). The three serine residues known to be phosphorylated have been identified and are indicated by asterisks in Fig. 1.

3.6 Cycles of Phosphorylation and Dephosphorylation Govern the Reversible Association of MARCKS with the Plasma Membrane

The subcellular distribution of MARCKS is unusual in that the vast majority of the myristic acid-labeled protein is associated with the plasma membrane in quiescent cells, while most of the phosphorylated protein is found in the cytosol of activated cells (ADEREM et al. 1988a). These data suggested a model in which myristic acid targets MARCKS to the membrane where it comes in close apposition with PKC. Upon activation, PKC phosphorylates MARCKS, resulting in the release of the protein from the membrane (ADEREM 1988). This hypothesis has been confirmed experimentally. First, mutational analysis showed that the myristic acid moeity is required for the stable attachment of MARCKS with the membrane (GRAFF et al. 1989a). Second, activation of PKC in either macrophages or neutrophils results in the displacement of the myristoylated protein from the membrane, and this occurs without the deacylation of the protein (M. THELEN et al. 1991). This also occurs in an in vitro reconstituted system where phosphorylation of MARCKS by purified PKC promotes its release from isolated membranes (M. THELEN et al. 1991).

Since cytosolic MARCKS which has been released from the membrane by phosphorylation still contains its myristic acid, membrane-targeting, moeity, we investigated whether dephosphorylation is accompanied by the reassociation of the protein with the plasma membrane. Since phorbol myristate acetate (PMA) activates PKC irreversibly, studies on the reversibility of membrane binding of MARCKS were undertaken using a chemotactic peptide which stimulates transient PKC-dependent responses in human neutrophils via a receptor, G-protein coupled pathway (DEWALD et al. 1988). Treatment of neutrophils with f-Met-Leu-Phe results in the rapid, but transient, phosphorylation of MARCKS which is accompanied by its release from the plasma membrane and its accumulation in the cytosol. After 40 stimulation with f-Met-Leu-Phe the equilibrium between kinase and phosphatase shifts to favor the dephosphorylation of MARCKS. As dephosphorylation proceeds there is a concomitant reassociation of the protein with the membrane, such that most of the protein is membrane-bound when phosphorylation has returned to basal levels (M. THELEN et al. 1991). This cycle of membrane release and attachment is not influenced by cycloheximide, indicating that de novo synthesis of MARCKS does not account for the increase in the membrane-bound form of the protein observed upon dephosphorylation.

Agents which shift the equilibrium of the steady state level of phosphorylated MARCKS also shift the equilibrium of membrane binding of the protein. Okadaic acid, a specific phosphatase inhibitor (BIALOJAN and TAKAI 1988; HAYSTEAD et al. 1989), blocks both the dephosphorylation of MARCKS and its reassociation with the plasma membrane. When dephosphorylation is accelerated by addition of the receptor antagonist, boc-met-leu-phe, the shift in the equilibrium towards the dephosphorylated species is accompanied by increased membrane binding (M. THELEN et al. 1991). The cycle of membrane attachment and detachment of MARCKS is illustrated in Fig. 2.

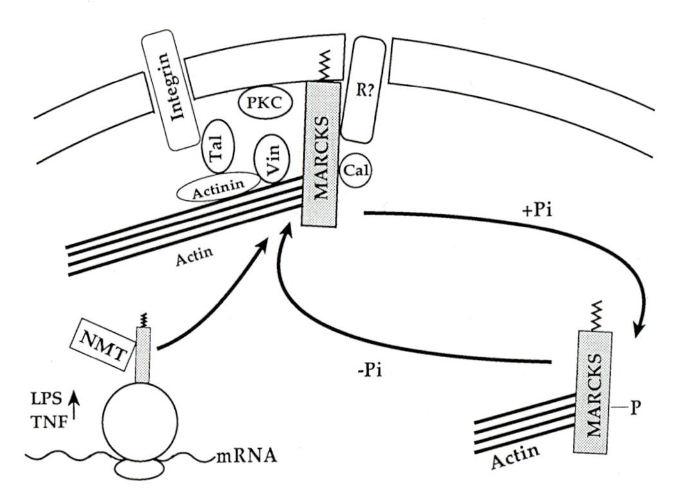

Fig. 2. A model indicating a possible role for MARCKS in macrophages. LPS and TNF-α induce the transcription and translation of MARCKS, following which it is myristoylated by the *N*-myristoyl transferase (*NMT*). MARCKS binds to the cytoplasmic face of the substrate-adherent plasma membrane, where it colocalizes with vinculin (*Vin*), talin (*Tal*), and PKC. MARCKS must be myristoylated for effective membrane binding and it may physically associate with a "myristoyl-protein" receptor (*R?*). MARCKS binds both actin and calmodulin/calcium (*Cal*). Upon activation of PKC, MARCKS is phosphorylated and this results in its translocation from the membrane to the cytoplasm, where it remains associated with F-actin. When MARCKS is dephosphorylated, it reassociates with the plasma membrane, and it may therefore provide a regulated cross-bridge between F-actin and the substrate-adherent plasma membrane

3.7 A Myristoyl-MARCKS Receptor at the Plasma Membrane?

The capacity of MARCKS to shuttle to and from the plasma membrane, together with the observation that MARCKS is located in discrete punctate structures at the substrate-adherent surface of macrophage filopodia (ROSEN et al. 1990) (see below), suggests that the protein associates with a receptor at the cytoplasmic surface of the plasma membrane, rather than through the nonspecific insertion of a hydrophobic moiety in the lipid bilayer. Some insight into the mechanism of membrane attachment of MARCKS might be gleaned by comparing it to the proto-oncogenic tyrosine kinase, $p60^{src}$, which also requires covalent modification with myristic acid for effective targeting to focal adhesions at the plasma membrane (PELLMAN et al. 1985; SHRIVER and ROHRSCHNEIDER 1981; ROHRSCHNEIDER and ROSOK 1983). While myristic acid is necessary for membrane attachment of $p60^{src}$, it is not sufficient, since a transformation-defective mutant of $p60^{src}$ has been described which does not associate with the membrane despite its myristoylation. Direct evidence for a "myristoyl-*src*" receptor was obtained recently in experiments which demonstrated specific and saturable binding of $p60^{v\text{-}src}$ to plasma membranes in vitro (RESH 1989; GODDARD et al. 1989). Binding was dependent on myristoylation and was inhibited competitively by a myristoylated peptide corresponding to the first 11 amino acids of $p60^{v\text{-}src}$ but not by the nonmyristoylated peptide or by myristoylated peptides derived from the sequences of other myristoylated proteins. The myristoylated *src* peptide can be crosslinked to a 32-kDa membrane protein which may well prove to be a "myristoyl-*src*" receptor (RESH and LING 1990). Further evidence for a "myristoylated protein receptor" comes from another member of the *src* family of proto- oncogenes, $p56^{lck}$, which has been shown to associate with the CD4 and CD8 T-lymphocyte surface glycoproteins (BARBER et al. 1989; VEILLETTE et al. 1988). The myristic acid-moiety is necessary (but not sufficient) for its attachment to CD4, and in this sense, CD4 represents the first myristoyl-protein receptor to be defined.

The cycle of membrane attachment and detachment described above for MARCKS may also extend to other myristoylated PKC substrates such as $p60^{src}$ and $p56^{lck}$, and it will be interesting to determine whether their association with the membrane is regulated in a similar way. The recent observation that phosphorylated $p56^{lck}$ has a lower affinity for CD4 than its nonphosphorylated counterpart is consistent with this proposal (HURLEY et al. 1989).

3.8 MARCKS Localizes to Points of Focal Adhesion in Macrophage Filopodia

We have investigated the subcellular location of MARCKS in macrophages by immunofluorescence microscopy. The protein has a punctate distribution at the substrate-adherent surface of macrophage pseudopodia and filopodia (ROSEN et al. 1990). When the cells are mechanically dislodged from the substratum, MARCKS continues to stain with a punctate pattern in residual patches of substrate-adherent membrane (ROSEN et al. 1990). Further evidence that the punctate staining pattern of MARCKS correlates with cellular adherence derives from an examination of murine resident peritoneal macrophages which are morphologically heterogeneous. MARCKS stains diffusely in rounded cells, while in elongated cells, punctate staining in membrane extensions is prominent (ROSEN et al. 1990). Since the punctate staining is due to association of MARCKS with areas of the macrophage membrane that are tightly associated with the substratum, we examined whether MARCKS colocalized with any components of focal adhesions. Many of the structures containing MARCKS also stain for vinculin and talin (ROSEN et al. 1990). A punctate distribution for vinculin and talin is in contrast to their typical plaque-like distribution in adherent fibroblasts (BURRIDGE et al. 1987), and has previously been observed in macrophages (MARCHISIO et al. 1987) and at the active cell edge of fibroblasts during the early stages of fibroblast attachment (BERSHADSKY et al. 1985). Punctate adhesions also predominate in fibroblasts transformed by Rous sarcoma virus (RSV), where adhesion to the substratum is less well developed (BERSHADSKY et al. 1985). It seems likely, therefore, that MARCKS is located in the initial, more transient adhesion complex that is formed at the leading edge of the motile macrophage.

Immunoelectron microscopy has further increased the resolution of our light microscopic observations. MARCKS is found in clusters near the plasma membrane in Lowicryl-embedded sections of fixed macrophages. To determine whether the immunogold particles that are bound near the membrane in cell sections are also associated with cytoskeletal fibers, we localized MARCKS in mechanically unroofed cells (A. ROSEN et al., submitted for publication). This technique preserves both membrane and its associated cytoskeletal elements, thereby allowing connections between filaments and the cytoplasmic side of the plasma membrane to be visualized (HARTWIG et al. 1989). Anti-MARCKS gold label is found in small, widely spaced clusters at points where actin filaments interact with the cytoplasmic surface of the plasma membrane.

3.9 LPS Increases Filopodia Formation and Induces MARCKS Synthesis and Clustering

Lipopolysaccharide induces the net synthesis of MARCKS (ADEREM et al. 1988a) and greatly increases the adherence of the cultured mouse macrophages to surfaces. This is accompanied by a striking increase in the number and prominence of lamellipodia, filopodia, and membrane veils (A. ROSEN et al., submitted for publication). Stimulation of macrophages with LPS also results in a dramatic increase in the number of punctate structures containing MARCKS, which are seen over the entire substrate-adherent surface of lamellipodia as well as along filopodia and veils of membrane. Electron microscopy confirms the immunofluorescence data, showing that the total amount of membrane-associated MARCKS, as well as the concentration of MARCKS in the clusters, increases. The close correlation between synthesis and acylation of MARCKS and its enrichment in pseudopodia and filopodia suggest that MARCKS might serve a function in these plastic structures.

The pronounced staining of MARCKS in filopodia, and the presence in filopodia of a conspicuous, central actin bundle (MITCHISON and KIRSCHNER 1988), prompted us to examine the distribution of MARCKS and actin in the same cell. Double staining and electron microscopy experiments show MARCKS spotted alongside filopodial actin bundles in LPS-treated macrophages (A. ROSEN et al., submitted for publication). Furthermore, clusters of MARCKS are intimately associated with the cytoplasmic surface of the plasma membrane in unroofed cells at points where multiple filaments contact the cytoplasmic surface of the plasma membrane.

3.10 The Effect of PKC Activation on the Distribution of MARCKS

Immunofluorescence studies have indicated that PKC is also a component of the substrate-adherent punctate structures which contain MARCKS, vinculin, and talin (ROSEN et al. 1990). This is not surprising, since it is known that PKC is a component of focal adhesions in fibroblasts (JAKEN et al. 1986), and that PKC very rapidly phosphorylates MARCKS when the cell encounters an activating stimulus (ROSEN et al. 1989). Activation of PKC with phorbol esters causes marked cell spreading and rounding and the almost complete disappearance of filopodia. This is accompanied by disappearance of punctate staining of MARCKS, and by a modest increase in the level of diffuse staining observed. The kinetics of the morphologic changes and the disappearance of punctate staining of MARCKS (ROSEN et al. 1990) mirror the kinetics of phosphorylation of MARCKS by PKC in intact macrophages (ROSEN et al. 1989), and the translocation of MARCKS from the membrane to the cytosol (ROSEN et al. 1990) (see above). In contrast, the effect of PMA on the distribution of vinculin and talin is quite distinct. After PKC treatment, vinculin and talin still stain prominently with a patchy, punctate organization and are particularly obvious at the phase-dense cell edge (ROSEN et al. 1990).

Immunoelectron microscopy confirms that the amount of plasma membrane-associated MARCKS markedly diminishes upon activation of PKC while cytoplasmic levels of the protein increase proportionally. Unexpectedly, experiments in mechanically unroofed macrophages show that MARCKS remains associated with the sides of actin filaments in PMA-treated cells, but that the filaments to which the gold label is attached are now spatially displaced from the membrane surface. MARCKS thus appears to be a component of the membrane-associated cytoskeleton, and is closely associated with actin filaments in this structure. Phosphorylation of MARCKS results in its translocation from the plasma membrane into the cytosol, where it appears still to be closely associated with actin filaments. This phosphorylation-regulated translocation from the plasma membrane is accompanied by the marked reorganization of the actin cytoskeleton that occurs upon activation of PKC (PHAIRE-WASHINGTON et al. 1980; SCHLIWA et al. 1984). The phosphorylation-dependent regulation of membrane binding of MARCKS might therefore serve to modify the attachment of actin filaments to the membrane, thereby influencing cytoskeletal organization and morphology in response to signals that activate PKC.

3.11 MARCKS Is an Actin-Binding Protein

The data described above suggested a role for MARCKS in actin-based motility. It was therefore important to determine whether MARCKS binds actin in vitro and whether this binding is influenced by phosphorylation. Both dephosphorylated MARCKS, and MARCKS phosphorylated in vitro with PKC, bind to F-actin in a cosedimentation assay (J. HARTWIG et al. 1992). Scatchard analysis indicates that the binding affinity of both phospho- and dephospho-MARCKS for actin is in the micromolar range. The MARCKS protein also appears to interact with actin in situ since actin copurifies with MARCKS through numerous chromatography and centrifugation steps and the two proteins can only be separated by reverse-phase chromatography.

3.12 Dephosphorylated MARCKS Is an Actin Filament Crosslinking Protein While Phosphorylated MARCKS Is Not

The interaction of MARCKS with actin filaments was studied in the electron microscope after negative staining. When actin is incubated at a ratio of 1 molecule of dephospho-MARCKS to 10 actin subunits, filaments become aggregated and are decorated with rod-shaped MARCKS molecules which attach along the actin fibers near the midpoints of the rods. At molar excess of MARCKS to actin subunits, the actin filaments become highly entangled and are now periodically decorated every—20 nm with dephospho-MARCKS molecules. Actin filaments in the absence of MARCKS or incubated in the presence of phospho-MARCKS appear bare and unaggregated. Thus, while phosphorylation does not diminish the binding of MARCKS to the actin filament, it does change the manner in which the MARCKS molecule binds to the actin filament (J. HARTWIG et al. 1992).

The effect of MARCKS binding on the function of F-actin, and alterations mediated by phosphorylation, were further evaluated by morphologic, optical, and hydrodynamic methods. As discussed above, dephosphorylated MARCKS crosslinked actin filaments into loose aggregates which were easily discernible by electron microscopy. The addition of molar ratios of $\leqslant$ 1 MARCKS to 20 actin filament subunits also increases the viscosity of actin, consistent with filament crosslinking. Higher ratios of dephosphorylated MARCKS to actin, however, decrease the viscosity of actin solutions relative to actin alone. These results suggest that low ratios of MARCKS link actin into a network and that higher concentrations aggregate filaments into bundles. Dynamic light scattering confirms that MARCKS causes the lateral alignment of actin filaments into larger structures. In contrast, phosphorylated MARCKS has minimal effects on the viscosity or light scatter intensity of F-actin solutions. These studies demonstrate that MARCKS binds to the sides of actin filaments and crosslinks them. High concentrations of the phospho- or dephosphoprotein do not affect the extent of linear actin assembly, making it unlikely that the MARCKS interacts with the ends of actin polymers. The stoichiometry of binding also indicates that the interaction of MARCKS with actin is not limited to the ends of polymers (J. HARTWIG et al. 1992).

3.13 Calmodulin Inhibits the Actin Crosslinking Activity of MARCKS

Nonphosphorylated MARCKS binds calmodulin in the presence of calcium while the phosphorylated protein does not (GRAFF et al. 1989d). In addition, calmodulin prevents PKC-dependent phosphorylation of MARCKS (ALBERT et al. 1984). Taken together, these data suggest that the calmodulin-binding site and the phosphorylation sites of MARCKS are closely linked. Since the phosphorylation of MARCKS inhibits its actin-crosslinking activity, it was of interest to determine whether calmodulin prevented the actin-crosslinking activity of nonphosphorylated MARCKS. This proves to be the case: stoichiometric amounts of calmodulin completely inhibit the actin-crosslinking activity of nonphosphorylated MARCKS (J. HARTWIG et al. 1992).

3.14 The Synthetic Phosphorylation Domain of MARCKS Crosslinks Actin in a PKC- and Calmodulin-Regulated Manner

As discussed above, the MARCKS protein contains two highly conserved domains: an N-terminal membrane-binding domain which is myristoylated, and a highly charged phosphorylation domain which also contains the calmodulin-binding site (GRAFF et al. 1989d) (See Fig. 1). A synthetic peptide which contains all the serine residues known to be phosphorylated, and corresponds to residues 146–163 of murine MARCKS (see Fig. 1) (SEYKORA et al. 1991), increases both the light scatter intensity of the F-actin in solution and aggregates actin filaments into tight bundles which are observable in the electron microscope. Phosphorylation of the serine residues in this peptide, however, prevents the peptide from crosslinking actin filaments, as determined by light scattering and viscosity measurements and by electron microscopy (J. HARTWIG et al. 1992). Phosphorylation, therefore, modulates the ability of this peptide to crosslink actin. Since calmodulin inhibits the actin-crosslinking activity of nonphosphorylated MARCKS, and since the synthetic peptide contains the calmodulin-binding site (GRAFF et al. 1989d), the effect of calmodulin on the actin-bundling activity of the peptide was determined. Calmodulin, at a molar ratio to the peptide of 1:1, completely inhibits the actin-bundling activity of the nonphosphorylated peptide.

The synthetic peptide containing the phosphorylation sites, the calmodulin-binding site, and the actin-binding site is predicted to form an α-helix in aqueous solution. A helical wheel representation of this domain of the murine protein is depicted in Fig. 1. The lysine residues (indicated with @) positioned on one side of the helix form a consensus calmodulin-binding site (O'NEIL and DEGRADO 1990) and are similarly positioned to the lysine residues found in known actin-binding domains (VANDEKERCKHOVE 1989). The phosphorylatable serines (indicated with *) are positioned on the opposite side of the helix. It is likely that the lysine residues comprise both the actin- and calmodulin-binding sites of MARCKS and this hypothesis is supported by the observation that calmodulin binding prevents actin-crosslinking activity. The crosslinking of actin filaments would require monomeric molecules to express a second actin-binding site or to dimerize. It is not yet clear whether the synthetic peptide has a second actin-binding site or whether it forms multimers.

The interaction of MARCKS with actin filaments is altered by phosphorylation or by calmodulin/calcium; although binding is not eliminated, filament crosslinking is. Phosphorylation or calmodulin/calcium, therefore, appears to inactivate one of the actin-binding site(s) contained between residues 146 and 163 of the murine protein.

During neurosecretion, leukocyte activation, and growth factor-dependent mitogenesis, cells undergo cytoskeletal rearrangements while MARCKS is both rapidly phosphorylated and redistributed to the cytoplasm from the cytoplasmic surface of the plasma membrane (ADEREM et al. 1988a; ROSEN et al. 1990; WANG et al. 1989). Upon dephosphorylation of MARCKS, it reassociates with the plasma membrane, demonstrating it to reversibly cycle to and from the plasma membrane (M. THELEN et al. 1991). Stimulation of leukocytes with chemotactic peptide results in the activation of PKC (THELEN et al. 1990) and an increase in cytosolic calcium levels. This generates two signals which can modify the interaction of MARCKS with actin and the attachment of MARCKS to the plasma membrane. Since MARCKS binds to points of plasma membrane-substrate adherence, it may be important in physically coupling membrane stimulation to cell movement.

4 The 48-kDa Myristoylated PKC Substrate; a Candidate Effector of IFN-γ-Induced Responses in Macrophages

During cell-mediated immunity, macrophages acquire an increased capacity to secrete reactive oxygen intermediates and to kill microbes and tumor cells. Treatment of macrophages with IFN-γ, and with other cytokines such as interleukin-4 and granulocyte-macrophage colony stimulating factor, induces properties that are similar to those of macrophages activated in situ (NATHAN et al. 1983; SCHREIBER et al. 1983). However, the intracellular events leading to the activation of macrophages by antigen-specific T cells are largely unknown, although it is clear that PKC has a role in these events [reviewed in (ADAMS and HAMILTON 1987)]. We have shown that physiologic concentrations of IFN-γ greatly enhance the induction of a 48-kDa myristoylated PKC substrate (ADEREM et al. 1988b). The protein contains the fatty acid in an amide linkage to an N-terminal glycine and myristoylation appears to occur cotranslationally since it is rapidly blocked by inhibitors of protein synthesis. Myristoylation of the 48-kDa PKC substrate is observed within 1 h and reaches a maximum by 3–4 h (ADEREM et al. 1988b). These kinetics are relatively rapid when compared with the majority of IFN-γ-induced responses in macrophages, which are generally observed 8–48 h after exposure to the cytokine (ADAMS and HAMILTON 1987). The kinetics are similar to those observed for IFN-γ-dependent potentiation of PKC activity in macrophages (HAMILTON et al. 1985), and it is possible that the 48-kDa protein may be an effector of the PKC- dependent signaling pathway leading to macrophage activation.

Interferon-α, -β, and -γ share a number of properties, including the induction of antiviral and antiproliferative states in a number of cell types, but IFN-γ is far more effective than IFN-α or -β for activation of macrophage oxidative metabolism and antiprotozoal activity (NATHAN et al. 1984). The induction of the 48-kDa PKC substrate is quite specific for IFN-γ, suggesting that this protein is related to the macrophage-activating capacity of the lymphokine and not to its antiviral or antiproliferative activity.

Consistent with it having a role in macrophage activation, the 48-kDa protein is constitutively induced in macrophages activated in vivo by intraperitoneal injection of formalin-killed *Corynebacterium parvum*. Indeed, we have been able to use the induction of the 48-kDa protein as an index of macrophage activation; an increase in the apparent "basal" level of this protein always correlates with increased numbers of activated macrophages in the peritoneal cavity, as evidenced by an increase in the number of cells expressing surface Ia molecules and by a decrease in the capacity of the macrophages to secrete leukotriene C_4 (ADEREM et al. 1988c).

The LPS-primed macrophage has proved an excellent model system in which to study the convergence of three distinct signal transduction systems mediated by protein myristoylation, PKC-mediated phosphorylation, and calcium/calmodulin. The biologic readouts are clear and distinct, and since the cells are terminally differentiated, the signals are not obscured by those involved in the regulation of the cell cycle. A number of myristoylated PKC substrates that are likely to have a role in LPS-, TNF-α-, and IFN-γ-induced signal transduction pathways have been identified. They include the MARCKS protein, which appears to regulate the reversible attachment of actin filaments to the substrate-adherent plasma membrane, the 48-kDa protein, which appears to have a role in immune activation of macrophages, and the macrophage-specific 42-, 40-kDa proteins which are induced by LPS. Future studies will concentrate on the identity and function of these proteins.

Acknowledgements. The work described above owes much to many. The members of my research group, past and present, did all the work, were dedicated, and persistent. More importantly, they knew how to have a good time. The group includes Antony Rosen, Marcus Thelen, John Seykora, Jianxun Li, Karen Keenan, Thoru Sato, Doug Marratta, Min Keum, and more recently Jennifer Darnell and Lee-Anne Allen. I also wish to acknowledge and thank Angus Nairn, Jeffrey Ravitch, John Hartwig, and Paul Janmey, who were essential collaborators and good friends. Thanks go to Kathy Barker for critically reading this manuscript. This work was supported by National Institutes of Health AI 25032 and by grants-in-Aid from the Squibb Institute for Medical Research and the Cancer Research Institute. Alan Aderem is an Established Investigator of the American Heart Association.

References

Adams DO, Hamilton TA (1987) Molecular transductional mechanisms by which IFN-γ and other signals regulate macrophage development. Immunol Rev 97: 5–27

Aderem AA (1988) Protein myristoylation as an intermediate step during signal transduction in macrophages: its role in arachidonic acid metabolism and in responses to interferon. J Cell Sci [Suppl] 9: 151–167

Aderem AA, Cohn ZA (1988) Calcium ionophore synergizes with bacterial lipopolysaccharides in activating macrophage arachidonic acid metabolism. J Exp Med 167: 623–631

Aderem AA, Cohen DS, Wright SD, Cohn ZA (1986a) Bacterial lipopolysaccharides prime macrophages for enhanced release of arachidonic acid metabolites. J Exp Med 164: 165–179

Aderem AA, Keum MM, Pure E, Cohn ZA (1986b) Bacterial lipopolysaccharides, phorbol myristate acetate, and zymosan induce the myristoylation of specific macrophage proteins. Proc Natl Acad Sci USA 83: 5817–5821

Aderem AA, Scott WA, Cohn ZA (1986c) Evidence for sequential signals in the induction of the arachidonic acid cascade in macrophages. J Exp Med 163: 139–154

Aderem AA, Albert KA, Keum MM, Wang JKT, Greengard P, Cohn ZA (1988a) Stimulus-dependent myristoylation of a major substrate for protein kinase C. Nature 332: 362–364

Aderem AA, Marratta DE, Cohn ZA (1988b) Interferon-gamma induces the myristoylation of a 48 k protein in macrophages. Proc Natl Acad Sci USA 85: 6310–6313

Aderem AA, Rosen A, Barker KA (1988c) Modulation of prostaglandin and leukotriene biosynthesis. Current Opinion in Immunology 1: 56–62

Aikten A, Cohen P, Santikarn S, Williams DH, Calder AG, Smith A, Klee CB (1982) Identification of the NH2-terminal blocking group of calcineurin B as myristic acid. FEBS Lett 150: 314–318

Albert KA, Wu WS, Nairn AC, Greengard P (1984) Inhibition by calmodulin of calcium/phospholipid-dependent protein phosphorylation. Proc Natl Acad Sci USA 81: 3622–3625

Albert KA, Nairn AC, Greengard P (1987) The 87 k protein, a major specific substrate for protein kinase C: purification from bovine brain and characterization. Proc Natl Acad Sci USA 84: 7046–7050

Barber ER, Dasgupta JD, Schlossman SF, Trevillyan JM, Rudd CE (1989) The CD4 and CD8 antigens are coupled to a protein-tyrosine kinase (p56 ck) that phosphorylates the CD3 complex. Proc Natl Acad Sci USA 86: 3277–3281

Bershadsky AD, Tint IS, Neyfakh AA Jr, Vasiliev JM (1985) Focal contacts of normal and RSV-transformed quail cells. Exp Cell Res 158: 433–444

Bialojan C, Takai A (1988) Inhibitory effect of a marine-sponge toxin, okadaic acid, on protein phosphatases. Biochem J 256: 283–290

Blackshear PJ, Wen L, Glynn BP, Witters LA (1986) Protein kinase C-stimulated phosphorylation in vitro of a Mr 80,000 protein phosphorylated in response to phorbol esters and growth factors in intact fibroblasts. Distinction from protein kinase C and prominence in brain. J Biol Chem 261: 1459–1469

Burridge K, Molony L, Kelly T (1987) Adhesion plaques: sites of transmembrane interaction between the extracellular matrix and the actin cytoskeleton. J Cell Sci 8: 211–229

Buss JE, Sefton BM (1985) Myristic acid, a rare fatty acid, is the lipid attached to the transforming protein of Rous sarcoma virus and its cellular homolog. J Virol 53: 7–12

Buss JE, Mumby SM, Casey PJ, Gilman AG, Sefton BM (1987) Myristylated alpha subunits of guanine nucleotide-binding regulatory proteins. Proc Natl Acad Sci USA 84: 7493–7497

Carr SA, Biemann K, Shoji S, Parmelee DC, Titani K (1982) *n*-Tetradecanoyl is the NH2-terminal blocking group of the catalytic subunit of cyclic AMP-dependent protein kinase from bovine cardiac muscle. Proc Natl Acad Sci USA 79: 6128–6131

Clark RA (1990) The human neutrophil respiratory burst oxidase. J Infect Dis 161: 1140–1147

Dewald B, Thelen M, Baggiolini M (1988) Two transduction sequences are necessary for neutrophil activation by receptor agonists. J Biol Chem 263: 16179–16184

Duronio RJ, Towler DA, Heuckeroth RO, Gordon JI (1989) Disruption of the yeast *N*-myristoyl transferase gene causes recessive lethality. Science 243: 796–800

Goddard C, Arnold ST, Felsted RL (1989) High affinity binding of an N-terminal myristoylated $p60^{src}$ peptide. J Biol Chem 264: 15173–15176

Graff JM, Gordon JI, Blackshear PJ (1989a) Myristoylated and nonmyristoylated forms of a protein are phosphorylated by protein kinase C. Science 246: 503–506

Graff JM, Stumpo DJ, Blackshear PJ (1989b) Molecular cloning, sequence, and expression of a cDNA encoding the chicken *m*yristoylated *a*lanine-*r*ich *C* *k*inase *s*ubstrate (MARCKS). Mol Endocrinol 3: 1903–1906

Graff JM, Stumpo DJ, Blackshear PJ (1989c) Characterization of the phosphorylation sites in the chicken and bovine myristoylated alanine-rich C kinase substrate protein, a prominent cellular substrate for protein kinase C. J Biol Chem 264: 11912–11919

Graff JM, Young TM, Johnson JD, Blackshear PJ (1989d) Phosphorylation- regulated calmodulin binding to a prominent cellular substrate for protein kinase C. J Biol Chem 264: 21818–21823

Hamilton TA, Adams DO (1987) Molecular mechanisms of signal transduction in macrophages. Immunol Today 8: 151–158

Hamilton TA, Becton DL, Somers SD, Gray PW, Adams DO (1985) Interferon-gamma modulates protein kinase C activity in murine peritoneal macrophages. J Biol Chem 260: 1378–1381

Hartwig JH, Chambers KA, Stossel TP (1989) Association of gelsolin with actin filaments and cell membranes of macrophages and platelets. J Cell Biol 109: 467–479

Hartwig JH, Thelen M, Rosen A, Janmey PA, Nairn AC, and Aderem A (1992) MARCKS is an actin filament crosslinking protein regulated by protein kinase C and calcium-calmodulin. Nature 356:618–622

Haystead TAJ, Sim ATR, Carling D, Honnor RC, Tsukitani Y, Cohen P, Hardie DG (1989) Effects of tumor promoter okadaic acid on intracellular protein phosphorylation and metabolism. Nature 337: 78–81

Hurley TR, Luo K, Sefton BM (1989) Activators of protein kinase C induce dissociation of CD4, but not CD8, from $p56^{lck}$. Science 245: 407–409

Jaken S, Leach K, Klauck T (1989) Association of type 3 protein kinase C with focal contacts in rat embryo fibroblasts. J Cell Biol 109: 697–704

James G, Olson EN (1989) Myristoylation, phosphorylation, and subcellular distribution of the 80-kDa protein kinase C substrate in BC3H1 myocytes. J Biol Chem 264: 20928–20933

Jones TLZ, Simonds WF, Merendino JJ Jr, Brann MR, Spiegel AM (1990) Myristoylation of an inhibitory GTP-binding protein α subunit is essential for its membrane attachment. Proc Natl Acad Sci USA 87: 568–572

Kikkawa U, Kishimoto A, Nishizuka Y (1989) The protein kinase C family: heterogeneity and its implications. Annu Rev Biochem 58: 31–44

Marchildon GA, Casnellie JE, Walsh KA, Krebs EG (1984) Covalently bound myristate in a lymphoma tyrosine protein kinase. Proc Natl Acad Sci USA 81: 7679–7682

Marchisio PC, Cirillo D, Teti A, Zambonin-Zallone A, Tarone G (1987) Rous sarcoma virus-transformed fibroblasts and cells of monocytic cell origin display a peculiar dot-like organization of cytoskeletal proteins involved in microfilament-membrane interactions. Exp Cell Res 169: 202–214

Mitchison T, Kirschner M (1988) Cytoskeletal dynamics and nerve growth. Neuron 1: 761–772

Mumby SM, Heukeroth RO, Gordon JI, Gilman AG (1990) G-protein α-subunit expression, myristoylation, and membrane association in COS cells. Proc Natl Acad Sci USA 87: 728–732

Nathan CF, Murray HW, Wiebe ME, Rubin BY (1983) Identification of interferon-gamma as the lymphokine that activates human macrophage oxidative metabolism and antimicrobial activity. J Exp Med 158: 670–689

Nathan CF, Prendergast TJ, Wiebe ME, Stanley ER, Platzer E, Remold HG, Welte K, Rubin BY, Murray HW (1984) Activation of human macrophages. Comparison of other cytokines with interferon-gamma. J Exp Med 160: 600–605

O'Neil KT, Degrado WF (1990) How calmodulin binds its targets: sequence independent recognition of amphiphilic alpha-helices. TIBS 15: 59–64

Olson EN, Spizz G (1986) Fatty acylation of cellular proteins. Temporal and subcellular differences between palmitate and myristate acylation. J Biol Chem 261: 2458–2466

Patel J, Kligman D (1987) Purification and characterization of an Mr 87,000 protein kinase C substrate from rat brain. J Biol Chem 262: 16686–16691

Pellman D, Garber EA, Cross FR, Hanafusa H (1985) An N-terminal peptide from p60src can direct myristoylation and plasma membrane localization when fused to heterologous proteins. Nature 314: 374–377

Phaire-Washington L, Silverstein SC, Wang E (1980) Phorbol myristate acetate stimulates micro-tubule and 10-nm filament extension and lysosome redistribution in mouse macrophages. J Cell Biol 86: 641–655

Prockop DJ (1990) Mutations that alter the primary structure of type I collagen. J Biol Chem 265: 15349–15352

Resh MD (1989) Specific and saturable binding of pp60v-src to plasma membranes: evidence of a myristyl-src receptor. Cell 58: 281–286

Resh MD, Ling H (1990) Identification of a 32 K plasma membrane protein that binds to the myristoylated amino-terminal sequence of $p60^{v\text{-}src}$. Nature 346: 84–86

Rohrschneider LR, Rosok MJ (1983) Transformation parameters and pp60src localization in cells infected with partial transformation mutants of Rous Sarcoma virus. Mol Cell Biol 3: 731–746

Rosen A, Nairn AC, Greengard P, Cohn ZA, Aderem AA (1989) Bacterial lipopolysaccharide regulates the phosphorylation of the 68 K protein kinase C substrate in macrophages. J Biol Chem 264: 9118–9121

Rosen A, Keenan KF, Thelen M, Nairn AC, Aderem AA (1990) Activation of protein kinase C results in the displacement of its myristoylated, alanine-rich substrate from punctate structures in macrophage filopodia. J Exp Med 172: 1211–1215

Rozengurt E, Rodriguez-Pena M, Smith KA (1983) Phorbol esters, phospholipase C, and growth factors rapidly stimulate the phosphorylation of a Mr 80,000 protein in intact quiescent 3T3 cells. Proc Natl Acad Sci USA 80: 7244–7248

Schliwa M, Nakamura T, Porter KR, Euteneuer U (1984) A tumor promoter induces rapid and coordinated reorganization of actin and vinculin in cultured cells. J Cell Biol 99: 1045–1059

Schmidt MFG (1989) Fatty acylation of proteins. Biochim Biophys Acta 988: 411–426

Schreiber RD, Pace JL, Russell SW, Altman A, Katz DH (1983) Macrophage-activating factor produced by a T cell hybridoma: physiochemical and biosynthetic resemblance to gamma-interferon. J Immunol 131: 826–832

Schultz AM, Henderson LE, Oroszlan S, Garber EA, Hanafusa H (1985) Amino terminal myristylation of the protein kinase p60*src*, a retroviral transforming protein. Science 227: 427–429

Schultz AM, Tsai SC, Kung HF, Oroszlan S, Moss J, Vaughan M (1987) Hydroxylamine-stable covalent linkage of myristic acid in G0 alpha, a guanine nucleotide-binding protein of bovine brain. Biochem Biophys Res Commun 146: 1234–1239

Schultz AM, Henderson LE, Oroszlan S (1988) Fatty acylation of proteins. Annu Rev Cell Biol 4: 611–647

Seykora JT, Ravetch JV, Aderem A (1991) Cloning and molecular characterization of the murine macrophage "68-kDa" protein kinase C substrate, and its regulation by bacterial lipopolysaccharide. Proc Natl Acad Sci USA (in press)

Shriver K, Rohrschneider LR (1981) Organization of pp60src and selected cytoskeletal proteins within adhesion plaques and junctions of Rous sarcoma virus-transformed rat cells. J Cell Biol 89: 525–535

Stumpo DJ, Graff JM, Albert KA, Greengard P, Blackshear PJ (1989) Molecular cloning, characterization, and expression of a cDNA encoding the "80- to 87-kDa" myristoylated alanine-rich C kinase substrate: a major cellular substrate for protein kinase C. Proc Natl Acad Sci USA 86: 4012–4016

Thelen M, Rosen A, Nairn AC, Aderem A (1990) Tumor necrosis factor α modifies agonist-dependent responses in human neutrophils by inducing the synthesis and myristoylation of a specific protein kinase C substrate. Proc Natl Acad Sci USA 87: 5603–5607

Thelen M, Rosen A, Nairn AC, and Aderem A (1991) Regulation by phosphorylation of reversible association of a myristoylated protein kinase C substrate with the plasma membrane. Nature 351:320–322

Towler DA, Adams SP, Eubanks SR, Towery DS, Jackson-Machelski E, Glaser L, Gordon JI (1987) Purification and characterization of yeast myristoyl CoA: protein *N*-myristoyl transferase. Proc Natl Acad Sci USA 84: 2708–2712

Towler DA, Gordon JI, Adams SP, Glaser L (1988) The biology and enzymology of eukaryotic protein acylation. Annu Rev Biochem 57: 69–99

Vandekerckhove J (1989) Structural principles of actin-binding proteins. Curr Opin Cell Biol 1: 15–22

Veillette A, Bookman MA, Horak EM, Bolen JB (1988) The CD4 and CD8 T cell surface antigens are associated with the internal membrane tyrosine-protein kinase p56lck. Cell 55: 301–308

Wang JKT, Walaas SI, Sihra TS, Aderem AA, Greengard P (1989) Phosphorylation and associated translocation of the 87-kDa protein, a major protein kinase C substrate, in isolated nerve terminals. Proc Natl Acad Sci USA 86: 2253–2256

Wilcox C, Hu JS, Olson EN (1987) Acylation of proteins with myristic acid occurs cotranslationally. Science 238: 1275–1278

Wu WS, Walaas SI, Nairn AC, Greengard P (1982) Calcium/phospholipid regulates phosphorylation of a Mr "87k" substrate protein in brain synaptosomes. Proc Natl Acad Sci USA 79: 5249–5253

Ribosomal RNA Metabolism in Macrophages

L. VARESIO[1], D. RADZIOCH[2], B. BOTTAZZI[3], and G. L. GUSELLA[4]

1 Introduction

Macrophages are ubiquitous cells critical for host defense (WILTROUT and VARESIO 1987; VARESIO 1985b). Metchnikoff's original report in the late nineteenth century first suggested that phagocytes are the body's prime detectors of foreign invaders; the functions of macrophages have since been delineated, initially at a cellular level and more recently at a molecular level.

Macrophages are active effector cells capable of tumoricidal and bactericidal activities and are integral regulatory components of the immune system, on which they exert stimulatory and inhibitory activities (VARESIO 1983, 1985b). The list of macrophage secretory products is impressive (NATHAN 1987) and clearly indicates that these cells learned to interact and communicate with the

[1] Immunobiology Section, Laboratory of Molecular Immunoregulation, Biological Response Modifiers Program, National Cancer Institute, Frederick Cancer Research Development Center, Frederick, MD 21702–1201, USA
[2] McGill Center for the Study of Host Resistance, Montreal General Hospital, Montreal, Quebec, Canada
[3] Instituto di Ricerche Farmacologiche Mario Negri, Milan, Italy
[4] Biological Carcinogenesis and Development Program, PRI/DynCorp, Frederick Cancer Research Development Center, Frederick, MD 21702–1201, USA

Current Topics in Microbiology and Immunology, Vol. 181

specialized and sophisticated structures that evolved around the ancestral ameba (VARESIO et al. 1980). In fact, there are no other cells as developmentally complex as macrophages that can adapt to so many different environments, including the secluded central nervous system, the variable lung, and the complex liver. As a result of the adaptation to different environments, of the contact with surrounding cells, and of the various stages of differentiation in which these cells may exist in the tissues, monocytic cells are a heterogeneous population with respect to morphologic and functional characteristics (LEUNG et al. 1985; GORDON et al. 1988; CROCKER et al. 1987).

One important feature of macrophage biology is that proliferation is not required and is not part of the activation program of the cell. Unlike activation of B or T lymphocytes, macrophage activation is not dependent upon clonal expansion or mitotic activity. The dissociation between macrophage activation and proliferation has been further documented by the demonstration that proliferating macrophage cell lines, immortalized by expression of various oncogenes, retain the biologic activities and characteristics of peritoneal macrophages (BLASI et al. 1985, 1987–1989; GANDINO and VARESIO 1990; COX et al. 1989). Under physiologic conditions, the expression of the activated phenotype is a transient event that is followed by a return of macrophages to a resting state (TAFFET et al. 1981). The magnitude of the macrophage-mediated response is determined mainly by number of responding cells and/or the accumulation of the macrophages into the inflammatory region, and it is modulated by the cellular feedback mechanisms and by the extent and nature of the environmental stimulation (ROSEN and GORDON 1990 a, b; CROCKER et al. 1988; PERRY and GORDON 1988). Proliferation is important to maintain an adequate supply of circulating monocytes ready to extravasate into the tissues in response to chemoattractants released during the inflammatory response. Monocytes can undergo limited proliferation in the circulation and in the tissues. However, the majority of circulating monocytes originate in the bone marrow from the proliferation of stem cells and their differentiation along the monocytic lineage. The proliferative ability of the cells decreases progressively during the differentiation (METCALF 1986). Thus, the majority of the mitotic activity is restricted to the bone marrow, and the balance of peripheral cells is dictated by cell death and replenishment from the bone marrow.

The recognition that proliferation is not part of the genetic program of a macrophage poses interesting questions of cell biology. Macrophages express receptors for a variety of growth factors specific for monocytic cells, such as the colony stimulating factor-1 (CSF-1) receptor (GUSELLA et al. 1990; BLASI et al. 1987), or typical of other lineages, such as the interleukin-2 (IL-2) receptor (COX et al. 1990; ESPINOZA-DELGADO et al. 1990a). However, macrophages respond to stimulation by growth factors with functional changes. For example, CSF-1 receptor is present in bone marrow monocyte precursors, circulating monocytes, and tissue macrophages. In bone marrow monocyte precursors, CSF-1 induces proliferation, while in monocytes and macrophages it induces secretory activity (MOORE et al. 1984; WARREN and RALPH 1986), resistance to viral infection and sustains cell viability and cytotoxic response (RALPH and NAKOINZ 1987; ESPINOZA-DELGADO et al. 1990b; LEE and WARREN 1987; SHERR 1990). Interestingly, in monocytes but not in macrophages, the CSF-1 receptor can be induced by IL-2 (ESPINOZA-DELGADO et al. 1990b), whereas in both cell types it is down-regulated by the ligand, phorbol esters, and endotoxins (GUSELLA et al. 1990; DOWNING et al. 1989). Thus, it appears that the differentiation from monocyte to macrophage is associated with a differential regulation of CSF-1 receptor expression. These considerations raise many questions on the nature of the molecular events initiated by CSF-1 in monocytic cells at different stages of differentiation. The signal transducing mechanisms of CSF-1 receptor may be different in the various monocytic populations. Alternatively, the same transducing signal initiated by the CSF-1 receptor could be filtered by a different genetic makeup typical of cells at different stages of differentiation and may initiate distinct programs of gene expression. Thus, the understanding of the molecular basis of growth factors and cytokine actions on macrophages involves answers to the following questions:

1. How is the proliferative potential of a cell gradually reduced?
2. What happens to the complex cellular machinery devoted to cell growth?

These problems are of major interest not only in understanding macrophage biology and cell differentiation, but also in developing and applying strategies to control neoplastic cell proliferation. Identification of the genes and mechanisms that induce the proliferative potential of a cell may lead to strategies to reduce the proliferative potential of tumor cells. An interesting aspect of this problem relates to the ribosomal genes, whose expression is readily activated during the proliferative response of a cell but is minimal in resting cells. The close association between proliferation and ribosomal RNA (rRNA) gene expression is a general phenomenon in nature occurring in normal and neoplastic cells, as well as in eukaryotic and prokaryotic cells. There are indications that ribosome biosynthesis can be induced in nonmitotic secretory cells (SCHMIT et al. 1985). Is the expression of ribosomal genes totally uncoupled to the response of macrophages to a growth factor or to an activating signal because the cells are not programmed to divide? If so, how do macrophages deal with the major changes in total protein synthesis occurring during the activation process? The finding of major changes in the metabolism of rRNA during macrophage activation suggested new and unexpected conclusions on the function of this molecule, and this will be one of the topics discussed in the present chapter.

The expression of a biologic function by macrophages is the net result of many events that include synthesis and secretion of biologically active mediators, expression of receptors, antigens, and ectoenzymes on the membrane, and activation of enzymatic pathways (ADAMS and HAMILTON 1987; VARESIO 1985b). The goal of the study of macrophage activation is to understand the sequence of molecular events initiated by the stimulus that result in the expression of the activated phenotype. We will focus our discussion on the pathways of macrophage activation that lead to the expression of tumoricidal activity by murine macrophages; hence, unless otherwise specified, the term "activated macrophages" will indicate a macrophage population endowed with cytotoxic activity against tumor target cells.

In many instances it has been possible to define a precise combination of stimuli that induces the expression of tumoricidal activity, and extensive information exists on the biochemical and molecular events induced by macrophage activators, such as interferon (IFN) and endotoxins (CHEN et al. 1990; ADAMS and HAMILTON 1987; VARESIO 1985b). The progressive characterization of the responses elicited by different signals has shown that the biochemical pathways triggered by various activators are different with regard to the utilization of signal transducing mechanisms and modalities of gene expression. The existence of multiple pathways of activation of cytotoxic macrophages has been further documented by differential sensitivity to metabolic inhibitors of macrophage activation initiated by different agents (RADZIOCH and VARESIO 1988; BLASI and VARESIO 1984; KOVACS et al. 1988; RADZIOCH et al. 1987b). It is still difficult to distinguish the sequence of events responsible for the activation of tumoricidal macrophages from those events leading to other functional activities. If different activating agents, although utilizing different biochemical pathways, generate a similar functional response, e.g., cytotoxicity, there should be common events elicited by every activator. Indeed, events correlated with the manifestation of cytotoxic activity have been identified, and some of them will be extensively discussed here. The task still to be performed is to tie together and organize into a mechanistic model the growing information on both the broad cascade of events initiated by various activating agents (stimulus-specific responses) and the more limited and selective molecular manifestation responsible exclusively for the tumoricidal activity (function-specific response). In this chapter, we will focus on the changes in RNA metabolism that are associated with the expression of tumoricidal activity by murine macrophages, elicited by different macrophage activators.

2 Control of RNA Metabolism

Analysis of the response of macrophages to specific stimuli has revealed changes in the levels of mRNA coding for structural and secretory proteins (HAMILTON et al. 1989). Run-off experiments and evaluation of the half-life of mRNA have revealed that transcriptional and post-transcriptional mechanisms control macrophage gene expression. For example, lipopolysaccharide (LPS) induces the expression of TNF and JE mRNA in peritoneal macrophages, and dibutyryl cAMP inhibits the inducibility of both mRNAs. However, the modulation of TNF is associated with transcriptional changes, whereas the modulation of JE mRNA seems to involve mainly a post-transcriptional control (KOERNER et al. 1987). Moreover, LPS induces transcriptional activation of TNF (TANNENBAUM and HAMILTON 1989) and inhibits the transcription of c-fms (GUSELLA et al. 1990).

These results indicate that distinct mechanisms may modulate the expression of different genes in response to the same macrophage stimulus. Finally, transcriptional and post-transcriptional mechanisms contribute to control the levels of mRNA expression, as shown for IL-1 (LEE et al. 1988), TNF (SARIBAN et al. 1988), CSF-1 (HORIGUCHI et al. 1988). Information on the levels at which mRNA expression is controlled derives mainly from run-off experiments assessing RNA synthesis/elongation in isolated nuclei, or from RNA stability experiments in which the decay of mRNA is evaluated following block of RNA synthesis by specific inhibitors. Both approaches are indicative but not conclusive since they are often affected by artifacts, and the results may be difficult to interpret. A clear understanding of the control of RNA metabolism will require precise information on the molecular events responsible for the transcriptional activation of a gene and definition of the parameters determining the stability of mRNA. It is beyond the scope of this chapter to review the changes in various RNA levels associated with the response of macrophages to stimulatory signals, or the growing literature on the molecular biology of gene expression. We will briefly mention the general concepts in RNA metabolism as background to a detailed discussion of the changes in ribosomal RNA expression and the relationship with the process of activation of cytotoxic macrophages.

Transcriptional activation plays a fundamental role in the control of gene expression in every cell type (CLEVELAND and YEN 1989). The mechanisms controlling gene transcription are very complex, and a large body of literature points toward the fundamental role played by the interactions among nuclear *trans*-activating factors that, by binding to the promoter region of the gene, modulate its expression in a positive or a negative way. However, it is difficult to predict the transcriptional regulatory properties of promoter elements from the binding activity of *trans*-activating factors. In fact, the transcription of many genes is cell type specific or cell activation dependent, and it may involve multiple promoter regions and different DNA binding proteins. An interesting example is the expression of the tumor necrosis factor-α (TNF-α) (GOLDFELD et al. 1990) or of the IFN-β gene (MANIATIS 1918; TANIGUCHI 1988). It was found that the promoter sequences required for virus-induced TNF-α expression were different in L929 fibroblasts and in the P388D1 macrophage cell line, indicating the presence of cell type-specific sequences. These differences are likely to reflect lineage-specific differences in the types or amounts of transcription factors that interact with TNF-α or IFN-β promoters. However, these conclusions are based on the expression of reporter genes driven by cytokine promoters in transfected cell lines. The availability and relative proportion of *trans*-activating factors may be different in tissue macrophages. Ideally, the promoter activity should be tested in fresh macrophages. Unfortunately, many attempts at establishing a transient expression system of transfected constructs in fresh macrophages have so far been unsuccessful. Presently, there is no information on the changes in nuclear *trans*-activating factors associated with the activation of macrophages.

Post-transcriptional events, including processing, transport, and stability of RNA, also play an important role in determining the final levels of gene expression at the mRNA levels, as demonstrated by the wide range of half-life values of mRNAs (BRAWERMAN 1987; COSMAN 1987; RAGOW 1987). In mammalian cells, the mRNA for transiently expressed genes, such as c-*fos* and c-*myc*, has half-life values as low as 15 min, while β-globin mRNA, for example, appears to be fully stable (BRAWERMAN 1987). The stability of mRNA can be controlled by specific nucleotide sequences and/or by the secondary conformation of the molecule. Evidence indicating that distinct regions of the mRNA molecule are important for its stability is provided by the demonstration that deletions of part of the gene alter the stability of the truncated transcript. For example, deletion of the 3′ end of c-*fos* untranslated region (MEIJLINK et al. 1985) or loss of the first noncoding exon of c-*myc* (RABBITTS et al. 1985) increases mRNA stability. A sequence stabilizing the mRNA molecule is the polyA tail. Degradation of mRNA is associated with shortening of the polyA tail, and removal of the poly A sequence results in rapid degradation of the RNA (BERGMANN and BRAWERMAN 1977). PolyA binding proteins have been implicated in the protection of mRNA from attack by ribonucleases (BERGMANN and BRAWERMAN 1977). AU-rich motifs at the 3′ end of many untranslated regions of mRNAs appear to be a destabilizing sequence for many mRNA species including some coding for oncogenes and cytokines (SHAW and KAMEN 1986; JONES and COLE 1987; KABNICK and HOUSMAN 1988; WILSON and TREISMAN 1988; WRESHNER and RECHAVI 1988). The relationship between these regions and mRNA instability has been suggested by experiments in which insertion or deletion of such regions caused a decrease or an increase in the half-life of the mRNA (PEPPEL et al. 1991).

The mechanism by which specific sequences control mRNA stability is unclear. One possibility is that RNA-binding proteins recognize such regions and modulate the susceptibility of RNA to ribonucleases (MALTER 1989). Like transcription factors, stability factors would differentially regulate mRNAs based on the presence of target RNA sequences. In addition to specific sequences, the secondary conformation of the RNA molecule may be involved in the control of RNA stability. Single-stranded RNA is capable of forming secondary structures, and the protein–RNA interaction may be conformational rather than, or in addition to, sequence specific (KLAUSNER and HARFORD 1989; LEVINE et al. 1986; MEIJLINK et al. 1985; BELASCO and HIGGINS 1988). It has been shown that RNA-binding proteins can bind to stem loops caused by folding of the RNA molecule, and that the configuration as well as the sequences within the loop is important in determining specific protein–RNA interaction (LAZINSKI et al. 1989). Proteins capable of interacting with RNA stem loop have been demonstrated (LAZINSKI et al. 1989). Moreover, the HIV-*tat* protein, which shares an arginine-rich motif with other RNA loop binding proteins, can interact with stem loops in the RNA of the HIV virus (CHANG and SHARP 1989; DINGWALL et al. 1989; GATIGNOL et al. 1989; OLSEN et al. 1990; ZAPP and GREEN 1989).

We are still far from being able to predict the stability of a given mRNA on the basis of the primary sequence or the potential secondary structure. The complexity of the mechanisms regulating RNA stability is evidenced by the fact that multiple factors are involved, even within the same cell, in controlling the stability of a given mRNA. Therefore, the relevance of specific sequences in determining the stability may vary substantially. This fact is exemplified by the observation of a differential post-transcriptional stabilization in the same cell of genes containing similar AU-rich regions at the 3′ end. For example, post-transcriptional stabilization of GM-CSF mRNA in a monocytic tumor was not associated with a similar stabilization of c-*myc* or c-*fos* mRNA, despite the fact these mRNAs share AU-rich motifs at the 3′ end (SHULER and COLE 1988).

In conclusion, the emerging information about the multiple mechanisms that may control the cellular content of RNA indicates that a major role is played by the interaction between nucleic acids and binding proteins. The evidence of differences among cell types in the control of mRNA expression indicates the need to extend the study of protein–nucleic acid interaction to the macrophage system in order to fully understand the mechanism of activation of these cells. In the following discussion, we will analyze the changes in ribosomal RNA (rRNA) occurring in macrophages and the potential role of the secondary structures of rRNA in the activation process.

3 Ribosomal RNA

Although a eukaryotic cell produces more than 10000 different RNA species, about half of its transcriptional capacity is devoted to the synthesis of one kind of RNA, namely rRNA. In virtually all eukaryotes, the 18S, 5.8S, and 28S mature rRNAs are initially transcribed as a large precursor encoded by genes in tandem arrays in the genome (rDNA) (SOLLNER-WEBB and TOWER 1986). In the mouse, the rDNA repeat unit is about 45 kb, and the largest detected rRNA precursor is approximately 47S or 14 kb (TIOLLAIS et al. 1971; GURNEY 1985). It extends from the initiation site (MILLER and SOLLNER-WEBB 1981) to a position 570 nucleotides downstream of the 28S coding region (GRUMMT et al. 1985b; GURNEY 1985). The presumptive primary transcript is rapidly processed, first at the 5′ end at residue +650 (MILLER and SOLLNER-WEBB 1981; GURNEY 1985) and then in the vicinity of the 3′ end of the 28S coding region (KOMIAMI et al. 1982; GRUMMT et al. 1985a), to yield the most prevalent precursor, the 45S rRNA. Subsequent processing gives rise to mature 18S, 5.8S, and 28S rRNAs (BOWMAN et al. 1983; CROUCH 1984). In general, more than 90% of the total cellular RNA is mature rRNA, and its half-life has been calculated as approximately 45 h (WEBER 1972). Because of its high synthetic rate and stability, rRNA is the predominant RNA species labeled by pulsing cells with radioactive uridine. The transcription of rDNA, the sole function of RNA polymerase I, is not only very efficient but also highly regulated, in large part reflecting the cellular need to produce more than a million new ribosomes per generation which, in turn, will be required to support protein synthesis in the daughter cells. Indeed, the biogenesis of ribosome particles is clearly coupled to the rate of cell proliferation. Quiescent cells accumulate ribosomes at a lower rate than do rapidly dividing cells (PERRY 1972; HOSICK and STROHMAN 1971). Terminal differentiation of myoblasts to myotubes represents an example of rapidly proliferating cells being converted into a nondividing fused population, resulting in an approximately 80% decline in the rate of ribosome accumulation (KRAUTER et al. 1979). Nevertheless, even in resting cells rRNA represents the major RNA species that continues to be actively synthesized. It has been reported that in lymphocytes, rRNA is synthesized at a rate higher than is needed for ribosome formation, resulting in a "wastage" of rRNA that is degraded before forming a ribosome (COOPER 1970). Modulation of rRNA synthesis has been associated mainly with alterations in cell growth. A close association between these two events is clearly evident in prokaryotic cells and has been observed in the eukaryotic system as well (NIERHOUS 1982; HADJIOLOV and NIKOLAEV 1976). However, if the proliferative stage were the only modulator of rRNA metabolism, one would have expected rRNA not to change in terminally differentiated cells not committed to proliferation, such as tissue macrophages or granulocytes. Therefore, it was intriguing to observe that the process of activation of resting peritoneal macrophages was associated with major changes in rRNA metabolism. The nature of the changes, their association with the stages of macrophage activation, and their possible biologic roles will be discussed.

4 rRNA Metabolism in Activated Murine Macrophages

The first evidence of changes in rRNA during macrophage activation was provided by studies of ^{3}H-uridine incorporation into acid-precipitable material (VARESIO et al. 1983, 1984b), a method that mainly detects changes in rRNA synthesis. It was found that murine macrophages activated to express tumoricidal activity exhibited a marked decrease in RNA synthesis. This decrease in RNA synthesis was evident in cytotoxic peritoneal murine macrophages activated in vitro by lymphokines, endotoxins, or synthetic compounds, as well as in vivo by injection into the mice of killed bacteria or endotoxins (VARESIO 1986). Normal levels of rRNA synthesis were found in macrophages stimulated, for example, to express immunosuppressive but not cytotoxic functions (VARESIO 1984). It was also established that the decrease in ^{3}H-uridine incorporation was a late event in the process of macrophage activation, being detectable in vitro after 8–10 h of exposure of macrophages to the activating agent (VARESIO et al. 1983, 1984b). These observations suggested that the down-regulation of RNA synthesis was a functional marker of macrophage activation that could be of relevance in the physiology of the activation process.

Direct evidence that rRNA metabolism was altered during the process of macrophage activation was obtained by the analysis of metabolically-labeled rRNA and by Northern blotting utilizing rRNA-specific probes. Size fractionation on denaturing agarose gels of RNA metabolically labeled by a short pulse with ^{3}H-uridine allows the detection of changes in rRNA synthesized de novo during the pulse with the radioactive tracer. Comparison of the radioactivity profile in resting and activated macrophages revealed an imbalanced accumulation of mature rRNA in activated cells (VARESIO et al. 1987; VARESIO 1985a). In cytotoxic macrophages, a major decrease in the accumulation of 28S rRNA relative to 18S rRNA was detected. 28S and 18S are the most prominent species of mature rRNA which are derived from a common high molecular weight 45S precursor. Pulse chase experiments, designed to follow the maturation of the 45S precursor, confirmed the existence of a selective block in the processing of the 28S rRNA in activated macrophages and provided clear evidence of a post-transcriptional level of regulation affecting the maturation of rRNA in cytotoxic macrophages (VARESIO 1985a). Since 18S and 28S rRNA derive from the same 45S precursor, these results indicated that the 28S rRNA was selectively degraded in activated macrophages. Alternatively, the maturation of the 45S precursor to 18S was normal, but the formation of the 28S rRNA was blocked (VARESIO et al. 1987). If this was the case, one would expect that rRNA precursors containing the 28S moiety would accumulate in activated macrophages.

Indeed, experiments in which total macrophage RNA was studied by Northern blotting followed by hybridization with probes specific for rRNA precursors showed that cytotoxic macrophages expressed significantly higher levels of 45S, 41S, and 36S rRNA precursors relative to resting peritoneal macrophages (RADZIOCH et al. 1987a). Northern blotting measures the steady state level of cellular RNA; therefore, these results demonstrated that 45S, 41S, and 36S rRNA precursors accumulate in cytotoxic macrophages.

The primary transcript of rRNA includes sequences for 18S and 28S mature rRNA. Formation of the final products occurs through a series of processing events which follow a general temporal sequence varying slightly from organism to organism and sometimes within a given organism (PERRY 1976). Generally, cleavage of the 41S by endonuclease separates the 18S and the 28S sequences in distinct precursor molecules (21S and 36S respectively). The 21S is coverted to 18S in a simple step; the 36S is sequentially cleaved to 32S and 28S species. In cytotoxic macrophages the levels of the 36S, but not of the 32S, are elevated. These results suggest that the process of activation inhibits the conversion of the 36S into 32S, thereby causing accumulation of the upstream precursors 36S, 41S, and 45S.

The steady state levels of mature 28S and 18S rRNA, however, do not change during the activation process, as shown by Northern blotting analysis (RADZIOCH et al. 1987a). There is no contradiction between these results and the reported decrease of the accumulation of de novo synthesized 28S rRNA detected in the analysis of metabolically labeled rRNA. The half-life of mature 28S rRNA is about 45 h (WEBER 1972a), and the changes in accumulation of 28S rRNA occurred during the last 6–8 h of the 18 h of activation. A decreased rate of 28S rRNA accumulation over 6–8 h is not sufficient to decrease the steady state level of 28S rRNA to such an extent as to be detected by Northern blotting.

In conclusion, the major changes in rRNA metabolism in cytotoxic macrophages can be summarized as being a decrease in rRNA synthesis and a selective decrease in the accumulation of de novo synthesized 28S but not of 18S rRNA. The major impact of these metabolic changes on the cellular rRNA content is represented by an accumulation of rRNA precursors (45S, 41S, and 36S) in cytotoxic macrophages. The precise relationship among these changes in rRNA is speculative. Changes in the methylation of rRNA precursors may be involved in the alteration of rRNA metabolism since it has been shown that IFN-γ augments the intracellular content of *S*-adenosylmethionine in macrophages, probably by inhibiting the transmethylation reactions (BONVINI et al. 1986). Little is known about the mechanisms of rRNA processing, the requisite protein species, or the nucleic acid sequences involved, and in vitro systems which produce fully processed rRNA have not yet been devised. It has been proposed that the key event is the block of the conversion of 36S to 32S pre-RNA. This event would affect the formation of the 28S but not the 18S rRNA, accounting for the imbalanced accumulation of de novo synthesized rRNA. Accumulation of the rRNA precursor would then feedback and inhibit the ribosomal DNA transcription. Evidence that rRNA can feedback and repress rDNA gene expression has been reported (GOURSE et al. 1985).

5 Modulation of rRNA in Different Cell Types

An imbalance between 28S and 18S rRNA accumulation is not a unique event occurring in macrophages; rather it has been observed in various cell systems (BIZZOZZERO et al. 1985; TONIOLO and BASILICO 1975; BOWMAN and EMERSON 1977; JOHNSON et al. 1976; MAUCH and GREEN 1973; TONIOLO et al. 1973; ABELSON et al. 1974; EMERSON 1971; WEBER 1972b; COOPER 1973). Imbalanced accumulation of mature rRNA species can be induced by inhibiting the growth of fibroblasts by serum starvation (BIZZOZZERO et al. 1985) or by contact inhibition (EMERSON 1971), by shifting temperature-sensitive cell lines to temperatures non-permissive for cell growth (OUELLETTE et al. 1976; TONIOLO and BASILICO 1975), or by inducing myoblast differentiation (BOWMAN and EMERSON 1977). These studies demonstrated that imbalanced accumulation of rRNA occurs during the transition of proliferating cell lines from a growing to a resting state. Moreover, it was reported that picolinic acid, a metabolite of tryptophan, can induce selective down-regulation of 28S rRNA, but not of 18S rRNA, in AKR normal rat kidney cells (COSTANTINI and JOHNSON 1981). These changes in rRNA metabolism have been attributed to the decrease in ribosome requirement associated with the decrease in proliferative activity. However, the mechanisms responsible for the occurrence of disproportionate accumulation of mature rRNA, RNA wastage, and alterations in rRNA metabolism characteristic of resting and rested cells have never been defined. The results obtained with macrophages indicate that rRNA metabolism can be modulated in the absence of changes in cell proliferation, since peritoneal macrophages are resting cells and the process of activation does not require or involve cell proliferation. Moreover, the dissociation between changes in rRNA metabolism and the cellular content of mature 28S and 18S rRNA indicates that the altered rRNA metabolism in macrophages is not directly related to the ribosome content. It is possible, instead, that the changes in rRNA metabolism detected in macrophages may reflect and be associated with the functional activity of the cells expressed in their lytic activity against tumor target cells. However, the occurrence of imbalanced accumulation of rRNA in other cell types suggests that such events may represent a more general mechanism of control of cellular functions.

6 Association Between Altered rRNA Metabolism and Cytotoxic Activity

Murine macrophages can be activated in vivo and in vitro to express tumoricidal activity. The effector mechanisms of macrophage-mediated killing are still undefined, and probably more than one mediator may contribute to the expression of tumoricidal activity. Therefore, different protocols to measure the tumoricidal activity of macrophages may generate somewhat different results that can be reconciled only by a careful side by side comparison (RUSSELL et al. 1986; PACE et al. 1985b). In any case, a reproducible tumoricidal assay in vitro is a good marker of a given stage of macrophage activation, and the changes in rRNA metabolism correlate very well with the expression of tumoricidal activity measured in an 18-h ^{51}Cr or ^{111}In release assay (RUSSELL et al. 1986).

The strongest correlation between changes in rRNA metabolism was found in studies of murine macrophages activated in vivo by injection of *C. parvum* (VARESIO 1984). Depending upon the time of the harvest of macrophages, it is possible to recover peritoneal macrophages expressing (a) immune suppressive activity but not tumoricidal activity or (b) both functions. However, only the macrophages expressing tumoricidal activity have altered rRNA metabolism (VARESIO 1984).

Activation of macrophages in vitro requires multiple signals, for example IFN-γ and traces of LPS (PACE et al. 1983). However, it was found that C57BL/6 macrophages can be activated by IFN-γ alone without the need for LPS as a second signal (RUSSELL et al. 1986; VARESIO et al. 1984a). Moreover, IFN-γ, IFN-α/β, endotoxins, or synthetic dsRNA (poly I:C) can activate macrophages in vitro alone or in the presence of a second signal (TARAMELLI and VARESIO 1981; BLASI et al. 1984; PACE et al. 1985a). The changes in rRNA followed exactly the pattern of expression of cytotoxic activity, since they were elicited by IFN-γ alone in C57BL/6 mice (RADZIOCH et al. 1987a) but required IFN-γ and traces of LPS in macrophages from C3H/HeN mice (VARESIO et al. 1990). In addition, changes in rRNA were also elicited by poly I:C (VARESIO et al. 1983), high levels of endotoxins (VARESIO 1986), and IFN-α/β (RADZIOCH et al. 1987a) under conditions in which these stimuli induced cytotoxic macrophages. Further evidence of an association between changes in rRNA metabolism and macrophage activation was provided by studies with macrophages from C3H/HeJ mice that can be activated in vitro by IFN-γ plus a strong second signal, such as *Listeria monocytogenes*, but not by IFN-γ alone or supplemented with traces of LPS (HOGAN and VOGEL 1988). As for the cytotoxic activity, only the combination of IFN-γ plus *Listeria monocytogenes* was able to modify the rRNA metabolism in C3H/HeJ macrophages (VERESIO et al. 1990).

In conclusion, a strong association exists between altered rRNA metabolism and expression of tumoricidal activity. It is important to note that the changes in rRNA metabolism are not stimulus specific since they can be elicited by different activating agents. Moreover, the changes in rRNA metabolism are not a constant response to an activating agent. For example, macrophages from many strains of mouse, including C57BL/6, C3H/HeN, and C3H/HeJ, respond to IFN-γ alone with induction of Ia antigens, of 2–5 A synthetase, etc. However, IFN-γ alone triggers changes in rRNA only in C57BL/6 macrophages, in which it also induces tumoricidal activity.

Changes in rRNA metabolism are not induced by costimulatory signals alone such as traces of LPS for macrophages from C3H/HeN or *Listeria monocytogenes* for macrophages from C3H/HeJ mice (VARESIO et al. 1990). Rather, only the combined action of two signals in conjunction with tumoricidal activity was able to induce changes in rRNA metabolism (VARESIO et al. 1990). Thus, the changes in rRNA are not a direct response to the first or the second signal but occur under conditions in which the stimulation is sufficient to induce tumoricidal activity.

Although results outlined above demonstrated the existence of a strong correlation between expression of tumoricidal activity and changes in rRNA metabolism, it was difficult to accept a causal relationship between these two distant phenomena since there is no obvious biochemical pathway that could theoretically connect these apparently unrelated events. Recent studies on the effects of picolinic acid on macrophages lend support to the existence of a causal connection between altered rRNA metabolism and cytotoxicity, and have led to the suggestion that there is a biochemical pathway going from rRNA to cytotoxic activity via tryptophan metabolism.

7 Picolinic Acid and the Tryptophan Connection

Picolinic acid is one end product of tryptophan catabolism, synthesized from the 2-amino-3-carboxymuconic semialdehyde by the enzyme picolinic decarboxylase. It has been shown that picolinic acid can inhibit the in vitro growth of various cell lines and that transformed cells are more sensitive to picolinic acid (FERNANDEZ-POL et al. 1977; FERNANDEZ-POL and JOHNSON 1977). The mechanism of action of picolinic acid is not known although it has been reported that picolinic acid interferes with iron uptake (FERNANDEZ-POL 1977) and ribosomal RNA metabolism (COSTANTINI and JOHNSON 1981; COLLINS et al. 1979). In macrophages, picolinic acid affects rRNA metabolism by inhibiting the accumulation of de novo synthesized 28S but not 18S rRNA and by inducing accumulation of rRNA precursors (VARESIO et al. 1990). These changes in rRNA metabolism are remarkably similar to those occurring in activated murine macrophages. Thus, the analysis of the macrophage-activating properties of picolinic acid could provide the tool for determining whether the changes in rRNA metabolism are causally related to the tumoricidal stage. Experiments in which macrophages were exposed to IFN-γ plus picolinic acid demonstrated that picolinic acid is a potent costimulator of macrophage activation. Specifically, picolinic acid acted synergistically with IFN-γ in activating macrophages from C57BL/6 mice (VARESIO et al. 1990). Moreover, it was found that macrophages from C3H/HeJ and C3H/HeN mice, which do not become cytotoxic in response to IFN-γ alone, could be fully activated by picolinic acid plus IFN-γ (VARESIO et al. 1990). Therefore, picolinic acid functions as a second signal in macrophage activation. Since altered rRNA metabolism, whether induced by picolinic acid or by another appropriate stimulus, was necessary for the expression of cytotoxic activity, a causal connection between these two events is supported by these results. Picolinic acid alone altered the rRNA metabolism but did not elicit the cytotoxic response even in C57BL/6 macrophages, which can be activated by IFN-γ alone (VARESIO et al. 1990), indicating that the accumulation of pre-RNA was a necessary but not sufficient event to trigger tumoricidal activity.

One difference between classical second signals and picolinic acid is that picolinic acid alone augments pre-rRNA, whereas low amounts of LPS or *Listeria monocytogenes* require the costimulatory activity of IFN-γ to affect rRNA metabolism (VARESIO et al. 1990). This difference in the mechanism of action can be reconciled by the consideration that picolinic acid is a metabolite of tryptophan and in itself may be a macrophage product, acting in an autocrine manner. In this context, only the combined action of IFN-γ plus a second signal may induce sufficient picolinic acid, or may deliver it to the correct intracellular compartment, to modify the rRNA metabolism. The possibility that picolinic acid is produced by activated macrophages is supported by several pieces of evidence. IFN-γ can induce, in vivo and in vitro, indoleamine 2,3-dioxygenase (IDO), the first enzyme in a major pathway for degradation of tryptophan (WERNER-FELMAYER et al. 1989; CARLIN et al. 1989a, b, d; WERNER et al. 1987a, b, 1988; BYRNE et al. 1986a; BIANCHI et al. 1988; YASUI et al. 1986), and some biologic effects of IFN-γ involve tryptophan degradation. (OZAKI et al. 1987, 1988; MAZA and PETERSON 1988; PFEFFERKORN 1984, 1986; PFEFFERKORN et al. 1986; AUNE and POGUE 1989; CARLIN et al. 1989c. MURRAY et al. 1989; BYRNE et al. 1986b). In the catabolism of tryptophan there is a branch point that leads to either picolinic or quinolinic acid. Picolinic acid is an end product whereas quinolinic acid is further metabolized to nicotinate mononucleotide, an important enzymatic cofactor. Thus IFN-γ initiates a pathway of tryptophan metabolism in which picolinic acid may be one of the by-products. The ability of the body to catabolize tryptophan to picolinic acid is shown by the identification of picolinic acid in human milk (EVANS and JOHNSON 1980). Moreover, there is evidence that picolinic acid is active in vivo in animals and humans (RUFFMANN et al. 1984; LEUTHAUSER et al. 1982; KRIEGER and STATTER 1987; KRIEGER 1980; SEAL and HEATON 1985; EVANS and JOHNSON 1980; MENARD and COUSINS 1983; HURLEY and LONNERDAL 1982) and that it may be involved in vivo in the process of macrophage activation. In fact, intraperitoneal injection of picolinic acid into C57BL/6 mice induced antitumor activity in macrophages without affecting the levels of natural killer cell (NK) activity (RUFFMANN et al. 1984, 1987). These results suggest that this compound may enhance the sensitivity of macrophages to activation by low levels of cytokines present in the peritoneal cavity and may account for the observed antitumor effect of picolinic acid in tumor-bearing mice (LEUTHAUSER et al. 1982). In conclusion, the demonstration that picolinic acid is a macrophage costimulator in vitro, taken together with the evidence that picolinic acid is produced and active in vivo (RUFFMANN et al. 1984; LEUTHAUSER et al. 1982; KRIEGER and STATTER 1987; KRIEGER 1980), supports the hypothesis that picolinic acid can function as an autocrine signal for the activation of macrophages to a cytotoxic stage.

The process of macrophage activation could be depicted as follows (Table 1). IFN-γ induces IDO, initiates the catabolism of tryptophan, and generates tryptophan degradation products. This process occurs in many cell types but does not necessarily lead to production of picolinic acid. In macrophages, however, the second signal activates the pathway generating picolinic acid and causes the build up of sufficient and/or available picolinic acid to alter the rRNA metabolism. The accumulation of the rRNA precursor, together with other IFN-γ-induced proteins, will lead to the expression of tumoricidal activity (VARESIO 1986; VARESIO et al. 1990). As previously discussed, the second signal alone will not induce rRNA changes, because picolinic acid cannot be formed in the absence of tryptophan degradation induced by IFN-γ. The reason why IFN-γ alone causes changes in RNA metabolism and activation to a tumoricidal stage of C57BL/6 macrophages may be that IFN-γ stimulates macrophages of this strain of mice to produce enough endogenous picolinic acid that an autocrine costimulatory signal is generated. This model is supported by indirect evidence, and the crucial piece of information is the identification of the source(s) of picolinic acid in the

Table 1. rRNA pathway in macrophage activation: a model

1. Stimulation of macrophages by one signal or a combination of signals sufficient to trigger tumoricidal activity
2. Induction of dsRNA-dependent enzymes
3. Induction of tryptophan catabolism with formation of picolinic acid
4. Alteration of rRNA metabolism by picolinic acid:
   a) inhibition of 28S rRNA accumulation leading to
   b) inhibition of rRNA synthesis and accumulation of rRNA precursors
5. Activation of dsRNA-dependent enzymes by the dsRNA structures present in the rRNA precursors and in their degradation products
6. Progression of macrophages in the activation process through phosphorylation of relevant substrates by the dsRNA-dependent kinase modulation of RNA levels by the 2–5 A synthetase

body and the evaluation of contributing macrophages and macrophage activation.

8 rRNA and Cytotoxic Activity: the Double-Stranded RNA Pathway

Although the correlation between changes in rRNA metabolism and expression of cytotoxic activity and the identification of picolinic acid as a second signal in macrophage activation provide compelling evidence for a causal association between rRNA and cytotoxic activity, it is difficult to transpose such a relationship to a biochemical/molecular level. As discussed previously, the alteration in the rRNA metabolism does not have a detectable impact on the content of mature rRNA during the time required to activate macrophages to a cytotoxic stage. Therefore, it seems unlikely that the altered rRNA metabolism will affect macrophage protein synthesis via a compromised ribosomal function. In contrast, the most prominent quantitative change affecting the cellular content of rRNA is the increase in rRNA precursors associated with the expression of cytotoxic activity. Is there a way by which rRNA precursors can affect cell biology independently of ribosome formation?

Ribosomal RNA precursors are very rich in cytidine and guanosine and the potential secondary structures, on the basis of computer modeling, predict the formation of very stable double-stranded structures (CROUCH 1984; WOESE et al. 1990; MICHOT et al. 1983). Thus, accumulation of rRNA precursors could be interpreted as a possible source of increase in the cellular content of double-stranded RNA (dsRNA). Changes in intracellular dsRNA are potentially relevant due to the existence of enzymes that require dsRNA as a cofactor to be active (SAMUEL 1987; FALTYNEK and KUNG 1988; LENGYEL 1982). We will briefly discuss the possibility that the rRNA precursors can function as an intracellular source of dsRNA to activate dsRNA-dependent enzymes and that such activation is needed for the expression of cytotoxic activity.

Two dsRNA-dependent enzymes have been described that are present in virtually every cell type and are induced by IFN (SAMUEL 1987; FALTYNEK and KUNG 1988; LENGYEL 1982). One of these is 2–5 A synthetase, which, when activated by dsRNA, synthesizes small 2′–5′ linked oligomers of adenosine with the general formula pppA (2′p5′A)n, where n is 2 or more. For convenience this mixture of oligonucleotides is referred to as 2–5 A. The function of 2–5 A is to bind and activate a latent endoribonuclease that degrades RNA. The other well-described enzyme is the dsRNA-dependent protein kinase, which, when activated by dsRNA, phosphorylates the alpha subunit of protein synthesis initiation factor 2 (eIF2) and inhibits its recycling. The dsRNA protein kinase has been purified and characterized (MEURS et al. 1990). Activation of the kinase with dsRNA is accompanied by autophosphorylation of a 65-kDa protein (p65) in mouse and in rabbit cells or of a 68-kDa protein (p68) in human cells. The p65 and p68 kinases are phosphorylated on several serine and threonine residues (HOVANESSIAN 1989; KURST et al. 1984), giving phosphate-saturated molecules that are more acidic and show higher molecular weights than those that are partially phosphorylated or unphosphorylated (KURST et al. 1984). p68 human kinase activity is independent of cAMP or cGMP, is markedly stimulated by manganese, and has two distinct activities: autophosphorylation and phosphorylation of exogenous substrates. The former is dependent on dsRNA, divalent cations, and ATP. The latter is not dependent on dsRNA and occurs as long as the p68 remains phosphorylated (GALBRU and HOVANESSIAN 1985). Dephosphorylation can be catalyzed by a manganese iondependent class I phosphatase (SZYSZKA et al. 1989).

Three main forms of 2–5 A synthetase have been described, corresponding to proteins of 40–46, 69, and 100 kDa (HOVANESSIAN et al. 1988). The small forms of 2–5 A synthetase are derived from the same gene by differential splicing between the fifth and an additional sixth exon of this gene (BENECH et al. 1985; SAUNDERS et al. 1985), and they thus differ in their C-terminus. The large forms of 2–5 A synthetase are probably encoded by another gene or genes. The small and large forms of 2–5 A synthetase differ substantially in their kinetics of expression and response to inducers. p100 seems to be a monomer found in a diffuse state in the cytoplasm whereas p69 is partially myristilated, seems to exist in a dimeric form, and is partly concentrated around the nucleus and partly distributed in the cytoplasm in a somewhat specific pattern (MARIE et al. 1990). p69 has the capacity to synthesize longer oligomers, whereas p100 has a tendency to mainly synthesize the dimeric form of 2–5 A (HOVANESSIAN et al. 1988). As the dimeric form has no known function and does not activate the latent ribonuclease, p100 may be involved in catalyzing other reactions. It is, therefore, possible that the different forms of 2–5 A synthetase may be activated under different physiologic conditions and in different cell types, and that they may have different functions. dsRNA-dependent enzymes were originally implicated in the antiviral effects of interferons since the kinase, by inactivating the eIF2, could block the translation of the viral proteins, and the 2–5 A synthetase could cause degradation of the viral mRNA through activation of the RNase activity. The viral RNA would provide the dsRNA needed for the activation.

dsRNA-dependent enzymes are induced during the process of macrophage activation since most of the activators are either IFNs or IFN-inducers (ADAMS and HAMILTON 1984, 1987). IFN-γ is among the most effective activators of macrophages. IFN-α/β can also induce cytotoxic macrophages, although the mechanisms of activation are different (BLASI et al. 1984; PACE et al. 1983). Poly I:C, LPS, and pyran copolymer among the many macrophage activators that are inducers of IFN, and the endogenous production of interferon could contribute to the activation process. Therefore, under conditions in which macrophages are activated and rRNA precursor accumulates, dsRNA-dependent enzymes are also induced in macrophages (GUSELLA et al., manuscript in preparation). Any biologic role of these dsRNA-dependent enzymes in macrophages requires that

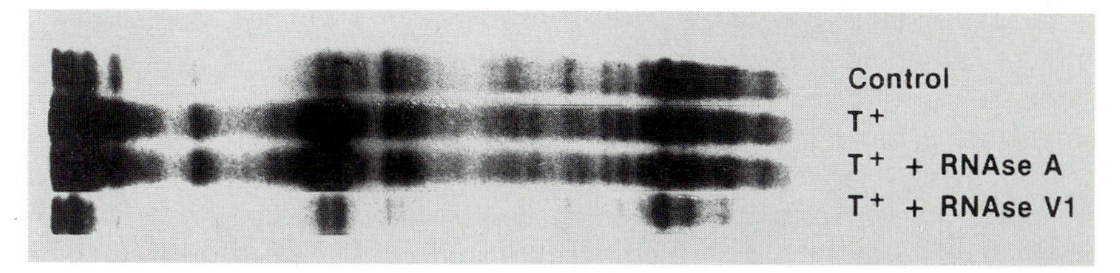

Fig. 1. Ribosomal RNA precursors were generated by in vitro transcription of the 3.7-kb *Eco*RI-*Bam*HI fragment of the mouse rDNA containing the 3′ terminal domain of the 18S rRNA, internal transcribe spacers, 5.8 rRNA, and the 5′ terminal domain of the 28S rRNA (Michot et al. 1983) cloned in the pGEM expression vector. Following in vitro transcription with the T7 polymerase, the rRNA transcript was purified and assayed for its ability to induced phosphorylation of histones by activating the dsRNA-dependent kinase. S100 pellets of IFN-β-treated fibroblasts were used as the source of dsRNA-dependent kinase, and the assay was performed as previously described (Minks et al. 1979). The "*control*" *lane* depicts the background phosphorylation in the absence of rRNA; the *lane* T^+ depicts the phosphorylation of histones induced by rRNA (0.1 μg/ml); the *lane* T^+ + *RNase A* depicts the phosphorylation induced by 0.1 μg/ml of rRNA pretreated with RNase A for 30 min at 37 °C to digest single-stranded RNA; the *lane* T^+ + *RNase V1* depicts the phosphorylation induced by 0.1 μg/ml of rRNA pretreated for 30 min at 37 °C with RNase V1 to digest the double-stranded RNA

cellular RNA can function as a source of dsRNA structures. It is postulated that the rRNA precursors or their byproducts generated during the rRNA precursor maturation contain double-stranded structures capable of activating dsRNA enzymes. Indeed, it has been observed that pre-rRNA generated by in vitro transcription of cloned murine rRNA genes could serve as a source of dsRNA in the activation of 2–5 A synthetase and kinase (Bottazzi et al., manuscript in preparation; Varesio 1986). Activation of murine dsRNA-dependent kinase by rRNA precursors is shown in Fig. 1. Experiments in which the rRNA transcripts were digested with RNAse A (degrading single- stranded RNA) or RNAse VI (degrading double-stranded RNA) demonstrated that RNAse VI, not RNAse A, could abrogate the ability of the rRNA transcripts to activate dsRNA-dependent protein kinase (Fig. 1). These experiments indicated that the double-stranded secondary structure of the rRNA transcripts was responsible for the activation of dsRNA-dependent enzymes. It is difficult, however, to know the secondary conformation of rRNA precursors in the intact cell, where the RNA is rapidly processed and is coupled to protein. Therefore, these results indicate that in vitro rRNA precursor folds into secondary structures capable of activating dsRNA, but the extent to which these double- stranded regions exist and are available for the dsRNA-dependent enzymes in vivo remains to be established. On the other hand, if we think that dsRNA is biologically relevant in nonvirally infected cells, we must accept the notion that cellular RNA can fold into double-stranded structures available for the activation of dsRNA-dependent enzymes, and rRNA is a good candidate.

A potential problem of such model is that rRNA precursors are mainly present in the nucleolus and dsRNA-dependent enzymes have so far been described in the cytoplasm and the perinuclear region. It cannot be excluded, however, that dsRNA-dependent enzymes exist in the nucleus since the components of this class of enzymes is growing, and a thorough analysis of the intracellular compartmentation has not been possible for lack of specific reagents. Moreover, it must be considered that degradation products of rRNA precursors generated during its maturation could reach the cytoplasm and activate dsRNA-dependent enzymes, particularly if folded in double-stranded structures. In fact, enzymes degrading dsRNA have not been described in eukaryotic cells and the double-stranded regions of rRNA precursors could be quite stable. Reports of small cytoplasmic RNA with various degrees of affinity for rRNA support this possibility (KING and GOULD 1970; BUSCH et al. 1982, MCLURE and PERRAULT 1986; PERKINS et al. 1986). However, a thorough investigation of the fate of the rRNA spacers, as well as of the introns of mRNA, has never been performed. We have evidence that discrete regions of the rRNA precursor spacers can activate dsRNA-dependent enzymes (BOTTAZZI et al., submitted for publication). In conclusion, the proposed model of activation of dsRNA-dependent enzymes by rRNA precursors or their degradation products needs further experimental support but it contains a provocative working hypothesis on one activation pathway of macrophages.

9 Double-Stranded RNA-Dependent Enzymes in Cell Biology

dsRNA-dependent enzymes may have important biologic functions in non-virally infected cells in mediating the cellular response to IFN. Evidence that dsRNA-dependent enzymes are also important in the biology of nonvirally infected cells is provided by studies on the effects of the kinase inhibitor 2-aminopurine (2AP). The two kinases that are known to be inhibited by 2AP in cell-free systems are the heme-regulated and the dsRNA-dependent eukaryotic initiation factor (eIF)-2α kinases (DE BENEDETTI and BAGLIONI 1983; FARRELL et al. 1977). Although the specificity of the inhibitory effects of 2AP has not been systematically studied in vivo, 2AP does not significantly alter the overall pattern of protein phosphorylation in HeLa cells. Thus, 2AP does not appear to be a general inhibitor of cellular kinases, although, like any other metabolic inhibitor, it may have unknown side-effects. It has been reported that 2AP specifically inhibits the induction of IFN-β, c-*fos*, and c-*myc* at the level of transcription in human osteosarcoma cells stimulated with poly I:C or virus, suggesting that dsRNA-dependent kinase could be involved in controlling gene expression (ZINN et al. 1988). Similar conclusions were reached in studies of the effects of 2AP on gene induction by interferons and double-stranded RNA in HeLa cells (TIWARY et al. 1988). Finally, 2AP has been shown to inhibit the activation of murine macrophages by high doses of LPS, providing preliminary evidence of the involvement of dsRNA-dependent kinase in the process of activation of macrophages to a tumoricidal stage (GUSELLA et al., manuscript in preparation). Macrophage activation is associated with expression of new proteins (MacKay et al. 1989) and phosphorylation by dsRNA-dependent kinase could modulate their activity.

The potential role played by 2–5 A synthetase in the regulation of cell growth is supported by the demonstration of correlations between changes in enzymatic activities and the degree of confluence of the culture (Stark et al. 1979), the state of differentiation (Sokawa et al. 1981), and the phase of the cell cycle (Krishnan and Baglioni 1981; Wells and Mallucci 1985). In addition, agents such as hormones (Stark et al. 1979) and inducers of differentiation (Bourgeade and Besancon 1984; Ferbus et al. 1985) also influence the levels of 2–5 A synthetase. In general, growth-inhibited cells have higher levels of 2–5 A synthetase than rapidly proliferating cells. Thus, 2–5 A synthetase may serve as an important mediator of the antiproliferative actions of IFN-γ. Moreover, studies on the antiproliferative affects of IFN-γ revealed varying kinetics of expression of 2–5 A synthetase isoenzymes in the A431 human tumor cell line: the 100-kDa and 40-kDa isoforms were induced early with a subsequent decline whereas the 67-kDa was induced late and remained high up to 9 days after IFN-γ treatment (Kumar and Mendelson 1989). It is unclear, however, which is the source of the dsRNA activating the enzymes. It is possible that in these situations the rRNA precursors may also play a role as donors of dsRNA, because it has been shown that the transition of cell lines from a growing to a resting state is associated with altered accumulation of rRNA (Ouellette et al. 1976; Bizzozzero et al. 1985; Toniolo and Basilico 1975; Bowman and Emerson 1977; Emerson 1971) and that the eIF-4F and eIF-2 activities are inhibited by serum deprivation (Duncan and Hershey 1985). One can speculate that the accumulation of rRNA precursors is involved in the activation of the dsRNA-dependent kinase responsible for the phosphorylation and inactivation of the initiation factors for protein synthesis. Also, in nonmacrophagic cell types the imbalanced accumulation of rRNA appears to be linked with a functional state rather than a response to IFN. In fact, growth-arrested fibroblasts do not alter rRNA metabolism in response to IFN-γ (Radzioch et al. 1987a). Degradation of rRNA but not of viral mRNA has been observed in IFN-treated L cells infected with DNA-containing viruses (Goswami and Sharma 1984). Since the degradation pattern of rRNA is consistent with the action of the latent ribonuclease induced by 2–5 A, it can be speculated that activation of 2–5 A synthetase by dsRNA structures of rRNA results in rRNA degradation via a feedback type of mechanism.

Every RNA molecule, including mRNA, can fold into a secondary structure containing a certain degree of double-stranded structure. If these double-stranded secondary structures are sufficiently stable and accessible, they can activate the dsRNA-dependent enzymes in a localized and compartmentalized fashion. Indeed, there are indications that double-stranded loops of mRNA can control its expression via activation of dsRNA-dependent enzymes. For example, the activation of dsRNA-dependent kinase is a mechanism that may control the translation of HIV-1 mRNA (Edery et al. 1989), although this possibility is still controversial (Gunnery et al. 1990). It was found that the TAR region of the 5′ end of the HIV-1mRNA, which responds to the *tat-trans*-activating protein, assumes a stable secondary structure responsible for the ability of TAR to inhibit *in trans* the translation of other mRNAs in a cell-free system. This mechanism of translation inhibition involves the activation of dsRNA-dependent kinase and phosphorylation of the protein synthesis initiation factor 2 (eIF-2). Mutations in the TAR region that diminish the stability of the secondary structure cause a significant inhibition of the *trans*-activation (ROY et al. 1991).

Indications of a potential role of dsRNA-dependent kinase in the control of mRNA expression derive from experiments with 2AP. If local activation of dsRNA-dependent kinase is an inhibitory signal for mRNA expression, one would expect 2AP to augment selectively the expression of mRNAs endowed with a stable double-stranded structure. Indeed, it has been shown that 2AP increases the translation efficiency of mRNA molecules from plasmid DNA transfected into COS-1 cells, presumably via alteration of the functional levels of eIF-2 (KAUFMAN and MURTHA 1987; KALVAKOLANU et al. 1991). These results confirm the pioneering work by DE BENEDETTI and BAGLIONI (1984), who demonstrated the specific inability to initiate translation of vesicular stomatitis virus mRNA containing a poly(U) tail hybridized to the poly(A) tail. Thus, the activation of the kinase would be bound specifically to the mRNA containing double-stranded features, resulting in specific, localized inhibition of protein synthesis. In conclusion, there is evidence that the dsRNA-dependent enzyme system may be involved in important cellular functions including macrophage activation. In order to delineate the role of dsRNA-dependent enzymes in cell biology it is crucial to define the endogenous dsRNA structures needed to activate these enzymes and the regulation of expression of these dsRNA species. The rRNAs are potential donor of double-stranded RNA capable of activating dsRNA-dependent enzymes, and rRNA is synthesized at a high enough rate to allow diversion of some of it to function as an enzyme activator rather than as a building block for the formation of ribosomes.

10 Conclusions

The goal of this article has been to review the changes in RNA metabolism that occur in macrophages and their connection with the process of activation. Evidence was presented that imbalanced accumulation of rRNA and accumulation of rRNA precursors in macrophages are directly associated with the activation of these cells to a cytotoxic stage. A metabolite of tryptophan, picolinic acid, induces similar changes in rRNA metabolism and functions as a second signal for macrophage activation, suggesting that picolinic acid produced by macrophages or by surrounding tissues may be the mediator of the changes in rRNA. Stable double-stranded regions of the rRNA precursors that accumulate in cytotoxic macrophages activate, in a cell-free system, dsRNA-dependent enzymes and could be a physiologic source of dsRNA structures in vivo. It is likely that dsRNA-dependent enzymes activated by rRNA precursors (and/or their degradation products) play a major role in the induction/expression of the activated phenotype.

Three players support the action in this scenario: rRNA, tryptophan metabolism, and dsRNA-dependent enzymes. The evidence implicating each of them in the process of macrophage activation is quite strong and the connection among the three of them is supported, although not yet unequivocally proven, by the results of ongoing research.

There is a consensus that the secondary structure of RNA is important for its stability and interactions with proteins. Although mRNA is being more actively studied at the present time, rRNA remains the most abundant and transcribed cellular RNA and the mechanisms controlling rRNA maturation are still largely obscure. A cyclical trend in the interest in rRNA and ribosome has indeed been noted (MOORE 1988). We feel that the macrophage activation system and the associated changes in rRNA metabolism provide an interesting and perhaps unique model for exploring the molecular basis of rRNA control since it has been established that transcriptional and post-transcriptional events modulate the expression of ribosomal genes in a nonproliferative system. Thereby, it may be possible to explore other functions of rRNA that might have been hidden by the ribosome formation task in proliferating cellular systems.

The study of dsRNA-dependent enzymes in the process of macrophage activation is important since IFNs are powerful activators of macrophages. Although the existence of these enzymes has long been known, only recently have some members of this family been cloned and only now can a critical evaluation of their functions be performed. Perhaps some hesitation in pursuing studies on these enzymes was due to questions concerning the origin and nature of the dsRNA needed for their activation. Growing evidence indicates that the secondary structure of rRNA and mRNA can activate dsRNA-dependent enzymes, supporting the notion that these enzymes may be of major biologic relevance even in the absence of viral infection.

Studies on tryptophan metabolism fell away decades ago. Picolinic acid was known to exist but questions regarding its function elicited merely sporadic interest. The discovery that IFN augments tryptophan metabolism and that picolinic acid affects rRNA maturation and macrophage activation gave rise to renewed interest in roles that this amino acid may have in the biology of macrophages and of the entire organism.

Since we have had to discuss so many varied topics in this chapter, and due to space limitations, we have not been able to cite all the important relevant papers; our apologies are offered to the authors of such papers.

J.H. Huxley made reference to "the great tragedy of science: the slaying of a beautiful hypothesis by an ugly fact." Ugly facts may modify our current view of macrophage activation. However, the questions addressed by our hypothesis are of basic relevance for macrophage and cell biology. Should an ugly fact disprove our hypothesis but also provide a definitive explanation for the interplay among rRNA, tryptophan, and the cytotoxic activity of macrophages, a major achievement would still have been accomplished and our hypothesis would have had the merit of having set the stage for the exposure of such a fundamental ugly fact.

Acknowledgements. The authors would like to thank Dr. J.J. Oppenheim and Dr. D. Longo for the critical review of the manuscript.

References

Abelson HT, Johnson LF, Penman S, Green H (1974) Changes in RNA in relation to growth of the fibroblasts. II. The lifetime of mRNA, rRNA and tRNA in resting and growing cells. Cell 1: 151–161

Adams DO, Hamilton TA (1984) The cell biology of macrophage activation. Annu Rev Biochem 2: 283–302

Adams DO, Hamilton TA (1987) Molecular transductional mechanisms by which rIFN and other signals regulate macrophage development. Immunol Rev 97: 5–27

Aune TM, Pogue SL (1989) Inhibition of tumor cell growth by interferon-gamma is mediated by two distinct mechanisms dependent upon oxygen tension: induction of tryptophan degradation and depletion of intracellular nicotinamide adenine dinucleotide. J Clin Invest 84: 863–875

Belasco JG, Higgins CF (1988) Mechanisms of mRNA decay in bacteria: a perspective. Gene 72: 15–23

Benech P, Merlin G, Revel M, Chebath J (1985) 3′ and structure of the human (2–5) oligoA synthetase gene: prediction of two distinct proteins with cell type-specific expression. Nucleic Acids Res 13: 1267–1281

Bergmann IE, Brawerman G (1977) Control of breakdown of the polyadenylate sequence in mammalian polyribosomes: role of poly (adenylic acid)-protein interaction. Biochemistry 16: 259–264

Bianchi M, Bertini R, Ghezzi P (1988) Induction of indoleamine dioxygenase by interferon in mice: a study with different recombinant interferons and various cytokines. Biochem Biophys Res Commun 152: 237–242

Bizzozzero NP, Espinoza IS, Medrano EE, Fernandez MTF (1985) Transcription of ribosomal RNA is differentially controlled in resting and growing Balb/c 3T3 cells. J Cell Physiol 124: 160–164

Blasi E, Varesio L (1984) Role of protein synthesis in the activation of cytotoxic mouse macrophages by lymphokines. Cell Immunol 85: 15–24

Blasi E, Herberman RB, Varesio L (1984) Requirement for protein synthesis for induction of macrophage tumoricidal activity by IFN-α and IFN-β but not by IFN-γ. J Immunol 132: 3226–3228

Blasi E, Mathieson BJ, Varesio L, Cleveland JL, Borchert PA, Rapp UR (1985) Selective immortalization of murine macrophages from fresh bone marrow by a *raf/myc* recombinant murine retrovirus. Nature 318: 667–670

Blasi E, Radzioch D, Durum SK, Varesio L (1987) A murine macrophage cell line, immortalized by v-*raf* and v-*myc* oncogenes, exhibits normal macrophage functions. Eur J Immunol 17: 1491–1498

Blasi E, Radzioch D, Varesio L (1988) Inhibition of retroviral mRNA expression in the murine macrophage cell line GG2EE by biologic response modifiers. J Immunol 141: 2153–2157

Blasi E, Radzioch D, Merletti L, Varesio L (1989) Generation of macrophage cell line from fresh bone marrow cells with a *myc/raf* recombinant retrovirus. Cancer Biochem Biophys 10: 303–317

Bonvini E, Hoffman T, Herberman RB, Varesio L (1986) Selective augmentation by recombinant interferon-gamma of the intracellular content of *S*-adenosyl methionine in murine macrophages. J Immunol 136: 2596–2604

Bourgeade MF, Besancon F (1984) Induction of 2′–5′-oligoadenylate synthetase by retinoic acid in two transformed human cell lines. Cancer Res 44: 5355–5360

Bowman LH, Emerson CP (1977) Post transcriptional regulation of ribosome accumulation during myoblast differentiation. Cell 10: 581–588

Bowman LH, Golman WE, Golberg GI, Herbert MB, Schlessinger D (1983) Location of the initial cleavage sites in mouse pre-rRNA Mol Cell Biol 3: 1501–1510

Brawerman G (1987) Determinants of messenger RNA stability. Cell 48: 5–6

Busch H, Reddy R, Rothblum L, Choi YC (1982) SnRNAs, SnRNPs, and RNA processing, Annu Rev Biochem 51: 617–654

Byrne GI, Lehmann LK, Kirschbaum JG, Borden EC, Lee CM, Brown RR (1986a) Induction of tryptophan degradation in vitro and in vivo: a gamma-interferon-stimulated activity. J Interferon Res 6: 389–396

Byrne GI, Lehmann LK, Landry GJ (1986b) Induction of tryptophan catabolism is the mechanism for gamma-interferon-mediated inhibition of intracellular *Chlamydia psittaci* replication in T24 cells. Infect Immun 53: 347–351

Carlin JM, Borden EC, Sondel PM, Byrne GI (1989a) Interferon-induced indoleamine 2,3-dioxygenase activity in human mononuclear phagocytes. J Leukocyte Biol 45: 29–34

Carlin JM, Borden EC, Byrne GI (1989b) Interferon-induced indoleamine 2,3-dioxygenase activity inhibits *Chlamydia psittaci* replication in human macrophages. J Interferon Res 9: 329–337

Carlin JM, Ozaki Y, Byrne GI, Brown RR, Borden EC (1989c) Interferons and indoleamine 2,3-dioxygenase: role in antimicrobial and antitumor effects. Experientia 45: 535–541

Carlin JM, Borden EC, Byrne GI (1989d) Enhancement of indoleamine 2,3-dioxygenase activity in cancer patients receiving interferon-beta Ser. J Interferon Res 9: 167–173

Chang DD, Sharp PA (1989) Regulation by HIV *rev* depends upon recognition of splice sites. Cell 59: 789–795

Chen TY, Bright SW, Pace JL, Russell SW, Morrison DC (1990) Induction of macrophage-mediated tumor cytotoxicity by a hamster monoclonal antibody with specificity for lipopolysaccharide receptor. J Immunol 145: 8–12

Cleveland DW, Yen TJ (1989) Multiple determinants of eukaryotic mRNA stability. New Biologist 1: 121–126

Collins JJ, Adler CA, Fernandez-Pol JA, Court D, Johnson GS (1979) Transient growth inhibition of *Escherichia coli* K-12 by ion chelators: "in vivo" inhibition of ribonucleic acid synthesis. J Bacteriol 138: 923–932

Cooper H (1970) Control of synthesis and wastage of ribosomal RNA in lymphocytes. Nature 227: 1105–1107

Cooper H (1973) Degradation of 28S RNA rate in ribosomal RNA maturation in nongrowing lymphocytes and its reversal after growth stimulation. J Cell Biol 59: 250–258

Cosman D (1987) Control of messenger RNA stability. Immunol Today 8: 16–17

Costantini MG, Johnson GS (1981) Disproportionate accumulation of 18S and 28S ribosomal RNA in cultured normal rat kidney cells treated with picolinic acid or 5-methylnicotinamide. Exp Cell Res 132: 443–451

Cox GW, Mathieson BJ, Gandino L, Blasi E, Radzioch D, Varesio L (1989) Heterogeneity of hematopoietic cells immortalized by v-*myc*/v-*raf* recombinant retrovirus infection of bone marrow or fetal liver. J Natl Cancer Inst 81: 1492–1496

Cox GW, Mathieson BJ, Giardina SL, Varesio L (1990) Characterization of IL-2 receptor expression and function on murine macrophages. J Immunol 145: 1719–1726

Crocker PR, Jefferies WA, Clark SJ, Chung PL, Gordon S (1987) Species heterogeneity in macrophage expression of the CD4 antigen. J Exp Med 166: 613–618

Crocker PR, Morris L, Gordon S (1988) Novel cell surface adhesion receptors involved in interactions between stromal macrophages and haematopoietic cells. J Cell Sci [Suppl] 9: 185–206

Crouch R (1984) Ribosomal RNA processing in eukaryotes. In Apirion D (ed) Processing of RNA. CRC, New York, pp 214–226

De Benedetti A, Baglioni C (1983) Phosphorylation of initiation factor eIF-2, binding of mRNA to 48S complexes, and its reutilization in initiation of protein synthesis. J Biol Chem 258: 14556–14562

De Benedetti A, Baglioni C (1984) Inhibition of mRNA binding to ribosomes by localized action of dsRNA-dependent protein kinase. Nature 311: 79–81

Dingwall C, Gait MJ, Karn J, Skinner M, Valerio R (1989) Human immuno-deficiency virus 1 *tat* protein binds *trans*-activation-response region (TAR) RNA in vitro. Proc Natl Acad Sci USA 86: 6925–6929

Downing R, Roussel MF, Sherr CJ (1989) Ligand and protein kinase C downmodulate the colony stimulating factor 1 receptor by independent mechanisms. Mol Cell Biol 9: 2890–2896

Duncan R, Hershey JWB (1985) Regulation of initiation factors during translational repression caused by serum depletion: covalent modification. J Biol Chem 260: 5493–5497

Edery I, Petryshyn R, Sonenberg N (1989) Activation of double stranded RNA-dependent kinase (dsl) by the TAR region of HIV-1 mRNA: a novel translational control mechanism. Cell 56: 303–312

Emerson CP (1971) Regulation of the synthesis and the stability of ribosomal RNA during contact inhibition of growth. Nature 232: 101

Espinoza-Delgado I, Ortaldo JR, Winkler-Pickett R, Sugamura K (1990a) Expression and role of p75 interleukin 2 receptor on human monocytes. J Exp Med 171: 1821–1826

Espinoza-Delgado I, Longo DL, Gusella GL, Varesio L (1990b) Interleukin 2 enhances c-*fms* expression in human monocytes. J Immunol 145: 1719–1726
Evans GW, Johnson EC (1980) Zinc absorption in rats fed a low-protein diet and supplemented with tryptophan or picolinic acid. J Nutr 110: 1076–1080
Evans GW, Johnson PE (1980) Characterization and quantitation of a zinc binding ligand in human milk. Pediatr Res 14: 867–880
Faltynek CR, Kung H (1988) The biochemical mechanisms of action of the interferons. BioFactors 1: 227–235
Farrell PJ, Balkow T, Hunt T, Jackson RJ, Trachsel H (1977) Phosphorylation of initiation factor eIF-2 and the control of reticulocyte protein synthesis. Cell 11: 187–200
Ferbus D, Testa U, Titeux M, Louache F, Thang MN (1985) Induction of 2′–5′-oligoadenylate synthetase activity during granulocyte and monocyte differentiation. Mol Cell Biochem 67: 125–133
Fernandez-Pol JA (1977) Iron: possible cause of the G1 arrest induced by picolinic acid. Biochem Biophys Res Commun 78: 136–143
Fernandez-Pol JA, Johnson GS (1977) Selective toxicity induced by picolinic acid in simian virus 40-transformed cells in tissue culture. Cancer Res 37: 4276–4279
Fernandez-Pol JA, Bono VH, Johnson GS (1977) Control of growth by picolinic acid: differential response of normal and transformed cells. Proc Natl Acad Sci USA 7: 2889–2893
Galbru J, Hovanessian AG (1985) Two interferon-induced proteins are involved in the protein kinase complex dependent on double stranded RNA. Cell 43: 685–694
Gandino L, Varesio L (1990) Immortalization of macrophages from mouse bone marrow and fetal liver. Exp Cell Res 188: 192–198
Gatignol A, Kumar A, Rabson A, Jeang KT (1989) Identification of cellular proteins that bind to the human immunodeficiency virus type 1 *trans*-activation responsive TER element RNA. Proc Natl Acad Sci USA 86: 7828–7832
Goldfeld EA, Doyle C, Maniatis T (1990) Human tumor necrosis factor alpha gene regulation by virus and lipopolysaccharide. Proc Natl Acad Sci USA 87: 9769–9773
Gordon S, Keshav S, Chung LP (1988) Mononuclear phagocytes: tissue distribution and functional heterogeneity. Curr Opin Immunol 1: 26–35
Goswami BB, Shrma OK (1984) Degradation of rRNA in interferon-treated vaccinia virus-infected cells. J Biol Chem 259: 1371–1374
Gourse L, Takebe Y, Sharrock RA, Nomura M (1985) Feedback regulation of rRNA and tRNA synthesis and accumulation of free ribosomes after conditional expression of rRNA genes. Proc Natl Acad Sci USA 82: 1069–1073
Grummt I, Sorbaz H, Hofmann A, Roth E (1985a) Spacer sequences downstream of the 28S RNA coding region are part of the mouse rDNA transcription unit. Nucleic Acids Res 13: 2293–2304
Grummt I, Maier U, Ohrlein A, Hassouna N, Bachellerie JP (1985b) Transcription of mouse rDNA terminates downstream of the 3′ end of 28S RNA and involves interaction of factors with repeated sequences in the 3′ spacer. Cell 43: 801–810
Gunnery S, Rice AP, Robertson HD, Mathews MB (1990) *tat*-Responsive region RNA of human immunodeficiency virus 1 can prevent activation of the double-stranded-RNA-activated protein kinase. Proc Natl Acad Sci USA 87: 8687-8691
Gurney T Jr (1985) Characterization of mouse 45S ribosomal RNA subspecies suggests that the first processing cleavage occurs 600 +/− 100 nucleotides from the 5′ end and the second 500 + − 100 nucleotides from the 3′ end of a 13.9 kb precursor. Nucleic Acids Res 13: 4905–4919
Gusella GL, Ayroldi E, Espinoza ID, Varesio L (1990) LPS but not IFN gamma down-regulates c-*fms* protooncogene expression in murine macrophages. J Immunol 145: 1137–1143
Hadjiolov AA, Nikolaev N (1976) Maturation of ribosomal ribonucleic acids and biogenesis of ribosomes. Prog Biophys Mol Biol 31: 95
Hamilton TA, Bredon N, Ohmori Y, Tannenbaum CS (1989) IFN-gamma and IFN-beta independently stimulate the expression of lipopolysaccharide-inducible genes in murine peritoneal macrophages. J Immunol 142: 2325–2331
Hogan MM, Vogel SN (1988) Production of tumor necrosis factor by rIFN-gamma-primed C3H/Hej macrophages requires the presence of lipid A-associated proteins. J Immunol 141: 4196–4202
Horiguchi E, Sariban E, Kufe D (1988) Transcriptional and posttranscriptional regulation of CSF-1 gene expression in human monocytes. Mol Cell Biol 8: 3951–3954
Hosick HL, Strohman RC (1971) Changes in ribosome-polysome balances in chick muscle cells. J Cell Physiol 77: 145–156

Hovanessian AG (1989) The double stranded RNA-activated protein kinase induced by interferon. J Interferon Res 9: 641–647

Hovanessian AG, Svab J, Marie I, Robert N, Chamaret S, Laurent A (1988) Characterization of 69- and 100 kd forms of 2–5 A-synthetase from interferon-treated human cells. J Biol Chem 263: 4959–4964

Hurley LS, Lonnerdal B (1982) Zinc binding in human milk: citrate versus picolinate. Nutr Rev 40: 65–71

Johnson LE, Levis R, Abelson HT, Green H, Penman S (1976) Changes in RNA in relation to growth of the fibroblasts. IV. Alterations in the production and processing of mRNA and rRNA in resting and growing cells. J Cell Biol 71: 933

Jones TR, Cole MD (1987) Rapid cytoplasmic turnover of c-*myc* specific mRNA: requirement of the 3′ untranslated sequences. Mol Cell Biol 7: 4513–4521

Kabnick KS, Housman DE (1988) Determinants that contribute to cytoplasmic stability of human c-*fos* and globin mRNA are located at several sites in each mRNA. Mol Cell Biol 8: 3244–3250

Kalvakolanu DV, Bandyopadhyay SK, Tiwary RK, Sen GS (1991) Enhancement of expression of exogenous genes by 2-aminopurine. J Biol Chem 266: 873–879

Kaufman RJ, Murtha P (1987) Translational control mediated by eucaryotic initiation factor-2 is restricted to specific mRNAs in transfected cells. Mol Cell Biol 7: 1568–1571

King HWS, Gould H (1970) Low molecular weight ribonucleic acid in rabbit reticulocyte ribosomes. Mol Biol 51: 687–702

Klausner RD, Harford JB (1989) *cis-trans* models for post-transcriptional gene regulation. Science 246: 870–872

Koerner TJ, Hamilton TA, Introna M, Tannenbaum CS, Bast C, Adams DO (1987) The early competence genes JE and KC are differently regulated in murine peritoneal macrophages in response to lipopolysaccharide. Biochem Biophys Res Commun 149: 969

Komiani R, Mishima Y, Urano Y, Sakai M, Muramatsu M (1982) Cloning and determination of the transcriptiontermination site of ribosomal RNA gene of the mouse. Nucleic Acids Res 10: 1963–1979

Kovacs EJ, Radzioch D, Young HA, Varesio L (1988) Differential inhibition of IL-1 and TNF-alpha mRNA expression by agents which block second messenger pathways in murine macrophages. Immunol 141: 3101–3105

Krauter KS, Soiero R, Nadal-Ginard B (1979) Transcriptional regulation of ribosome RNA accumulation during myoblast differentiation. J Mol Biol 134: 727–741

Krieger I (1980) Picolinic acid in the treatment of disorders requiring zinc supplementation. Nutr Rev 38: 148–150

Krieger I, Statter M (1987) Tryptophan deficiency and picolinic acid: effect on zinc metabolism and clinical manifestation of pellagra. Am J Clin Nutr 46: 511–517

Krishnan I, Baglioni C (1981) Elevated levels of 2′–5′-oligoadenylic acid polymerase activity in growth-arrested human lymphoblastoid Namalva cells. Mol Cell Biol 1: 932–938

Kumar R, Mendelson J (1989) Role of 2′–5′-oligoadenylate synthetase in interferon-gamma mediated growth inhibition of A431 cells. Cancer Res 49: 5180–5184

Kurst B, Galbru J, Hovanessian AG (1984) Further characterization of the protein kinase activity mediated by interferon in mouse and human cells. J Biol Chem 259: 8494–8498

Lazinski D, Grzadzielska E, Das A (1989) Sequence specific recognition of RNA hairpins by bacteriophage antiterminators requires a conserved arginine rich motif. Cell 59: 207–218

Lee MT, Warren K (1987) CSF-1 induced resistance to viral infection in murine macrophages. J Immunol 138: 3019–3022

Lee SW, Chan H, Petrie K, Allison AC (1988) Glucocorticoids selectively inhibit the transcription of the interleukin 1 gene and decrease the stability of interleukin 1 beta mRNA. Proc Natl Acad Sci USA 58: 1024

Lengyel P (1982) Biochemistry of interferons and their actions. Annu Rev Biochem 51: 251–282

Leung KP, Russell SW, LeBlanc PA, Caballero S (1985) Heterogeneity among macrophages cultured from mouse bone marrow. Morphologic, cytochemical and flow cytometric analyses. Cell Tissue Res 239: 693–701

Leuthauser SWC, Oberley LW, Oberley TD (1982) Antitumor activity of picolinic acid in CBA/J mice. J Natl Cancer Inst 68: 123–125

Levine RA, McCormack JE, Buckler A, Sononshein GE (1986) Transcriptional and posttranscriptional control of c-*myc* gene expression in WEHI231 cells. Mol Cell Biol 6: 4112–4116

MacKay RJ, Pace JL, Parpe MA, Russell SW (1989) Macrophage activation- associated proteins. Characterization of stimuli and conditions needed for expression of proteins 47b, 71/73, and 120. J Immunol 142: 1639–1645

Malter JS (1989) Identification of AUUUA-specific messenger RNA binding protein. Science 246: 664–666

Maniatis T (1988) Control of gene expression in eukaryotes. Harvey Lect 82: 71–104

Marie I, Svab J, Robert N, Galbru J, Hovanessian AG (1990) Differential expression and distinct structure of 69- and 100-kDa forms of 2–5 A synthetase in human cells treated with interferon. J Biol Chem 265: 18601–18607

Mauch JC, Green TC (1973) Regulation of RNA synthesis in fibroblasts during the transition from resting to growing state. Proc Natl Acad Sci USA 70: 2819

Maza LM, Peterson EM (1988) Dependence of the in vitro antiproliferative activity of recombinant human gamma-interferon on the concentration of tryptophan in culture media. Cancer Res 48: 346–350

McLure MA, Perrault J (1986) RNA virus genomes hybridize to cellular rRNA and to each other. J Virol 57: 917–921

Meijlink F, Curran T, Miller AD, Verma LM (1985) Removal of 67-base-pair sequence in the noncoding region of protooncogene *fos* converts it to a transforming gene. Proc Natl Acad Sci USA 82: 4987–4991

Menard MP, Cousins RJ (1983) Effect of citrate, glutathione and picolinate on zinc transport by brush border membrane vesicles from rat intestine. J Nutr 113: 1653–1656

Metcalf D (1986) The molecular biology and functions of the granulocyte-macrophage colony-stimulating factor. Blood 67: 257

Meurs E, Chong K, Galbru J, Thomas NSB, Kerr IM, Williams BRG, Hovanessian AG (1990) Molecular cloning and characterization of the human double stranded RNA-activated protein kinase induced by interferon. Cell 62: 379–390

Michot B, Bachellerie J, Raynal F (1983) Structure of the mouse rRNA precursor. Complete sequence and potential folding of the spacer regions between 18S and 28S rRNA. Nucleic Acids Res 11: 3375–3390

Miller KG, Sollner-Webb B (1981) Transcription of mouse rRNA genes by RNA polymerase I: in vitro and in vivo initiation and processing sites. Cell 27: 165–174

Minks MA, Benvin S, Maroney PA, Baglioni C (1979) Synthesis of 2′–5′-oligo (A) in extracts of interferon-treated HeLa cells. J Biol Chem 245: 5058–5064

Moore PB (1988) The ribosome returns. Nature 331: 223–227

Moore RN, Larsen HS, Horohov DW, Rouse BT (1984) Endogenous regulation of macrophage proliferative expansion by colony stimulating factor induced interferon. Science 223: 178

Murray HW, Szuro-Sudol A, Wellner D et al. (1989) Role of tryptophan degradation in respiratory burst-independent antimicrobial activity of gamma interferon-stimulated human macrophages. Infect Immun 57: 845–849

Nathan CF (1987) Secretory products of macrophages. J Clin Invest 79: 319–326

Nierhous KN (1982) Structure, assembly and function of ribosomes. Curr Top Microbiol Immunol 97: 81–157

Olsen HS, Nelbolock P, Cochrane AW, Rosen CA (1990) Secondary structure is the major determinant for interaction of HIV *rev* protein with RNA. Science 247: 845–848

Ouellette AJ, Bandman E, Kumar A (1976) Regulation of ribosomal RNA methylation in a temperature sensitive mutant of BHK cells. Nature 262: 619–621

Ozaki Y, Edelstein MP, Duch DS (1987) The actions of interferon and antiinflammatory agents of induction of indoleamine 2, 3-dioxygenase in human peripheral blood monocytes. Biochem Biophys Res Commun 144: 1147–1153

Ozaki Y, Edelstein MP, Duch DS (1988) Induction of indoleamine 2, 3- dioxygenase: a mechanism of the antitumor activity of interferon gamma. Proc Natl Acad Sci USA 85: 1242–1246

Pace JL, Russell SW, Torres BA, Johnson HM, Gray PW (1983) Recombinant mouse gamma interferon induces the priming step in macrophage activation for tumor cell killing. J Immunol 130: 2011–2013

Pace JL, Russell SW, LeBlanc PA, Murasko DM (1985a) Comparative effects of various classes of mouse interferons on macrophage activation for fumor cell killing. J Immunol 134: 977–981

Pace JL, Varesio L, Russell SW, Blasi E (1985b) The strain of mouse and assay conditions influence whether MuIFN-gamma primes or activates macrophages for tumor cell killing. J Leukocyte Biol 37: 475–479

Peppel K, Vinci JM, Baglinoi C (1991) The AU-rich sequences in the 3′ untranslated region mediate the increased turnover of interferon mRNA induced by glucocorticoids. J Exp Med 173: 349–355
Perkins KK, Furneaux HM, Hurwitz J (1986) RNA splicing products formed with isolated fractions from HeLa cells are associated with fast-sedimenting complexes. Proc Natl Acad Sci USA 83: 887–891
Perry RP (1972) Ribosomal RNA synthesis and processing. Biochem Soc Symp 37: 105–116
Perry RP (1976) Processing of RNA. Annu Rev Biochem 45: 605–629
Perry VH, Gordon S (1988) Macrophages and microglia in the nervous system. Trends Neurosci 11: 273–277
Pfefferkorn ER (1984) Interferon gamma blocks the growth of *Toxoplasma gondii* in human fibroblasts by inducing the host cells to degrade tryptophan. Proc Natl Acad Sci USA 81: 908–912
Pfefferkorn ER (1986) Interferon gamma and the growth of *Toxoplasma gondii* in fibroblasts. Ann Inst Pasteur Microbiol 137: 348–352
Pfefferkorn ER, Eckel M, Rebhun S (1986) Interferon-gemma suppresses the growth of *Toxoplasma gondii* in human fibroblasts through starvation for tryptophan. Mol Biochem Parasitol 20: 215–224
Rabbitts PH, Foster A, Stinson MA, Rabbitts TH (1985) Truncated exon 1 from the c-*myc* gene results in prolonged c-*myc* mRNA stability. EMBO J 4: 3727–3733
Radzioch D, Varesio L (1988) Protein Kinase C inhibitors block the activation of macrophages by IFN-beta but not by IFN-gamma. J Immunol 140: 1259–1263
Radzioch D, Clayton M, Varesio L (1987a) Interferon-alpha, -beta, and -gamma augment the levels of rRNA precursors in peritoneal macrophages but not in macrophage ce;; ;ines and fibroblasts. J Immunol 139: 805–812
Radzioch D, Bottazzi B, Varesio L (1987b) Augmentation of c-*fos* mRNA expression by activators of protein kinase C in fresh, terminally differentiated resting macrophages. Mol Cell Biol 7: 595–599
Ragow R (1987) Regulation of messenger RNA turnover. TIBS 12: 358–360
Ralph P, Nakoinz I (1987) Stimulation of macrophage tumoricidal activity by the growth and differentiation factor CSF-1. Cell Immunol 105: 270–279
Rosen H, Gordon S (1990a) the role of the type 3 complement receptor in the induced recruitment of myelomonocytic cells to inflammatory sites in the mouse. Am J Respir Cell Mol Biol 3: 3–10
Rosen H, Gordon S (1990b) Adoptive transfer of fluorescence-labeled cells shows that resident peritoneal microphages are able to migrate into specialized lymphoid organs and inflammatory sites in the mouse. Eur J Immunol 20: 1251–1258
Roy S, Agy M, Hovanessian AG, Sonenberg N, Katze MG (1991) The intergrity of the stem structure of human immunodeficiency virus type 1 *tat*-responsive sequence RNA is required for interaction with the interferon-induced 68000-*Mr* protein kinase. J Virol 65: 632–640
Ruffmann R, Welker RD, Saito T, Chirigos MA, Varesio L (1984) In vivo activation of macrophages but not natural killer cells by picolinic acid (PLA). J Immunopharmacol 6: 291–304
Ruffmann R, Schlick R, Chirigos MA, Budzynsky W, Varesio L (1987) Antiproliferative activity of picolinic acid due to macrophage activation. Drugs Exp Clin Res 13: 607–614
Russell SW, Pace JL, Varesio L et al. (1986) Comparison of five short-term assays that measure nonspecific cytotoxicity mediated to tumor cells by activated macrophages. J Leukocyte Biol 40: 801–813
Samuel CE (1987) Mechanisms of interferon actions. in: Pfeffer ML (ed) Mechanisms of interferon actions, Vol 1. CRC, Boca Raton, Fl, pp 111–130
Sariban EK, Imamura R, Luebbers R, Kufe D (1988) Transcriptional and post-transcriptional regulation of tumor necrosis factor gene expression in human monocytes. J Clin Invest 81: 1506
Saunders ME, Gewert DR, Tugwell ME, McMahon M, Williams BRG (1985) Human 2–5 A synthetase: characterization of a novel cDNA and corresponding structure. EMBO J 4: 1761–1768
Schmit T, Chen PS, Pellegrini M (1985) The induction of ribosome biosynthesis in a nonmitotic secretory tissue. J Biol Chem 12: 7645–7650
Seal CJ, Heaton FW (1985) Effect of dietary picolinic acid on the metabolism of exogenous and endogenous zinc in the rat. J Nutr 115: 986–993
Shaw G, Kamen R (1986) A conserved AU sequence from the 3′ untranslated region of GM-CSF mRNA mediates selective mRNA degradation. Cell 46: 659–667
Sherr CJ (1990) Colony-stimulating factor-1 receptor. Blood 75: 1–12
Shuler GD, Cole MD (1988) GM-CSF and oncogene mRNA stabilities are independently regulated in *trans* in a mouse monocytic tumor. Cell 55: 1115–1122
Sokawa Y, Nagata K, Ichikawa Y (1981) Induction and function of 2′–5′ oligoadenylate synthetase in differentiation of mouse myeloid leukemia cells. Exp Cell Res 135: 191–197

Sollner-Webb B, Tower J (1986) Transcription of cloned eukaryotic ribosomal RNA genes. Annu Rev Biochem 55: 801–830

Stark GR, Dower WJ, Schimke RT, Brown RE, Kerr IM (1979) 2–5 (A) synthetase: assay, distribution and variation with growth or hormone status. Nature 278: 471–473

Szyszka R, Kudicki W, Kramer G, Hardestty B, Galbru J, Hovanessian AG (1989) A type 1 phosphoprotein phosphatase active with phosphorylated M_r 68 000 initiation factor 2 kinase. J Biol Chem 264: 3827–3731

Taffet SM, Pace JL, Russell SW (1981) Lymphokine maintains macrophage activation for tumor cell killing by interfering with the negative regulatory effect of prostaglandin E_2. J Immunol 127: 121–127

Taniguchi T (1988) Regulation of cytokine gene expression. Annu Rev Immunol 6: 439–464

Tannenbaum CS, Hamilton TA (2989) Lipopolysaccharide-induced gene expression in murine peritoneal macrophage is selectively suppressed by agents that elevate intracellular cAMP. J Immunol 142: 1274–280

Taramelli D, Varesio L (1981) Activation of murine macrophage. I. Different pattern of activation by poly I: C than by lymphokine or LPS. J Immunol 127: 58–63

Tiollais P, Galibert F, Boiron M (1971) Evidence for the existence of several molecular species in the 45S fraction of mammalian ribosomal precursor RNA. Proc Natl Acad Sci USA 68: 1117–1120

Tiwary RK, Kusari J, Kumar R, Sen GS (1988) Gene induction by interferons and double-stranded RNA: selective inhibition by 2-aminopurine. Mol Cell Biol 8: 4289–4294

Toniolo D, Basilico C (1975) Processing of ribosomal RNA in a temperature sensitive mutant of BHK cells. Biochim Biophys Acta 425: 409

Toniolo D, Meiss HK, Basilico C (1973) A temperature sensitive mutation affecting 28S ribosomal RNA production in mammalian cells. Proc Natl Acad Sci USA 70: 1273

Varesio L (1983) Inhibition of immune functions by macrophages. In: Friedman H, Herberman RB (eds) The reticuloendothelial system. Plenum, New York, pp 217–252

Varesio L (1984) Down regulation of RNA labeling as a selective marker for cytotoxic but not suppressor macrophages. J Immunol 132: 2683–2685

Varesio L (1985a) Imbalanced accumulation of ribosomal RNA in macrophages activated in vivo or in vitro to a cytolytic stage. J Immunol 134: 1262–1267

Varesio L (1985b) Induction and expression of tumoricidal activity by macrophages. In: Dean RT, Jessup W (eds) Mononuclear phagocytes: physiology and pathology. Elsevier Science, Amsterdam, pp 381–407

Varesio L (1986) Molecular bases for macrophage activation. Ann Inst Pasteur Immunol 137: 235–240

Varesio L, Landolfo S, Forni G (1980) The macrophage as the social interconnection within the immune system. Dev Comp Immunol 4: 11–19

Varesio L, Issaq HJ, Taramelli D (1983) RNA synthesis in activated macrophages I. Poly(I) X poly(C)-induced triggering of cytolytic activity is associated with decrease in RNA synthesis. Eur J Immunol 13: 959–964

Varesio L, Blasi E, Thurman GB, Talmadge JE, Wiltrout RH, Herberman RB (1984a) Potent activation of mouse macrophages by recombinant interferon-gamma. Cancer Res 44: 4465–4469

Varesio L, Issaq HJ, Kowal R, Bonvini E, Taramelli D (1984b) Lymphokines inhibit macrophage RNA synthesis. Cell Immunol 84: 51–64

Varesio L, Clayton M, Radzioch D, Bonvini E (1987) Selective inhibition of 28S ribosomal RNA in macrophages activated by interferon-gamma or -beta. J Immunol 138: 2332–2337

Varesio L, Clayton M, Blasi E, Ruffmann R, Radzioch D (1990) Picolinic acid, a catabolite of tryptophan, as the second signal in the activation of IFN gamma primed macrophages. J Immunol 145: 4265–4271

Warren MK, Ralph P (1986) Macrophage growth factor CSF-1 stimulates human monocytes production of interferon, tumor necrosis factor, and colony stimulating activity. J Immunol 137: 2281

Weber MJ (1971b) Ribosomal RNA turnover in contact inhibited cells. Nature 235: 51

Wells V, Mallucci L (1985) Expression of the 2–5 A system during the cell cycle. Exp Cell Res 159: 27–36

Werner ER, Bitterlich G, Fuchs D et al. (1987a) Human macrophages degrade tryptophan upon induction by interferon-gamma. Life Sci 41: 273–280

Werner ER, Hirsch-Kauffmann M, Fachs D, Hausen A, Ribnegger G, Schweiger M, Wachter H (1987b) Interferon-gamma-induced degradation of tryptophan by human cells in vitro. Biol Chem Hoppe Seyler 368: 1407–1412

Werner ER, Werner-Felmayer G, Fuchs D, Hausen A, Reibnegger G (1988) Influence of interferon-gamma and extracellular tryptophan on indoleamine 2,3-dioxygenase activity in T24 cells as determined by a non-radiometric assay. Biochem J 256: 537–541

Werner-Felmayer G, Werner ER, Fuchs D, Hausen A, Reibnegger G (1989) Characteristics of interferon induced tryptophan metabolism in human cells in vitro. Biochim Biophys Acta 1012: 140–147

Wilson T, Treisman R (1988) Removal of poly(A) and consequent degradation of c-*fos* mRNA facilitated by 3′ AU rich sequences. Nature 336: 369–399

Wiltrout RH, Varesio L (1987) Activation of macrophages for cytotoxic and suppressive functions. In: Oppenheim JJ, Shevach E (eds) Textbook of immunophysiology: role of cells and cytokines in Immunity and inflammation. Oxford Press, New York

Woese CR, Winker S, Gutell RR (1990) Architecture or ribosomal RNA: constraints on the sequence of "tetra-loops". Proc Natl Acad Sci USA 87: 8467–8471

Wreshner DH, Rechavi G (1988) Differential mRNA stability to reticulocyte ribonucleases correlates with 3′ non-coding (U)nA sequences. Eur J Biochem 172: 333–340

Yasui H, Takai K, Yoshida R, Hayaishi O (1986) Interferon enhances tryptophan metabolism by inducing pulmonary indoleamine 2, 3-dioxygenase: its possible occurrence in cancer patients. Proc Natl Acad Sci USA 83: 6622–6626

Zapp ML, Green MR (1989) Sequence specific RNA binding by the HIV-1 Rev protein. Nature 342: 714–716

Zinn K, Keller A, Whittemore L, Maniatis T (1988) 2-Aminopurine selectively inhibits the induction of beta-interferon, c-*fos*, and c-*myc* gene expression. Science 240: 210–213

Mononuclear Phagocytes as Targets, Tissue Reservoirs, and Immunoregulatory Cells in Human Immunodeficiency Virus Disease

M. S. MELTZER and H. E. GENDELMAN

1 Mononuclear Phagocytes as a Tissue Reservoir for Virus in the HIV-Infected Patient

Infection by the human immunodeficiency virus (HIV) initiates a slowly progressive degenerative disease of the immune system termed the acquired immunodeficiency syndrome (AIDS). The primary immunologic defect in AIDS is an inexorable depletion of $CD4^+$ T cells, a depletion invariably associated with opportunistic infection, degenerative neurologic disease, a variety of neoplastic changes, and ultimately death (LIFSON et al. 1988). The fequency of infected cells in blood of asymptomatic HIV-seropositive subjects, as detected by polymerase chain reaction gene amplification of DNA from leukocyte lysates or by direct isolation of HIV from limiting dilutions of blood leukocytes, is about 1 in 40000 (0.0025%). This frequency increases about 1000-fold in patients with symptomatic disease: AIDS-related complex (ARC) and AIDS (SCHNITTMAN et al. 1989; PSALLIDOPOULOS et al. 1989; HO et al. 1989a). Studies further show that virtually all infected blood leukocytes throughout HIV disease are $CD4^+$ T cells (SCHNITTMAN et al. 1989; PSALLIDOPOULOS et al. 1989). Indeed, in late-stage disease about 1 in 40 $CD4^+$ T cells harbor virus. Attempts to detect viral protein or mRNA in blood leukocytes of seropositive patients, however, reveal a frequency of productively infected cells in early or late disease of no more than 0.001%

HIV Immunopathogenesis Program, Department of Cellular Immunology, Walter Reed Army Institute of Research, Washington, DC 20307-5100, USA

Current Topics in Microbiology and Immunology, Vol. 181

(HARPER et al. 1986). Thus, >99% of infected T cells are latently infected. But T cells are not the only target cell for HIV. In certain bodily tissues, such as those of the central nervous system, lymph nodes, or lung, the frequency of cells productively infected with HIV may be 10000-fold higher than that in blood. In each of these tissues, the predominant cell type infected with HIV and producing virus is not the CD4$^+$ T cell, but rather the macrophage (for reviews see GENDELMAN et al. 1989; MELTZER et al. 1990).

Neurologic disease is strongly associated with HIV infection: more than 60% of patients with AIDS show symptomatic CNS disease; 80%–90% have neuropathologic abnormalities at autopsy (HO et al. 1989b). Such abnormalities are characterized by typical pathologic changes in brain and spinal cord. HIV-induced changes in the brain are most evident within the white matter and include atrophic changes without inflammation but associated with microglial nodules (clusters of microglia and reactive fibrous astrocytes) and multinucleated giant cells. Virus isolation from cerebrospinal fluid or homogenates of brain tissue is successful in most patients with AIDS-associated encephalopathy by (LEVY et al. 1985). Indeed, virus isolation from cerebrospinal fluid of patients with acute aseptic meningoencephalitis during HIV infection can occur before seroconversion (HO et al. 1985b). The predominant infected cell (and in most studies, the only infected cell) is the macrophage (STOLER et al. 1986; KOENIG et al. 1986). In situ hybridization for HIV RNA in brain tissue of infected individuals shows a frequency of productively infected macrophages at 1%–10%. Brain macrophages (subarachnoid, perivascular, and parenchymal cells), microglia, and macrophage-derived, multinucleated giant cells possess 500–1500 copies of HIV RNA per cell. This amount of virus RNA per infected cell is at least ten fold higher than that found in blood leukocytes. Interestingly, almost all of the HIV-infected brain macrophages are negative for CD4 by immunocytochemistry (VAZEUX et al. 1987).

Pathologic changes induced in the spinal cord by HIV are different from those in the brain. Vacuolar myelopathy with macrophage infiltrates is found at autopsy in about 25% of patients with AIDS (EILBOTT et al. 1989). The high frequency of HIV-infected macrophages with myelin in phagocytic vacuoles suggests a proximate role for these cells in the pathogenesis of AIDS-associated myelopathy syndrome.

Transmission electron microscopic analysis of lymph nodes from HIV-infected individuals showed typical virions in virtually all specimens examined (26 of 30 lymph nodes), even in those patients with early asymptomatic infection (ARMSTRONG and HORNE 1984; LE TOURNEAU et al. 1986). Viral particles are found only in follicular dendritic cells with an approximate frequency of 10% (GYORKEY et al. 1985). Similarly, HIV can be isolated from cells in bronchoalveolar lavage fluids (>90% macrophages) (ZIZA et al. 1985). HIV proteins or nucleic acids are detected in 10%–50% of the macrophages in such fluids (CHAYT et al. 1986; PLATA et al. 1987). Epidermal Langerhans cells, the dendritic, CD4$^+$ antigen-presenting cell of skin, are also targets for HIV infection. In skin biopsies of 40 seropositive patients, HIV-infected Langerhans cells were identified in about 20%. However, such HIV-infected Langerhans cells were present at a frequency of infection much lower than that for macrophages of brain, lymph node, or lung (TSCHACHLER et al. 1987). Indeed, other investigators found immunocytochemical evidence for HIV infection in cells of the oral mucosa in only 2 of 26 seropositive patients (BECKER et al. 1988) or in 0 of 44 skin biopsies (KANITAKIS et al. 1989). These studies suggest that while Langerhans cells are susceptible targets for HIV infection (RAPPERSBERGER et al. 1988), the frequency of this event during HIV disease is much lower than that of other tissue macrophages.

The relatively high frequency of productively infected macrophages found in brain, lymph nodes, and lung (10%–50%) is not observed in all bodily tissues. For example, in the steady state about 60% of blood monocytes settle in the liver as Kupffer cells, yet Kupffer cell infection in HIV disease has not been described. Capacity to infect a tissue macrophages may be dependent upon cellular factors that change with differentiation. In visna-maedi disease of sheep, certain tissue macrophages are highly permissive for virus replication (macrophages of brain, lung, lymph nodes, and bone marrow), while other macrophages are resistant (connective tissue histiocytes, liver Kupffer cells) (GENDELMAN et al. 1985). Permissiveness of tissue macrophages to HIV and other lentivirus infection is dependent upon cell differentiation. Unlike many other cell types, however, changes in macrophage differentiation do not occur with cell cycle changes (tissue macrophages replicate at very low levels, if at all), but rather after exposure to any of a multitude of exogenous and endogenous stimuli. It is these stimuli that control the replication of HIV in mononuclear phagocytes.

2 Changes in Mononuclear Phagocyte Number, Phenotype, or Function During HIV Infection

Numerical, phenotypic, or functional changes that occur in macrophage subpopulations during HIV infection are not well defined. Observations made by different investigators using similar experimental techniques are often contradictory. Moreover, virtually all studies to date suffer two major interpretive problems. First, patient selection has been limited to individuals with late-stage, symptomatic HIV infection. The average time interval from seroconversion for HIV antibody to onset of AIDS may be > 10 years (LIFSON et al. 1988). Analysis of macrophage numbers, morphology, phenotype, or function in this asymptomatic time interval is almost non-existent. Second, the macrophage subpopulation most frequently analyzed has been the blood monocyte. This precursor cell to all tissue macrophages has a relatively short circulating half-life in blood of about 30 h. Migration of blood monocytes into tissue is unidirectional: unlike T cell traffic patterns, there is no evidence for tissue macrophage reentry into the blood. As previously stated, the reservoir for HIV in blood is the $CD4^+$ T cell. Thus, virtually no blood monocytes examined for any change during HIV disease are infected.

Most studies (but not all) document normal numbers of blood monocytes in HIV-infected patients even during late-stage disease where $CD4^{+}$ T cells may be undetectable (POLI et al. 1985; SIEGAL et al. 1986; ENK et al. 1986). Similarly, most studies document normal phenotypic expression of plasma membrane antigens by flow cytometric analysis with monoclonal antibodies. Expression of the class II major histocompatibility complex determinants (HLA-DP, HLA-DQ, HLA-DR) by monocytes from HIV seropositive individuals is indistinguishable from that of cells from seronegative donors (HAAS et al. 1987). Changes in class II major histocompatibility complex determinants induced in monocytes after treatment in vitro with interferon-γ (IFN-γ) or bacterial endotoxic lipopolysaccharides (LPS) were similar to those induced in control cells (HEAGY et al. 1984). Expression of several other plasma membrane determinants (CD4, CD11, CD14, CR3, transferrin receptor, Fc receptor I and II, or Mo3e) by monocytes from HIV-infected patients was also normal (DAVIDSON et al. 1988). Other studies performed with similar methodologies document significant changes in the expression of these monocyte membrane antigens (SEI et al. 1986; RIEBER and RIETHMULLER 1986; ROY et al. 1987; KOETHE et al. 1989). While it is difficult to reconcile such disparate observations, recent reports that the envelope glycoproteins of HIV, gp41 and gp120, both act directly on monocytes to induce phenotypic or functional change provide a potential mechanism to explain such variability (WAHL et al. 1989; WAHL LM et al. 1989; NAKAJIMA et al. 1989; TAS et al. 1988; MERRILL et al. 1989). Indeed, that expression of HLA-DR on monocytes was decreased only in those patients with detectable levels of p24 capsid protein in their blood supports this hypothesis (BRAUN et al. 1988).

Analysis of macrophage function during HIV infection has been approached both in vivo and in vitro. Several studies by BENDER and colleagues showed impaired clearance from circulation of particles that express Fc (spleen macrophage-mediated clearance) or C3 (hepatic Kupffer cell-mediated clearance) determinants in most patients with late-stage HIV infection (BENDER et al. 1988). In vitro assays that assess monocyte function suffer the identical interpretive problems previously mentioned for phenotypic changes. Monocyte chemotactic responses to any of several different chemoattractants were each depressed below normal levels (POLI et al. 1985; SMITH et al. 1984). This phenomenon can be duplicated with monocytes from seronegative donors after exposure of cells of purified gp41 or gp120 proteins (WAHL SM et al. 1989; TAS et al. 1988). Monocyte mircrobicidal activity against any of several unrelated pathogens (*Candida albicans*, *C. guelliermondi*, *C. neoformans*, *Aspergillus fumigatus*, *Thermoascus crustaceus*, *Toxoplasma gondii*, *Chlamydia psittaci*) was normal both in steady state and after further in vitro exposure to IFN-γ (MURRAY et al. 1984, 1987, 1988; NIELSEN et al. 1986; EALES et al. 1987; ESTEVEZ et al. 1986). Phagocytosis of latex beads or infectious microbes such as *Candida* or *Toxoplasma* was normal (POLI et al. 1985; MURRAY et al. 1984; ESTEVEZ et al. 1986). Release of toxic monocyte secretory products that serve as effector molecules in antimicrobial reactions such as H_2O_2, interleukin-1 (IL-1), or tumor necrosis factor (TNF-α) was normal with cells from HIV-infected donors and appropriately increased after further in vitro treatment with IFN-γ or LPS (HAAS et al. 1987; MURRAY et al. 1984, 1987; NIELSEN et al. 1986). Moreover, monocytes from HIV-infected patients treated with recombinant IFN-γ in vivo also showed increased secretion of H_2O_2 and microbicidal activity against *T. gondii* (MURRAY et al. 1987).

Analysis of tissue macrophages in HIV-infected patients suggests a very different picture from that of the relatively normal blood monocyte population. In skin, Langerhans cells undergo extensive morphologic change even in early disease. Up to 30% of epidermal Langerhans cells show condensation of cytoplasmic and nuclear chromatin, vacuole formation, and cytolysis in the absence of obvious HIV infection (TSCHACHLER et al. 1987). Detection of HIV virions by transmission electron microscopy or of HIV proteins by immunocytochemistry was rare. Other studies show profound phenotypic changes in epidermal Langerhans cells in otherwise unaffected, clinically normal skin: the numbers of cells that express HLA-DR or CD1 or show ATPase activity decrease to at least 50% of control levels with late-stage HIV disease (BELSITO et al. 1984; OXHOLM et al. 1986; DRENO et al. 1988).

Follicular dendritic cells in lymph nodes also show major degenerative changes early in HIV infection that increase in extent with disease progression (ARMSTRONG and HORNE 1984; LE TOURNEAU et al. 1986; GYORKEY et al. 1985). Indeed, in late-stage disease there can be complete loss of the follicular dendritic network so important for antigen presentation in lymph nodes (CAMERON et al. 1987). Unlike observations with the epidermal Langerhans cell, viral particles are easily detected in follicular dendritic cells (LE TOURNEAU et al. 1986; CAMERON et al. 1987). Interestingly, Langerhans cells of lymph nodes also do not show HIV virions even in close proximity to obviously infected follicular dendritic cells. It is possible that the degenerative changes observed in epidermal Langerhans cell and lymph node follicular dendritic cell populations represent a special event not apparent with macrophages of other tissues. Unlike other tissue macrophages, the Langerhans cell and the follicular dendritic cell have exceedingly high levels of expression of cell membrane CD4. Mechanisms (largely unknown) that induce depletion of $CD4^+$ T cells with time after HIV infection may also affect the Langerhans and follicular dendritic cells. Blood monocytes and tissue macrophages have low to undetectable levels of CD4 and may therefore not be susceptible to these degenerative or lytic events.

3 Mononuclear Phagocytes as Susceptible Target Cells for HIV In Vitro

Initial attempts to infect blood monocytes or alveolar macrophages with HIV were suggestive for productive infection but the results were inconclusive. Monocytes cultured with HTLV-IIIB, a strain of HIV passaged continuously in T cells or T cell lines, bound virus to the cell membrane and ingested viral particles into phagocytic vacuoles within 10 min. HIV was detected in such vacuoles by transmission electron microscopy through 3 days of culture, but no virions were observed budding from the plasma membrane. Assays for reverse transcriptase (RT) activity, p24 antigen (Ag), or other viral proteins by direct immunofluorescence were uniformly negative (NICHOLSON et al. 1986). Addition of mitogen-induced lymphoblasts from seronegative donors to such HIV-infected monocyte cultures 2–3 weeks after the initial virus exposure initiated a productive infection in the T cell targets. Thus, at the very least, monocytes were able to sequester viable HIV and to transmit these infectious particles to T cells (SALAHUDDIN et al. 1986). Other studies with HTLV-IIIB and cultured monocytes showed low levels of RT activity (twice background) 2 weeks after virus inoculation, but no cytopathic effects were induced in the cell monolayer and no cell-associated virions were observed by transmission electron microscopy. The frequency of HIV-infected cells in these cultures as quantified by direct immunofluorescence for HIV proteins was 1%–5% (HO et al. 1986).

The first evidence for productive HIV infection of monocytes in vitro developed from studies using primary cultures of brain tissue from patients with AIDS-associated encephalopathy (GARTNER et al. 1986). Primary brain explants, enriched for macrophages by repeated trypsin digestion of adherent cell monolayers, released RT activity through 6 weeks of culture. Virus budding from plasma membranes of cultured cells was evident by transmission electron microscopy in a small fraction of cells. This relatively low number of infected cells after 6 weeks of culture was confirmed by in situ hybridization for HIV RNA. Passage of progeny virus on blood monocytes from seronegative donors initiated a productive infection sustained in the monocyte target cells through 2 months. Such monocytes infected with HIV showed a frequency of infected cells in culture of 5%–20% by immuno-fluorescence with monoclonal anti-p17 capsid protein and developed profound HIV-associated cytopathic effects of multinucleated giant cells not present in the uninfected control cultures. Serial dilutions of virus inoculum derived from brain explant cultures were 10- to 100-fold more efficient for infection of other monocyte target cells than for T cells. Conversely, HTLV-IIIB was 10 000-fold more efficient in infection of T cell targets than monocytes.

Such distinct differences in target cell tropism for different HIV isolates were confirmed in a subsequent study (KOYANAGI et al. 1987). HIV isolated from cerebrospinal fluid of a patient with AIDS-associated encephalopathy replicated in phytohemagglutinin (PHA)-induced lymphoblasts from seronegative donors, but not in blood monocyte cultures. In contrast, virus isolated from brain tissue infected both lymphoblasts and monocyte target cells. Restriction endonuclease cleavage maps of these two HIV strains were different at only 4 of 26 restriction enzyme sites (15%), a difference much less than that found between two isolates from different patients (15 of 33 sites or 45%). Thus, biologically different but closely related strains of HIV with distinct target cell tropism coexist in various tissues within the same infected patient.

Detailed analysis of HIV–monocyte interaction was impeded by the inability to culture blood monocytes for extended intervals. Conventional culture of monocytes as an adherent monolayer in medium with fetal calf serum results in death of most of the initial cell population (80% loss of viable cells in 1 week). In contrast, monocytes cultured in medium with human serum and recombinant human macrophage colony stimulating factor (MCSF) survived for weeks with little or no loss of cell viability (BECKER et al. 1987; GENDELMAN et al. 1988a). MCSF is made by many cells of the body, including the macrophage itself (but not by T cells). Indeed, in the steady state, normal human blood has about 300–800 U/ml MCSF (HANAMURA et al. 1988). The MCSF receptor, a tyrosine kinase identical to the c-*fms* proto-oncogene product, is found only on mononuclear phagocytes; the number of receptors per cell increases with cell differentiation. In murine systems, MCSF is a potent growth factor. Bone marrow myeloid precursors and tissue macrophages respond to MCSF with a strong proliferative response and colony formation in growth medium: a single progenitor cell can yield $> 1 \times 10^9$ progeny (about 1 gram of macrophages). For human macrophages, MCSF is a survival and differentiation factor, not a growth factor (CLARK and KAMEN 1987). Low levels of ^{3}H-thymidine incorporation are observed in treated cultures, but the numbers of proliferating cells as quantified by counting cells with mitotic figures or nuclear grains on ^{3}H-thymidine autoradiography were $< 3\%$ of the total cell population. Phenotypic analysis of this monocyte population (blood leukocytes obtained through leukapheresis, and the monocytes purified by ficoll-diatrizoate

Table 1. Phenotypic characterization of fresh and cultured monocytes by monoclonal antibodies and flow cytometry

Antibody	Antigen cluster	% positive cells	
		Blood monocytes	MCSF-treated monocytes (10 days)
leukocyte			
HLe-1	CD45	99	99
transferrin receptor	CD71	2	82
myeloid			
M01	CD11b	90	96
Leu-15	CD11b	87	94
M02	CD14	85	96
Leu-M3	CD14	88	97
HLA-DR	—	64	73
B cell			
B4	CD19	5	7
T cell			
T11	CD2	3	5
OKT3	CD3	2	3
OKT4	CD4	4	5
Leu-3a	CD4	5	3
OKT8	CD8	6	8

density gradient centrifugation and counter-current centrifugal elutriation) at 2 weeks in culture documented a cell population >98% pure (Table 1). The CD4 determinant, the HIV receptor for T cells, was undetected in this cell population by any of several different monoclonal antibodies. Although CD4 is present at low concentrations in blood monocytes, the number of cells that display this plasma membrane determinant has been reported to range from <5% to 90% (HAAS et al. 1987; GARTNER et al. 1986; GENDELMAN et al. 1988a; CROWE et al. 1987; FALTYNEK et al. 1989; MCELRATH et al. 1989). Studies with radiolabeled recombinant gp120 are consistent with the lower estimates. The MOLT/4 T cell line has about 7000 specific gp120 binding sites/cell, and the U937 myeloid cell line about 4000 specific binding sites/cell. In both instances, specific binding is inhibited by monoclonal anti-CD4 and soluble recombinant CD4 (sCD4). In contrast, blood monocytes tested at the time of isolation or after 7 days in culture have <200 specific sites/cell. Almost all binding of radiolabeled gp120 to monocytes is nonspecific and not inhibited by cold gp120, by monoclonal anti-CD4, or by sCD4 (D.S. FINBLOOM, D.L. HOOVER and M.S. MELTZER, unpublished).

Repeated attempts to infect MCSF-treated monocytes with HTLV-IIIB were uniformly negative even at viral inoculum 100000-fold higher than that necessary to infect T cells. In contrast, virus isolation onto MCSF- treated monocytes from blood leukocytes of HIV-infected patients was much more successful. Peripheral blood mononuclear cells (PBMCs) from 33 individuals seropositive for HIV or at risk for HIV infection were cocultivated with MCSF-treated monocytes from seronegative donors (GENDELMAN et al. 1990b). Culture fluids were assayed at 2- to 3-day intervals for p24 Ag by ELISA. Significant levels of p24 Ag were detected in 31 of 33 cultures, an overall viral isolation frequency of 93%. HIV was detected in cultures of MCSF-treated monocyte target cells with PBMCs from patients independent of the subject's age, sex, numbers of $CD4^+$ T cells, or clinical stage (POPOVIC and GARTNER 1987; WEISS et al. 1988). This relatively high frequency of virus isolation was also unaffected by coincident 3′-azido-3′-deoxythymidine therapy. The average time interval to first detect p24 Ag in culture fluids was 20 ± 2 days (mean $\pm$ SEM for 25 patients), with a median time of 18 days (range of 7–45 days). The time interval necessary to first detect p24 Ag in cultures with PBMCs of patients with normal numbers of $CD4^+$ T cells (830 ± 160 cells/mm^3) and early disease was significantly longer than that required in cultures with PBMCs of patients with decreased numbers of $CD4^+$ T cells (160 ± 20 cells/mm^3) and later stages of disease: 29 ± 5 days for seven early-stage patients vs 17 ± 2 days for 17 later-stage patients (GENDELMAN et al. 1990b)

Human immunodeficiency virus isolates from patient PBMCs in MCSF-treated monocytes were serially passaged in MCSF-treated monocyte cultures. Passage was successful with 17 of 20 isolates, an efficiency of 84%. The average time interval necessary to first detect p24 Ag in these cultures was 7 ± 1 days (mean $\pm$ SEM for 16 isolates), with a median time of 7 days (range of 2–19 days). HIV-associated cytopathic changes in monocyte monolayers (multinucleated giant cells, cell syncytia, and lysis in about 20%–40% of the cell population) were

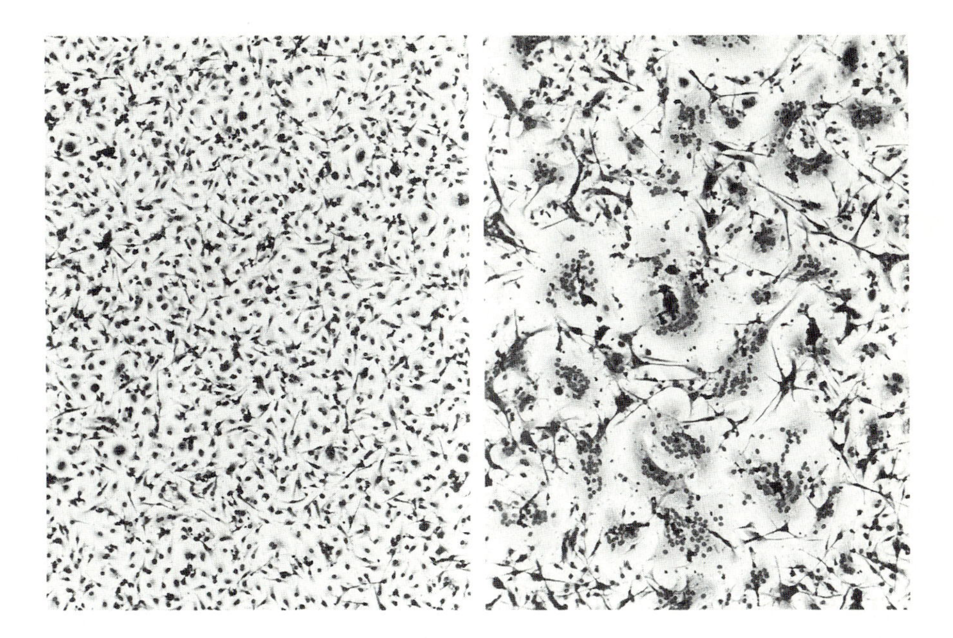

Fig. 1. HTV-1 induced cytopathic effects in MCSF-treated monocytes. PBMC from HIV-seronegative donors purified to >98% monocytes and cultured as abherent monolayers in medium with human serum and MCSF were exposed to ADA, a monocyte HIV isolate, at a multiplicity of infection of 0.01 infectious virus/cell. Cultures were refed with fresh medium every 2 to 3 days. Photomicrographs of adherent cells 15 days after infection are at 200 X original magnification: (*left*) uninfected MCSF-treated monocytes; (*right*) HIV-1 infected monocytes (GENDELMAN et al. 1988)

apparent at 2 weeks in all cultures (Fig. 1). There was no correlation between these cytopathic effects and the clinical stage at which the virus was isolated.

Human immunodeficiency virus isolates serially passaged three times in MCSF-treated macrophages were added to PHA/IL-2 treated lymphoblasts: levels of p24 Ag released into culture fluids were indistinguishable from those of HTLV-IIIB infected lymphoblasts through 2 weeks of infection. Analysis of such HIV-infected lymphoblasts by levels of RT activity, by in situ hybridization for HIV-specific RNA (5%–20% frequency of cells expressing HIV-specific mRNA), by formation of cell syncytia during infection (3%–10% of total cells), by down-modulation of T cell plasma membrane CD4 (60% $CD4^+$ PHA/IL-2 treated lymphoblasts prior to infection vs 10% $CD4^+$ cells 1 week after infection), and by transmission electron microscopy (progeny virions budding at the plasma membrane only, with no intracytoplasmic accumulation of viral particles) showed no qualitative or quantitative differences between HTLV-IIIB and the MCSF-treated monocyte-derived HIV isolates. These experiments document little or no target cell restriction in virus replication for HIV isolated from patient PBMCs into MCSF-treated macrophages; viral isolates grew equally well in macrophages or PHA/IL-2 treated lymphoblasts.

The preceding observations contrast previous reports of target cell restriction in the propagation of HIV isolates (GARTNER et al. 1986; KOYANAGI et al. 1987; GENDELMAN et al. 1988a). It is possible that target cell permissiveness to HIV infection may vary with different viral isolates. To clarify this point, we examined the serial passage of five clinical isolates of HIV in both MCSF-treated monocytes and PHA/IL-2 treated lymphoblasts: a representative isolate is shown in Fig. 2. PBMCs from five different patients seropositive for HIV were cocultivated with both MCSF-treated monocytes and PHA/IL-2 treated lymphoblasts from seronegative donors. In each of the five patients, an HIV primary isolate was recovered in both monocyte and lymphoblast culture systems. For each of the five patients, HIV isolated in PHA/IL-2 treated lymphoblasts or MCSF-treated monocytes were serially passaged into cultures of the homologous cell type. Furthermore, HIV isolated in MCSF-treated monocytes also infected PHA/IL-2 treated lymphoblasts. In marked contrast, five of five viral isolates recovered from PHA/IL-2 treated lymphoblasts showed no growth in the heterologous MCSF-treated monocytes by the criteria of p24 Ag release, RT levels, or infectious titer. Moreover, viral isolates initially recovered in MCSF-treated monocytes and then passaged in PHA/IL-2 treated lymphoblasts showed little or no evidence of virus growth when placed back into MCSF-treated monocytes. The preceding experiments document the existence of two distinct species of HIV: viruses isolated in MCSF-treated monocytes show dual tropism and infect monocytes and T cells equally; viruses isolated in PHA/IL-2 treated lymphoblasts replicate only in T cells.

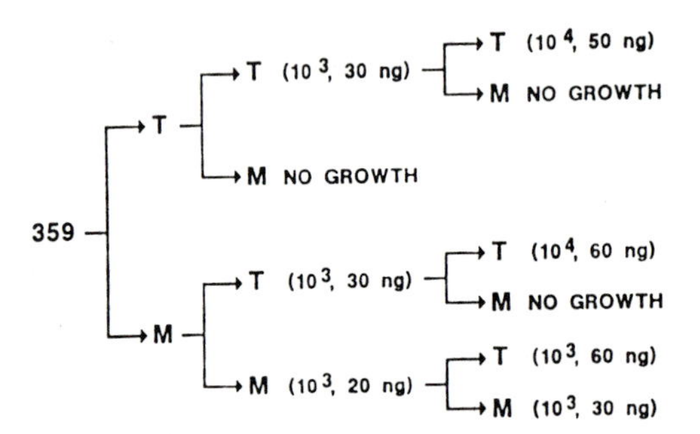

Fig. 2. Serial passage of HIV isolated from patient blood leukocytes in MCSF-treated monocyte and PHA/IL-2 treated lymphoblast cultures. PBMC from patient 359 were added to 7 day MCSF-treated monocytes or 3 day PHA/IL-2 treated lymphoblasts. Dilutions of culture fluids from the primary isolation and each successive passage were added to other monocyte or T lymphoblast cultures. All cultures were incubated for a 2 hour viral adsorption interval, then washed, and refed with fresh medium every 2 to 3 days through 2 months. HIV infection in monocyte cultures at 2 to 3 weeks and lymphoblast cultures at 1 to 2 weeks were estimated through 2 serial passages by infectious titer and p24 antigen (ng/ml) in pooled culture fluids shown in parenthesis (GENDELMAN et al. 1990b)

The phenomenon of distinct HIV variants with different target cell tropism is found both in vitro and in vivo. Whatever the mechanism for cell tropism, several lines of evidence suggest that the major determinant for this biologic feature of HIV resides in the envelope. HIV released from infected monocytes are very different from those of infected T cells. Radioimmunoprecipitation analysis of HIV proteins in infected monocytes and in the virions released from infected monocytes underscores this profound difference (Fig. 3). The predominant viral proteins synthesized by HIV-infected T cells are *env* gene products. Envelope glycoproteins and their breakdown products exceed those capsid proteins (*gag* gene products) and their breakdown products. This ratio (*env* > *gag* gene products) is maintained for both T cell synthesis of viral proteins and the proteins assembled into viral particles. In the virions released from HIV-infected T cells, > 35% of total viral proteins (gp160 + gp120)/(gp160 + gp120 + p55 + p24) are *env* gene products. In contrast, the dominant viral proteins synthesized in the HIV-infected monocyte and assembled into viral particles are capsid proteins: envelope glycoproteins in the virions released from HIV-infected monocytes represent < 10% of total virus protein. At equivalent levels of RT activity and infectious titer, HIV particles released from infected T cells have at least five times

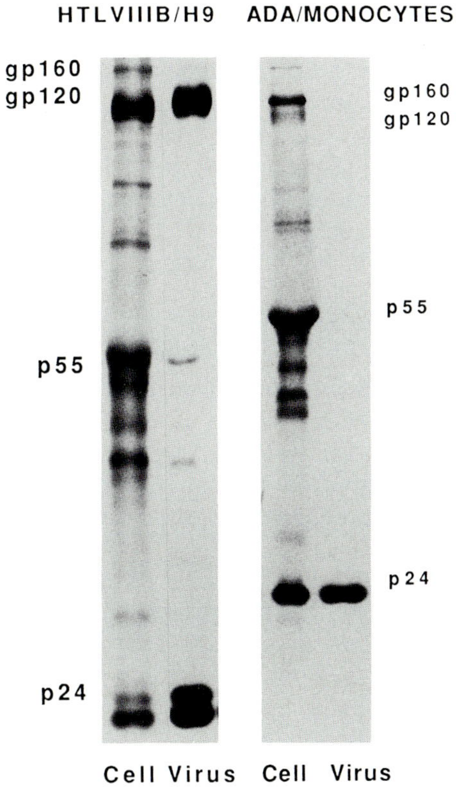

Fig. 3. Radioimmunoprecipitation analysis of HIV proteins in infected monocyte and T cell targets. HIV-specific proteins shown after gel electrophoresis and autoradiography were isolated by radioimmunoprecipitation with pooled HIV-seropositive sera of ^{35}S-methionine labeled cell lysates and ultracentrifuged culture fluids (virions) from MCSF treated monocyte infected with ADA, a monocyte tropic HIV isolate, or the H9 T cell line infected with HTLVIIIB. Infected cells were to the radiolabel for 30 minutes, then washed with medium and cultured for an additional 1, 2 or 4 h

more envelope glycoproteins than the virions released from infected monocytes. Transmission electron microscopic analysis of progeny virus from infected monocytes and T cells confirms the radioimmunoprecipitation analysis. Characteristic surface projections or envelope "spikes," the morphologic representation of gp120, are evident in the HIV released from infected T cells. Progeny virus released from HIV-infected monocytes show little or no "spikes" and are relatively bald. Perhaps the strongest evidence for the envelope as prime determinant of cell tropism derives from studies with hybrid virions. Proviral clones derived from T cell trophic HIV will not replicate in monocytes; clones derived from monocyte tropic HIV replicate in both T cells and monocytes. Hybrid constructs of T cell tropic clones with the *env* gene of the monocyte tropic HIV infect both T cells and monocytes (W.A. O'BRIAN and I.S.Y. CHEN, unpublished observations).

Functional consequences of the changes that produce envelope-deficient virions from HIV-infected monocytes are illustrated in experiments with sCD4 (GOMATOS et al. 1990). Soluble recombinant CD4 binds with high affinity to gp120, the envelope glycoprotein of HIV, and at relatively low concentration (0.1–1 μg/ml) completely inhibits infection of many HIV strains in T cells or T cell lines (SMITH et al. 1987; FISHER et al. 1988; HUSSEY et al. 1988; DEEN et al. 1988; CLAPHAM et al. 1989). HTLV-IIIB infection of the H9 T cell line is completely inhibited by prior treatment of virus with 10 μg/ml sCD4 (50% inhibitory dose at 1×10^5 $TCID_{50}$ HTLV-IIIB: <2 μg/ml sCD4). No p24 Ag or HIV-induced syncytia are detected in cultures of H9 cells exposed to 1×10^5 $TCID_{50}$ HTLV-IIIB in the presence of sCD4. Under identical conditions and at a 100-fold *lower* viral inoculum, 10 μg/ml sCD4 has little or no effect on infection of monocytes by any of six different monocyte tropic HIV isolates as assessed by three different criteria: levels of p24 Ag and RT activity, virus-induced cytopathic effects, and the frequency of infected cells that express HIV specific mRNA (GOMATOS et al. 1990). At 10- to 100-fold higher concentrations of sCD4, however, infection is completely inhibited. Monoclonal anti-CD4 (Leu-3a or OKT4a) also prevents infection of these same viral isolates in monocytes. The relative inefficiency of sCD4 for inhibition of HIV infection in monocytes (about 10 000-fold) is a property of the virion and not the target cell: HIV isolates that infect both monocytes and T cells require similarly high levels of sCD4 for inhibition of infection (50% inhibitory dose at 1×10^3 $TCID_{50}$ ADA: 100–200 μg/ml sCD4). These data suggest that the gp120 of HIV derived from macrophages interacts with sCD4 differently than that of virions derived from T cells. For both variants of HIV, however, the predominant mechanism of virus entry for infection is CD4 dependent.

All of the preceding data are consistent with the hypothesis that HIV interaction with CD4 is an obligate reaction for infection of both T cells and monocytes. None of these observations, however, preclude another, CD4-independent route of infection. HIV may enter macrophages through phagocytosis, FcR-mediated endocytosis (NICHOLSON et al. 1986; ROBINSON et al. 1988, 1989; HOMSY et al. 1988; TAKEDA et al. 1988), or interaction with receptors for mannosylated proteins (EZEKOWITZ et al. 1989). Infection of T cells or monocytes with HIV is markedly enhanced (five to ten fold increase in RT activity in culture fluids) by sera from certain HIV-infected patients. Such antibody-mediated enhancement is not inhibited by monoclonal anti-CD4 (HOMSY et al. 1989). Most impressively, HIV isolates that do not replicate in monocytes (T cell tropic HIV) will infect these cells after treatment with enhancing antibodies (HOMSY et al. 1989). In monocytes, antibody-mediated enhancement of HIV infection is inhibited by monoclonal anti-FcRIII (the predominant Fc receptor in tissue macrophages and monocytes in culture, but absent on circulating blood monocytes) but not monoclonal antibodies directed against FcRI or FcRII. Antibody-mediated enhancement of HIV infection in the myeloid cell line U937, which lacks FcRIII but does express FcRI and FcRII, is blocked by heat-aggregated IgG (TAKEDA et al. 1988). In these studies, the infection pattern of U937 stimulates more closely that of T cells rather than that of monocytes or macrophages. Interestingly, U937 is also a susceptible target cell for several T cell tropic HIV, but not monocyte tropic viruses (KOYANAGI et al. 1987; COLLMAN et al. 1989).

The numbers of monocytes infected with HIV in vitro as quantitated by in situ hybridization for HIV RNA is at least three-fold greater than in lymphoblast cultures. But the number of virions released into culture fluids of HIV-infected monocytes as quantified by infectious titer or RT activity is 10- to 100-fold less than that of infected lymphoblasts. The basis for this apparent dissociation

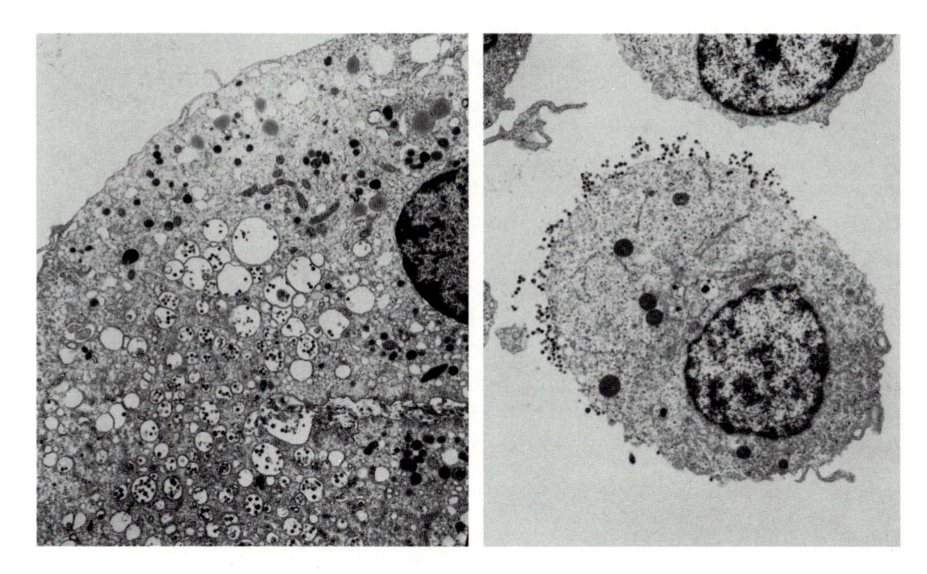

Fig. 4. Virion budding and release in HIV-infected MCSF-treated monocytes and PHA/IL-2 treated lymphoblasts. Transmission electron microscopy of an HIV infected MCSF-treated monocyte (*left*) (viral particles sequestered within intracytoplasmic vacuoles; few or no virions at the plasma membrane) and PHA/IL-2 treated lymphoblasts (*right*) (numerous viral particles budding at the plasma membrane; no intracellular virions), X 9400 (GENDELMAN et al. 1988; ORENSTEIN et al. 1988). (The authors thank DR. Jan M. Orenstein, Dept. of Pathology, George Washington University Medical Center, Washington, DC, for electron microscopy)

between high numbers of infected cells and low levels of infectious virus released into the culture is shown by transmission electron microscopic analysis (Fig. 4). HIV-infected T cells show hundreds of viral particles associated with the plasma membrane: HIV assembles and buds only from the plasma membrane of infected T cells; there is no intracellular accumulation of mature or even immature virions. HIV interaction with macrophages is quite different from that of T cells. Ultrastructural analysis of HIV-infected macrophages 2–6 weeks after infection (time intervals at which 60%–90% of cells express both HIV-specific mRNA and proteins) shows little or no virions at the plasma membrane. Yet these infected cells contain large numbers of viral particles. Virus is localized almost exclusively to intracellular vacuoles. Infected macrophages display numerous vacuolar structures not associated with the plasma membrane, each of which contains scores of mature and immature virions. Indeed, HIV not only accumulates within these intracellular vacuoles but also assembles and buds from the vacuolar membranes. Morphologic evidence strongly suggests that these vacuoles are not endosomes, but rather are derived from the Golgi complex (ORENSTEIN et al. 1988). The macrophage handles HIV much like any other secretory glycoprotein: HIV is assembled in the Golgi complex and transported in Golgi complex-derived vacuoles towards the plasma membrane. Significantly, the final step of secretion, exocytosis into the extracellular milieu, appears suppressed: the amount of virus released from HIV-infected macrophages, quantitated by RT activity or p24 Ag in culture fluids, is ten fold less than that released by an equal number of infected T cells. Thus, the HIV-infected macrophage represents a veritable virus factory, but a factory whose entire output remains hidden from the host. Experiments confirm that the intracellular virions of HIV-infected macrophages are infectious: release of these viral particles by freeze-thaw cycles increases the infectious titer of the culture fluids at least ten fold (GENDELMAN et al. 1988a). Most importantly, these in vitro observations have been confirmed in the AIDS patient. Macrophages in the brain of a seropositive individual also showed intracellular localization of virus particles within vacuoles; little or no virus was detected at the plasma membrane (ORENSTEIN et al. 1988). Such virus, sequestered from host immunity within cytoplasmic vacuoles, represents a true reservoir for continued infection. Release of infectious virus from this macrophage reservoir and dissemination of HIV into other macrophages or T cells could be initiated by any agent that perturbs macrophage function: factors released during inflammation, normal tissue remodeling, or host response to intercurrent infection.

4 Mononuclear Phagocytes as Regulatory Cells in the Pathophysiology of HIV Infection

The preceding observations clearly document major roles for macrophages as both target cell and tissue reservoir for infectious virus during HIV disease. HIV-infected macrophages are found in brain, lung, lymph node, skin, bone marrow, and blood of seropositive patients. It is probable that these infected cells directly participate in the pathogenesis of HIV-induced immunosuppression and CNS dysfunction. However, the means and mediators for this participation are not yet understood. A major role for macrophages in the steady state and during disease is regulation of tissue function. This regulatory role is mediated by the literally hundreds of secretory molecules released by the macrophage under a variety of pathophysiologic conditions (NATHAN 1987). Changes in the secretion or release of certain mediators occur during HIV infection and underlie the symptomatology of AIDS.

The paucity of virus-infected lymphocytes in AIDS and the absence of cytolytic infections of neurons or neuroglia suggest an indirect mechanism for immune and rervous system dysfunction in HIV infection (HO et al. 1989b). For example, macrophages release many secretory products that have direct effects on nerve growth, function, or repair of injury. Inappropriate secretion of these monokines (IL-1, IL-6, TNF-α, platelet-derived growth factor, apolipoprotein e) by HIV-infected macrophages in brain may induce both neurologic symptoms and tissue injury. Moreover, macrophages have receptors for and respond to several neuropeptides (ACTH, β-endorphins, somatotropin, neurotensin, substance P, and vasoactive intestinal peptide) to secrete toxic oxygen metabolites and other injurious monokines. Indeed, recent reports document induction of prostaglandins, IL-1, IL-6, and TNF-α by gp120, the HIV envelope glycoprotein (WAHL SM et al. 1989; WAHL LM et al. 1989; NAKAJIMA et al. 1989; MERRILL et al. 1989). These individual observations provide the basis for a regulatory network in which HIV-infected macrophages affect nerve cells through any of several monokine or virus-derived secretory factors; the injured neural tissue reciprocally affects the HIV-infected macrophage to release even more toxic secretory products (GENDELMAN et al. 1988b).

Similar cytokine networks regulate virus production and latency in HIV-infected macrophages and T cells. But such cytokine networks are very complicated. The sequela of any single cytokine treatment is often unpredictable. For example, HIV replication in monocytes pretreated with granulocyte-macrophage colony ctimulating factor (GMCSF) is markedly reduced, yet GMCSF added to monocytes after HIV infection increases virus production (Fig. 5) (HAMMER et al. 1986; MELTZER and GENDELMAN 1988; KOYANAGI et al. 1988; PERNO et al. 1989). Moreover, cytokines almost never operate alone. Most monokines are autocrine factors that induce other monokines (IL-1, IL-6, IL-8, TNF-α, MCSF) which in turn induce cytokines in lymphocytes, endothelial cells, and fibroblasts of adjacent tissue. Effects of cytokines on the replication of HIV are dependent upon the mixture of cytokines that the cell is exposed to, the time of exposure, and the state of differentiation of the responsive cell.

The long interval of clinical latency during HIV infection (50% of infected individuals develop AIDS 10 years after infection) (LIFSON et al. 1988), may be associated with true viral latency (perhaps intermittent) with no expression of HIV in infected cells. In the latent state, HIV exists in T cells as a provirus integrated within host genomic DNA without transcriptional activity. HIV becomes transcrip-

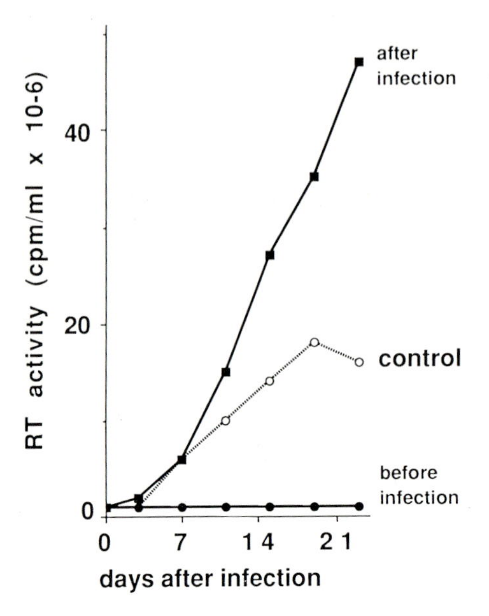

Fig. 5. Effect of GMSCF on replication of HIV in monocytes at various times of virus infection. PBMC from HIV-seronegative donors purified to > 98% monocytes, and cultured 7 days as adherent monolayers in medium with human serum and MSCF were exposed to ADA, a monocyte tropic HIV isolate, at a multiplicity of infection of 0.01 infectious virus/cell. In certain cultures, monocytes were treated with 200 U/ml GMCSF 24 h before viral inoculation, washed free of cytokine, and exposed to HIV. In other cultures, 200 U/ml GMCSF was added 2 h after virus inoculation and maintained at this concentration through 3 weeks. RT activity (cpm/ml 10^{-6}), was determined in culture fluids

tionally active and reenters the replicative cycle after exposure of the T cell to any of a variety of apparently unrelated stimuli that include cytokines, mitogens, phorbol esters, infection with herpesvirus, adenovirus, and exposure to sunlight. Control of HIV replication in such infected cells in mediated in large part by cellular rather than viral factors. An animal model that simulates latent HIV infection in man was produced to further identify factors that affect induction of viral gene synthesis in latently infected target cells. Transgenic mice that contain integrated copies of the HIV long terminal repeat (LTR) linked to the bacterial gene for chloramphenicol acetyltransferase (CAT) were constructed (LEONARD et al. 1989). Thus HIV LTR-directed expression of the CAT gene in cells of transgenic animals would be analogous to HIV gene expression for progeny virus production in infected patients whose cells harbor latent integrated proviruses. Under steady state conditions in blood and in tissue, neither macrophages nor lymphocytes express detectable levels of CAT activity. CAT activity in macrophage populations is increased by in vitro treatment with any of several recombinant murine cytokines (IL-1, IL-2, IL-4, MCSF, and GMCSF). A similar phenomenon was observed in vivo. Resident peritoneal macrophages or cells elicited by a sterile, chronic irritant (thioglycollate) showed no CAT activity. However, macrophages recovered from an intraperitoneal immune reaction to *Mycobacterium bovis*, strain BCG or *Proprionibacterium acnes* (*Corynebacterium parvum*) were strongly positive. Purified splenic T cells also showed no constitutive CAT activity. Significantly, cultivation of these T cells in murine IL-2 increased CAT expression 20-fold. Infection of both macrophages and T cells with certain DNA viruses (herpes simplex 1, adenovirus, murine cytomegalovirus) increased CAT activity by as much as 50-fold. Thus, HIV activation in this mouse model system is clearly regulated by endogenous cytokines released from both T cells and macrophages.

Human immunodeficiency virus subverts cellular transcriptional factors and other cellular gene products to direct virus gene expression. For example, the HIV LTR contains a number of *cis*- acting sequences that are targets for transactivators including the cellular DNA-binding proteins NF-κB and Sp1 (NABEL and BALTIMORE 1987; JONES et al. 1986). Activation of the HIV LTR in proliferating T cells after treatment with mitogens or phorbol esters is associated with synthesis of the endogenous DNA-binding protein, NF-κB. Monocytes exposed to LPS release IL-1 and INF-α, monokines that also induce NF-κB binding activity and thereby increase HIV expression in T cells (CLOUSE et al. 1989; GRIFFIN et al. 1989; OSBORN et al. (1989). HIV gene expression in the myeloid cell line U937 is similarly regulated by NF-κB: treatments that induce NF-κB binding activity in U937 (phorbol esters, TNF-α) markedly increase HIV replication. But the HIV LTR can also be activated by endogenous signals independent of NF-κB: GMCSF, which does not induce NF- κB binding activity, also activates HIV expression in U937. In contrast to observations with the U937 myeloid cell line, NF- κB binding activity is present constitutively in blood monocytes (GRIFFIN et al. 1989). The precise regulatory role for this DNA binding protein for HIV replication in monocytes is therefore less clear. MCSF increases HIV expression in monocytes at concentrations equivalent to that normally found in human blood (300–800 U/ml MCSF). Interestingly, levels of MCSF in the blood of patients with infectious disease (sepsis, pneumonia) can increase three to eight fold. Thus, as with the macrophage–neural tissue interactions, complex cytokine networks are formed in the steady state and during immune reactions that regulate HIV expression in macrophages and T cells.

A major player in these regulatory networks in interferon (IFN). Indeed, there is a well-established precedent for IFN as a dominant regulatory molecule in the pathogenesis of several retroviral diseases: murine and avian leukemia, and a number of lentiviral diseases (FRIEDMAN and PITHA 1984). In visna-maedi infection of sheep, virus replication in macrophages is reduced $\geqslant$1000-fold by IFN released from T cells (NARAYAN et al. 1985; KENNEDY et al. 1985). Levels of IFN released by ovine lung leukocytes directly correlate with lentivirus infection of alveolar macrophages (LAIRMORE et al. 1988). In man, IFN activity is found in sera of patients with late-stage HIV disease and is an index of poor prognosis (PREBLE et al. 1985; VADHAN et al. 1986). In a survey of 15 different HIV isolates in both monocytes and lymphoblasts, we found no IFN activity in culture fluids through 3 weeks of infection (GENDELMAN et al. 1990a). But antiviral activity is reported with addition of IFN-α, IFN-β, and IFN-γ to HIV-infected T cells and macrophages (HO et al. 1985; PUTNEY et al. 1986; YAMAMOTO et al. 1986; DOLEI et al. 1986; HARTSHORN et al. 1986, 1987; KOYANAGI et al. 1988; MACE et al. 1988; WONG et al. 1988; YAMADA et al. 1988; KORNBLUTH et al. 1989; MICHAELIS and LEVY 1989; CRESPI 1989; CROWE et al. 1989). Any effects of IFN on HIV replication in human traget cells must therefore depend upon exogenous sources for this cytokine. Possible sources for such INF include direct induction of IFN by coincident infection with another microorganism, IFN production as a consequence of immune reactions to foreign antigens, or administration through immunotherapeutic intervention.

Monocytes treated with recombinant human IFN-α at the time of virus challenge show no evidence of HIV infection 3 weeks later: no viral protein, no viral mRNA, and no proviral DNA (Fig. 6). IFN interrupts one or more early event in the virus replication cycle *before* the formation of proviral DNA: binding, uptake, uncoating, or reverse transcription. The exact mechanisms for this antiviral activity are conjectural and include: (a) changes in virus receptor number or distribution (IFN-treated monocytes show marked changes in CD4, FcR, CD11a, and mannosylated protein receptors, each of which is implicated in the uptake of HIV into monocytes); (b) changes in the monocyte plasma membrane that interrupt fusion or uptake of the virion into the cell (IFN-treated cells show alterations in membrane fluidity, microfilament organization, and membrane proteases that could damage bound virus); and (c) changes in subcellular compartments or cytosolic milieu that preclude reverse transcription [IFN- treated cells synthesize 2′-5′(A)oligonucleotides that induce RNases and directly inhibit reverse transcription].

The effect of IFN on monocytes infected with HIV prior to treatment is equally dramatic. Monocyte cultures infected with HIV 7 days before IFN treatment show a gradual decrease in levels of p24 Ag and reverse transcriptase activity to baseline by 3 weeks. HIV-induced cytopathic changes (multinucleated giant cells and cell lysis) are markedly reduced, and the frequency of productively infected cells as quantified by in situ hybridization for HIV mRNA is $\leqslant 1\%$ of total cells. In the interim, viral particles released from the IFN-treated, HIV-infected cells are 1000- to 100-fold less infectious than equal numbers of control virions. But,

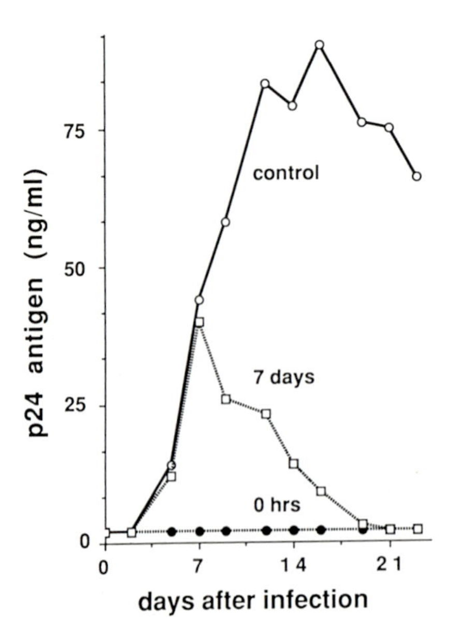

Fig. 6. Effect on IFN on replication of HIV in monocytes at various times after infection, PBMC from HIV-seronegative donors purified to >98% monocytes, and cultured 7 days as adherent monolayers in medium with human serum and MCSF were exposed to ADA, a monocyte tropic HIV isolate, at a multiplicity of infection of 0.01 infectious virus/cell. At the time of infection (0 h) and 7 days infection, 500 IU/ml rIFNα was added and maintained at this concentration throughout the culture interval. Levels of p24 antigen in culture fluids were determined by ELISA (GENDELMAN et al. 1990a)

unlike the outcome of IFN pretreatment, monocytes treated with IFN 7 days after HIV infection are not free of the retroviral pathogen: levels of proviral DNA in the IFN-treated and control HIV-infected cells were indistinguishable. The presence of large quantities of proviral DNA in cells with little or no evidence for active transcription suggests true microbiologic latency—and this in a nonreplicating cell with no direct evidence for integrated virus. Such transcriptional restriction of virus replication in the IFN-treated, HIV-infected monocytes has no precedent in previously described retroviral systems.

5 Summary

We have presented evidence in this review for the following:

1. Macrophages are likely the first cell infected by HIV. Studies document recovery of HIV into macrophages in the early stages of infection in which virus isolation in T cells is unsuccessful and detectable levels of antibodies against HIV are absent.
2. Macrophages are major tissue reservoirs for HIV during all stages of infection. Unlike the lytic infection of T cells, many HIV-infected macrophages show little or no virus-induced cytopathic effects. HIV-infected macrophages persist in tissue for extended periods of time (months) with large numbers of infectious particles contained within intracytoplasmic vacuoles.
3. Macrophages are a vector for the spread of infection to different tissues within the patient and between individuals. Several studies suggest a "Trojan horse" role for HIV-infected macrophages in dissemination of infectious particles. The predominant cell in most bodily fluids (alveolar fluid, colostrum, semen, vaginal secretions) is the macrophage. In semen, for example, the numbers of macrophages exceed those of lymphocytes by more than 20-fold (WOLF and ANDERSON 1988).
4. Macrophages are major regulatory cells that control the pace and intensity of disease progression in HIV infection. Macrophage secretory products are implicated in the pathogenesis of CNS disease and in control of viral latency in HIV-infected T cells.

This litany of events in which macrophages participate in HIV infection in man parallels similar observations in such animal lentivirus infections as visna-maedi or caprine arthritis-encephalitis viruses. HIV interacts with monocytes differently than with T cells. Understanding this interaction may more clearly define both the pathogenesis of HIV disease and strategies for therapeutic intervention.

Acknowledgments. The authors thank members of the Walter Reed Retroviral Research Group for excellent patient management. Dr. H.E. Gendelman is a Carter-Wallace fellow of The Johns Hopkins University School of Public Health and Hygiene in the Department of Immunology and infections Diseases. These studies were supported in part by the Henry M. Jackson Foundation for the Advancement of Military Medicine, Rockville, MD.

References

Armstrong JA, Horne R (1984) Follicular dendritic cells and virus-like particles in AIDS-related lymphadenopathy. Lancet II: 370–372

Becker J, Ulrich P, Kunze R, Gelderblom H, Langford A, Reichart P (1988) Immunohistochemical detection of HIV structural proteins and distribution of T-lymphocytes and Langerhans cells in the oral mucosa of HIV-infected patients. Virchows Arch [A] 412: 413–419

Becker S, Warren MK, Haskill S (1987) Colony-stimulating factor-induced monocyte survival and differentiation into macrophages in serum-free cultures. J Immunol 139: 3703–3709

Belsito DV, Sanchez MR, Baer RL, Valentine F, Thorbecke GL (1984) Reduced Langerhans' cell Ia antigen and ATPase activity in patients with the acquired immunodeficiency syndrome. N Engl J Med 310: 1279–1282

Bender BS, Davidson BL, Kline R, Brown C, Quinn TC (1988) Role of the mononuclear phagocyte system in the immunopathogenesis of human immunodeficiency virus infection and the acquired immunodeficiency syndrome. Rev Infect Dis 10: 1142–1154

Braun DP, Kessler H, Falk L, Paul D, Harris JE, Blaauw B, Landay A (1988) Monocyte functional studies in asymptomatic, human immunodeficiency disease virus (HIV)-infected individuals. J Clin Immunol 8: 486–494

Cameron PU, Dawkins RL, Armstrong JA, Bonifacio E (1987) Western blot profiles, lymph node ultrastructure and viral expression in HIV-infected patients: a correlative study. Clin Exp Immunol 68: 465–478

Chayt KJ, Harper ME, Marselle LM, Lewin EB, Rose RM, Oleske JM, Epstein LG, Wong-Staal F, Gallo RC (1986) Detection of HTLV-III RNA in lungs of patients with AIDS and pulmonary involvement. JAMA 256: 2356–2359

Clapham PR, Weber JN, Whitby D, McIntosh K, Dalgleish AG, Maddon PJ, Deen KC, Sweet RW, Weiss RA (1989) Soluble CD4 blocks the infectivity of diverse strains of HIV and SIV for T cells and monocytes but not for brain and muscle cells. Nature 337: 368–370

Clark SC, Kamen R (1987) The human hematopoietic colony-stimulating factors. Science 236: 1229–1234

Clouse KA, Powell D, Washington I, Poli G, Strebel K, Farrar W, Barstad P, Kovacs J, Fauci AS, Folks TM (1989) Monokine regulation of human immunodeficiency virus-1 expression in a chronically infected human T cell clone. J Immunol 142: 431–438

Collman R, Hassan N, Walker R, Godfrey B, Cutilli J, Hastings JC, Friedman H, Douglas SD, Nathanson N (1989) Infection of monocyte-derived macrophages with human immunodeficiency virus type 1. Monocyte tropic and lymphocyte tropic strains of HIV-I show distinctive patterns of replication in a panel of cell types. J Exp Med 170: 1149–1156

Crespi M (1989) The effect of interferon on cells persistently infected with HIV. AIDS 3: 33–36

Crowe SM, Mills J, McGrath MS (1987) Quantitative immunocytofluorographic analysis of CD4 antigen expression and HIV infection of human peripheral blood monocyte/macrophages. AIDS Res Hum Retroviruses 3: 135–138

Crowe SM, McGrath MS, Elbeik T, Kirihara J, Mills J (1989) Comparative assessment of antiretrovirals in human monocyte-macrophages and lymphoid cell lines acutely and chronically infected with the human immunodeficiency virus. J Med Virol 29: 176–180

Davidson BL, Kline RL, Rowland J, Quinn TC (1988) Surface markers of monocyte function and activation in AIDS. J Infect Dis 158: 483–486

Deen KC et al. (1988) A soluble form of CD4 (T4) protein inhibits AIDS virus infection. Nature 331: 82–84

Dolei A, Fattorossi A, D'Amelio R, Aiuti F, Dianzani F (1986) Direct and cell-mediated effects of interferon-α and -γ on cells chronically infected with HTLV-III. J Interferon Res 6: 543–549

Dreno B, Milpied B, Bignon JD, Stalder JF, Litoux P (1988) Prognostic value of Langerhans cells in the epidermis of HIV patients. Br J Dermatol 118: 481–486

Eales L-J, Moshtael O, Pinching AJ (1987) Microbicidal activity of monocyte derived macrophages in AIDS and related disorders. Clin Exp Immunol 67: 227–235

Eilbott DJ, Peress N, Burger H, LaNeve D, Orenstein J, Gendelman HE, Seidman R, Weiser B (1989) Human immunodeficiency virus type I in spinal cords of acquired immunodeficiency syndrome in patients with myelopathy: expression and replication in macrophages, Proc Natl Acad Sci USA 86: 3337–3341

Weiser B, Peress N, La Neve D, Eilbott DJ, Seidman R, Berger H (1990) Human immunodeficiency virus type 1 expression in the central nervous system correlates directly with extent of disease. Proc Natl Acad Sci USA 87, 3997

Enk C, Gerstoft J, Møller S, Remvig L (1986) Interleukin I activity in the acquired immunodeficiency syndrome. Scand J Immunol 23: 491–497

Estevez ME, Ballart IJ, Diez RA, Planes N, Scaglione C, Sen L (1986) Early defect of phagocytic cell function in subjects at risk for acquired immunodeficiency syndrome. Scand J Immunol 24: 215–221

Ezekowitz RAB, et al. (1989) A human serum mannose-binding protein inhibits in vitro infection by the human immunodeficiency virus. J Exp Med 169: 185–196

Faltynek CR, Finch LR, Miller P, Overton WR (1989) Treatment with recombinant IFN-γ decreases cell surface CD4 levels on peripheral blood monocytes and on myelomonocytic cell lines. J Immunol 142: 500–508

Fisher RA et al. (1988) HIV infection is blocked in vitro by recombinant soluble CD4. Nature 331: 76–78

Friedman RM, Pitha PM (1984) The effect of interferon on membrane-associated viruses. In: Friedman RM (ed) Interferon 3: Mechanisms of production and action. Elsevier Science, Amsterdam, pp 319–341

Gartner S, Markovits P, Markovitz DM, Kaplan MH, Gallo RC, Popovic M (1986) The role of mononuclear phagocytes in HTLV-III/LAV infection. Science 233: 215–219

Gendelman HE, et al. (1985) Slow persistent replication of lentiviruses: role of tissue macrophages and macrophage precursors in bone marrow. Proc Natl Acad Sci USA 82: 7086–7093

Gendelman HE, Orenstein JM, Martin MA, Ferrua C, Mitra R, Phipps T, Wahl LA, Lane HC, Fauci AS, Burke DS, Skillman D, Meltzer MS, (1988a) Efficient isolation and propagation of human immunodeficiency virus on recombinant colony stimulating factor 1-treated monocytes. J Exp Med 167: 1428–1441

Gendelman HE, Leonard JM, Dutko FJ, Koenig S, Khillan JS, Meltzer MS (1988b) Immunopathogenesis of human immunodeficiency virus infection in the central nervous system. Ann Neurol 23: S78–81

Gendelman HE, Orenstein JM, Baca L, Weiser B, Burger H, Kalter DC, Meltzer MS (1989) Editorial review: The macrophage in the persistence and pathogenesis of HIV infection AIDS 3: 475–495

Gendelman HE, Baca L, Turpin J, Kalter DC, Hansen B, Orenstein JM, Friedman RM, Meltzer MS (1990a) Regulation of HIV replication in infected T cells and monocytes by interferonα: mechanisms for viral restriction. AIDS Res Hum Retroviruses 6: 1045–1049

Gendelman HE, Baca L, Husayni H, Orenstein JM, Turpin JA, Skillman D, Hoover DL, Meltzer MS (1990b) Macrophage-human immunodeficiency virus interaction: viral isolation and target cell tropism AIDS 4: 221–228

Gomatos PJ, et al. (1990) Relative inefficiency of soluble rCD4 for inhibition of infection by monocyte tropic HIV in monocytes and T cells. J Immunol 144: 4183–4188

Griffin GE, Leung K, Folks TM, Kunkel S, Nabel GJ (1989) Activation of HIV gene expression during differentiation by induction of NF-κB. Nature 339: 70–73

Gyorkey F, Melnick JL, Sinkovics JG, Gyorkey P (1985) Retrovirus resembling HTLV in macrophages of patients with AIDS. Lancet i: 106

Haas JG, Riethmuller G, Ziegler-Heitbrock HWL (1987) Monocyte phenotype and function in patients with the acquired immunodeficiency syndrome (AIDS) and AIDS-related disorders. Scand J Immunol 26: 371–379

Hammer SM, Gillis JM, Groopman JE, Rose RM (1986) In vitro modification of human immunodeficiency virus infection by granulocyte-macrophage colony stimulating factor and g interferon. Proc Natl Acad Sci USA 83: 8734–8738

Hanamura T, Motoyoshi K, Yoshida K, Saito M, Miura Y, Kawashima T, Nishida M, Takaku F (1988) Quantitation and identification of human monocytic colony-stimulating factor in human serum by enzyme-linked immunosorbent assay. Blood 72: 886–892

Harper ME, Marselle LM, Gallo RC, Wong-Staal F (1986) Detection of lymphocytes expressing human T-lymphotropic virus type III in lymph nodes and peripheral blood from infected individuals by in situ hybridization. Proc Natl Acad Sci USA 83: 772–776

Hartshorn KL, Sandstorm EG, Neumeyer D, Paradis TJ, Chou TC, Schooley RT, Hirsch MS (1986) Synergistic inhibition of human T-cell lymphotropic virus type III replication in vitro by phosphonoformate and recombinant alpha-A interferon. Antimicrob Agents Chemother 30: 189–191

Hartshorn KL, Neumeyer D, Vogt MW, Schooley RT, Hirsch MS (1987) Activity of interferons alpha, beta, and gamma against human immunodeficiency virus replication in vitro. AIDS Res Human Retroviruses 3: 125–133

Heagy W, Kelley VE, Strom TB, Mayer K, Shapiro HM, Mandel R, Finberg R (1984) Decreased expression of human class II antigens on monocytes from patients with acquired immune deficiency syndrome: increased expression with interferon-γ. J Clin Invest 74: 2089–2096

Ho DD, Hartshorn KL, Rota TR, Andrews CA, Kaplan JC, Schooley RT, Hirsch MS (1985a) Recombinant human interferon alpha-A suppresses HTLV-III replication in vitro. Lancet i: 602

Ho DD, Rota TR, Schooley RT, Kaplan JC, Allan JD, Groopman JE, Resnick L, Felsenstein D, Andrews CA, Hirsch MS (1985b) Isolation of HTLV-III from cerebrospinal fluid and neural tissues of patients with neurologic syndromes related to the acquired immunodeficiency syndrome. N Engl J Med 313: 1493–1497

Ho DD, Rota TR, Hirsch MS (1986) Infection of monocyte/macrophages by human T lymphotropic virus type III. J Clin Invest 77: 1712–1715

Ho DD, Moudgil T, Alam M (1989a) Quantitation of human immunodeficiency virus type 1 in the blood of infected persons. N Engl J Med 321: 1621–1625

Ho DD, Bredesen DE, Vinters HV, Daar ES (1989b). The acquired immunodeficiency syndrome (AIDS) dementia complex. Ann Intern Med 111: 400–410

Homsy J, Tateno M, Levy JA (1988) Antibody-dependent enhancement of HIV infection. Lancet i: 1285–1286

Homsy J, Meyer M, Tateno M, Clarkson S, Levy JA (1989) The Fc and not CD4 receptor mediates antibody enhancement of HIV infection in human cells. Science 244: 1357–1360

Hussey RE et al. (1988) A soluble CD4 protein selectively inhibits HIV replication and syncytium formation. Nature 331: 78

Jones K et al. (1986) Activation of the AIDS retrovirus promotor by the cellular transcription factor, SP1. Science 232: 755–759

Kanitakis J, Marchand C, Su H, Thivolet J, Zambruno G, Schmitt D, Gazzolo L (1989) Immunohistochemical study of normal skin of HIV-1-infected patients shows no evidence of infection of epidermal Langerhans cells by HIV. AIDS Res Hum Retroviruses 5: 293–302

Kennedy PGE, Ghotbi A, Hopkins J, Gendelman HE, Clements JE, Narayan O (1985) Persistent expression of Ia antigen and viral genome in visna maedi virus- induced inflammatory cells: possible role of lentivirus induced interferon. J Exp Med 162: 1970–1982

Koenig S, Gendelman HE, Orenstein JM, Dal Canto MC, Pezeshkpour GM, Yungbluth M, Janotta F, Aksamit A, Martin MA, Fauci AS (1986) Detection of AIDS virus in macrophages in brain tissue from AIDS patients with encephalopathy. Science 233: 1089–1093

Koethe SM, Carrigan DR, Turner PA (1989) Increased density of HLA-DR antigen on monocytes of patients infected with the human immunodeficiency virus. J Med Virol 29: 82–87

Kornbluth RS, Oh PS, Munis JR, Cleveland PH, Richman DD (1989) Interferons and bacterial lipopolysaccharide protect macrophages from productive infection by human immunodeficiency virus in vitro. J Exp Med 169: 1137–1151

Koyanagi Y, Miles S, Mitsuyasu RT, Merrill JE, Vinters HV, Chen ISY (1987) Dual infection of the central nervous system by AIDS viruses with distinct cellular tropism. Science 236: 819–822

Koyanagi Y, O'Brian WA, Zhao JQ, Golde DW, Gasson JC, Chen ISY (1988) Cytokines alter production of HIV-1 from primary mononuclear phagocytes. Science 241: 1673–1675

Lairmore MD, Butera ST, Callahan GN, DeMartini JC (1988) Spontaneous interferon production by pulmonary leukocytes is associated with lentivirus-induced lymphoid interstitial pneumonia. J Immunol 140: 779–785

Leonard JM, Khillan JS, Gendelman HE, Adachi A, Lorenzo S, Westphal H, Martin MA, Meltzer MS (1989) The human immunodeficiency virus long terminal repeat is preferentially expressed in Langerhans cells in transgenic mice AIDS Res Hum Retroviruses 5: 421–430

Le Tourneau A, Audouin J, Diebold J, Marche C, Tricottet V, Reynes M, (1986) LAV-like particles in lymph node germinal centers in patients with the persistent lymphadenopathy syndrome and the acquired immunodeficiency syndrome-related complex. An ultrastructural study of 30 cases. Hum Pathol 17: 1047–1053

Levy JA, Shimabukuro J, Hollander H, Mills J, Kaminsky L (1985) Isolation of AIDS-associated retroviruses from cerebrospinal fluid and brain of patients with neurologic symptoms. Lancet ii: 586–588

Lifson AR, Rutherford GW, Jaffe HW (1988) The natural history of human immunodeficiency virus infection. J Infect Dis 158: 1360–1367

Mace K, Duc Dodon M, Gazzolo L (1988) Restriction of HIV-1 replication in promonocytic cells: a role IFN-α. Virology 168: 399–405

McElrath MJ, Pruett JE, Cohn ZA (1989) Mononuclear phagocytes of blood and bone marrow: comparative roles as viral reservoirs in human immunodeficiency virus type 1 infections. Proc Natl Acad Sci USA 86: 675–679

Meltzer MS, Gendelman HE (1988) Effects of colony stimulating factors on the interaction of monocytes and the human immunodeficiency virus. Immunol Letters 19: 193–198

Meltzer MS, Skillman DS, Gomatos PJ, Kalter DC, Gendelman HE (1990) Role of mononuclear phagocytes in the pathogenesis of human immunodeficiency virus infection. Ann Rev Immunol 8: 169–194

Merrill JE, Koyanagi Y, Chen ISY (1989) Interleukin-1 and tumor necrosis factor α can be induced from mononuclear phagocytes by human immunodeficiency virus type 1 binding to the CD4 receptor. J Virol 63: 4404–4410

Michaelis B, Levy JA (1989) HIV replication can be blocked by recombinant human interferon beta. AIDS 3: 27–31

Murray HW (1988) Macrophage activation in the acquired immunodeficiency syndrome. Adv Exp Med Biol 239: 297–307

Murray HW, Rubin BY, Masur H, Roberts RB (1984) Impaired production of lymphokines and immune (gamma) interferon in the acquired immunodeficiency syndrome. N Engl J Med 310: 883–889

Murray HW, Jacobs JL, Bovbjerg DH (1987) Accessory cell function of AIDS monocytes. J Infect Dis 156: 696

Nabel G, Baltimore D (1987) An inducible transcription factor activates expression of human immunodeficiency virus in T cells. Nature 326: 711–713

Nakajima K, Martinez-Maza O, Hirano T, Breen BC, Nishanian PG, Salazar- Gonzalez JF, Fahey JL, Kishimoto T (1989) Induction of IL-6 (B cell stimulatory factor-2/IFNβ2) production by HIV. J Immunol 142: 531–536

Narayan O, Sheffer D, Clements JE, Teenekoon G (1985) Restricted replication of lentivirus Visna viruses induce a unique interferon during interaction between lymphocytes and infected macrophages. J Exp Med 162: 1954–1969

Nathan CF (1987) Secretory products of macrophages. J Clin Invest 79: 319–326

Nicholson JKA, Cross GD, Callaway CS, McDougal JS (1986) In vitro infection of human monocytes with human T lymphotropic virus type III/lymphadenopathy-associated virus (HTLV-III/LAV). J Immunol 137: 323–329

Nielsen H, Kharazmi A, Faber V (1986) Blood monocytes and neutrophil functions in the acquired immune deficiency syndrome. Scand J Immunol 24: 291–296

Orenstein JM, Meltzer MS, Phipps T, Gendelman HE (1988) Cytoplasmic assembly and accumulation of human immunodeficiency virus types 1 and 2 in recombinant human colony stimulating factor-1 treated human monocytes: an ultrastructural study. J Virol 62: 2578–2586

Osborn L, Kunkel S, Nabel GJ (1989) Tumor necrosis factor α and interleukin 1 stimulate the human immunodeficiency virus enhancer by activation of the nuclear factor κB. Proc Natl Acad Sci USA 86: 2336–2340

Oxholm P, Helweg-Larsen S, Permin H (1986) Immunohistological skin investigations in patients with the acquired immune deficiency syndrome. Acta Path Microbiol Immunol Scand Sect A 94: 113–116

Perno C-F, Yarchoan R, Cooney DA, Hartman NR, Webb DSA, Hao Z, Mitsuya H, Johns DG, Broder S (1989) Replication of human immunodeficiency virus in monocytes. Granulocyte/macrophage colony-stimulating factor (GM-CSF) potentiates viral production yet enhances the antiviral effect mediated by 3′-azido '2′3′-dideoxythymidine (AZT) and other dideoxynucleoside congeners of thymidine. J Exp Med 169: 993: 951

Plata F, Autran B, Martins LP, Wain-Hobson S, Raphael M, Mayaud C, Denis M, Guillon J-M, Debre P (1987) AIDS virus-specific cytotoxic T lymphocytes in lung disorders Nature 328: 348–351

Poli G, Bottazzi B, Acero R, Bersani L, Rossi V, Introna M, Lazzarin A, Galli M, Mantovani A (1985) Monocyte function in intravenous drug abusers with lymphadenopathy syndrome and in patients with acquired immunodeficiency syndrome: selective impairment of chemotaxis. Clin Exp Immunol 62: 136–142

Popovic M, Gartner S (1987) Isolation of HIV-1 from monocytes but not T phocytes. Lancet ii: 916

Preble OT, Rook AH, Quinnan QV, Vilcek J, Friedman RM, Steis R, Gelmann EP, Sonnabend JA (1985) Role of interferon in AIDS. Ann NY Acad Sci., 437: 65–87

Psallidopoulos MC, Schnittman SM, Thompson LM, Baseler B, Fauci AS, Lane HC, Salzman NP (1989) Integrated proviral human immunodeficiency virus type I is present in CD4+ peripheral blood lymphocytes in healthy seropositive individuals. J Virol 63: 4626–4631

Putney SD, Matthews TJ, Robey WG, Lynn DL, Robert-Guroff M, Muller WT, Langlois AJ, Ghrayab J, Petteway SR, Weinhold KJ, Fischinger PJ, Wong-Staal F, Gallo RC, Bolognesi DP (1986) HTL VIIIB/LAV-neutralizing antibodies to an E. coli-produced fragment of the virus envelope. Science 234: 1392–1395

Rappersberger K, Gartner S, Schenk P, Stingl G, Groh V, Tschachler E, Mann DL, Wolff K, Konrad K, Popovic M (1988) Langerhans' cells are an actual site of HIV-1 replication. Intervirology 29: 185–194

Ratner L et al. (1977) Interferon, double-stranded RNA and RNA degradation: Characteristics of an endonuclease activity. Eur J Biochem 79: 565–577

Rieber P, Riethmuller G (1986) Loss of circulating T4+ monocytes in patients infected with HTLV-III. Lancet i: 270

Robinson WE, Montefiori DC, Mitchell WM (1988) Antibody-dependent enhancement of human immunodeficiency virus type-1 infection. Lancet i: 790–794

Roy G, Rojo N, Leyva-Cobian F (1987) Phenotypic changes in monocytes and alveolar macrophages in patients with acquired immunodeficiency syndrome (AIDS) and AIDS-related complex (ARC). J Clin Lab Immunol 23: 135–141

Salahuddin SZ, Rose RM, Groopman JE, Markham PD, Gallo RC (1986) Human T lymphotropic virus type III infection of human alveolar macrophages. Blood 68: 281–284

Schnittman SM, Psallidopoulos MC, Lane HC, Thompson L, Baseler B, Massari F, Fox CH, Salzman NP, Fauci AS (1989) The reservoir for HIV-1 in human peripheral blood is a T cell that maintains expression of CD4. Science 245: 305–308

Sei Y, Petrella RJ, Tsang P, Bekesi JG, Yokoyama MM (1986) Monocytes in AIDS. N Engl J Med 315: 1611–1612

Siegal FP, Lopez C, Fitzgerald PA, Shah K, Baron P, Leiderman IZ, Imperato D, Landesman S (1986) Opportunistic infections in acquired immune deficiency syndrome result from synergistic defects of both the natural and adaptive components of cellular immunity. J Clin Invest 78: 115–123

Smith DH et al. (1987) Blocking of HIV-I infectivity by a soluble, ecreted form of the CD4 antigen Science 238: 1704–1707

Smith PD, Ohura K, Masur H, Lane CH, Fauci AS, Wahl SM (1984) Monocyte function in the acquired immunodeficiency syndrome. Defective chemotaxis. J Clin Invest 74: 2121–2128

Stoler MH, Eskin TA, Benn S, Angerer RC, Angerer LM (1986) Human T-cell lymphotropic virus type III infection of the central nervous system. Preliminary in situ analysis. JAMA 256: 2360–2364

Szebeni J et al. (1989) Dipyridamole potentiates the inhibition by 3′-azido-3′-deoxythymidine and other dideoxynucleosides of human immunodeficiency virus replication in monocyte-macrophages. Proc Natl Acad Sci USA 86: 3842–3846

Takeda A, Tuazon CU, Ennis FA (1988) Antibody-enhanced infection by HIV-1 via Fc receptor mediated entry. Science 242: 580–583

Tas M, Drexhage HA, Goudsmit J (1988) A monocyte chemotaxis inhibiting factor in serum of HIV-infected men shares epitopes with the HIV transmembrane protein gp41. Clin Exp Immunol 71: 13–18

Tschachler E, Groh V, Popovic M, Mann DL, Konrad K, Safai B, Eron L, diMarzo- Veronese F, Wolff K, Stingl G (1987) Epidermal Langerhans cells-a traget for HTLV-III/LAV infection. J Invest Dermatol 88: 233–237

Vadhan R, Wong G, Gnecco C, Cunninham-Rundles S, Krim M, Real FX, Oettgen HF, Krown SE (1986) Immunological variables as predictors of prognosis in patients with Kaposi's sarcoma and the acquired immuno-deficiency syndrome. Cancer Res 46: 417–425

Vazeux R et al. (1987) AIDS subacute encephalitis. Identification of HIV-infected cells. AM J Pathol 126: 403–410

Wahl LM, Corcoran ML, Pyle SW, Arthur A, Harel-Bellan A, Farrar WL (1989) Human immunodeficiency virus glycoprotein (gp120) induction of monocyte arachidonic acid metabolites and interleukin-1. Proc Natl Acad Sci USA 86: 621–625

Wahl SM, Allen JB, Gartner S, Orenstein JM, Popovic M, Chenoweth DE, Arthur LO, Farrar WL, Wahl LM (1989) Human immunodeficiency virus-1 and its envelope glycoprotein down-regulate chemotactic ligand receptors and chemotactic function of peripheral blood monocytes. J Immunol 142: 3553–3559

Weiss SH, Goedert JJ, Gartner S, Popovic M, Waters D, Markham P, di Marzo Veronese F, Gail MH, Barkley WE, Gibbons J, Gill FA, Leuther M, Shaw GM, Gallo RC, Blattner WA (1988) Risk of human immunodeficiency virus (HIV-1) infection among laboratory workers. Science 239: 68–71

Wolff H, Anderson DJ (1988) Potential human immunodeficiency virus-host cells in human semen. AIDS Res Hum Retroviruses 4: 1–2

Wong GHW, Krowka JF, Sites DP, Goeddel DC (1988) In vitro anti-human immunodeficiencyvirus activities of tumor necrosis factor-α and interferon-γ. J Immunol 140: 120–124

Yamada O, Hattori N, Kurimura T, Kita M, Kishida TH (1988) Inhibition of growth of HIV by human natural interferon in vitro. Aids Res Human Retro viruses 4: 287–294

Yamamoto JK et al. (1986) Human alpha- and beta-interferon but not gamma-suppresses the in vitro replication of LAV, HTLV-III, and ARV-2. J Interferon Res 6: 143–152

Ziza J-M, Brun-Vezinet F, Venet A, Rouzioux CH, Traversat J, Israel-Biet B, Barre-Sinoussi F, Godeau P (1985) Lymphadenopathy-associated virus isolated from bronchoalveolar lavage fluid in AIDS-related complex with lymphoid interstitial pneumonitis. N Engl J Med 313: 183

Interactions Between Macrophages and Legionella pneumophila

M. A. HORWITZ

1 Introduction

Legionella pneumophila, the first member of the family Legionellaceae, is aerobic, motile, gram-negative bacterium that causes Legionnaires' disease and Pontiac fever (FRASER et al. 1977; MCDADE et al. 1977; GLICK et al. 1978). Although the family Legionellaceae now contains more than 40 species, *L. pneumophila* causes over 90% of human infections. *L. micdadei*, the second most frequently isolated species in human infection, evidently has lower virulence for humans than *L. pneumophila* as it appears to infect only immunocompromised hosts (MYEROWITZ et al. 1971). *L. pneumophila* normally inhabits aquatic environments,

Division of Infectious Diseases, Department of Medicine, UCLA School of Medicine, Center for the Health Sciences, 37-121, Los Angeles, CA 90024, USA

Current Topics in Microbiology and Immunology, Vol. 181

and thus humans are accidental, albeit frequent hosts. The bacterium is spread to humans by the airborne route in aerosols arising out of contaminated sources; possibly, the organism is also spread to humans by the waterborne route (MUDER et al. 1986).

L. pneumophila is a facultative intracellular parasite that parasitizes human mononuclear phagocytes (HORWITZ and SILVERSTEIN 1980). The bacterium has been shown to multiply in vitro in cultures of human monocytes and alveolar macrophages (HORWITZ and SILVERSTEIN 1980; NASH et al. 1984), monkey alveolar macrophages (KISHIMOTO et al. 1979; JACOBS et al. 1984), guinea pig alveolar macrophages (ELLIOTT and WINN 1986), guinea pig peritoneal macrophages (KISHIMOTO et al. 1981; YOSHIDA and MIZUGUCHI 1986; YAMAMOTO et al. 1987), hamster peritoneal macrophages (YOSHIDA and MIZUGUCHI 1986), rat alveolar macrophages (ELLIOTT and WINN 1986), rat peritoneal macrophages (YOSHIDA and MIZUGUCHI 1986), and thioglycolate-elicited but not resident peritoneal macrophages from A/J mice (YAMAMOTO et al. 1988). However, thioglycolate-elicited and resident peritoneal macrophages of BDF-1 mice and thioglycolate-elicited peritoneal macrophages of DBA2, C3H/HeN, C57BL/6, and Balb/c mice do not support growth of *L. pneumophila* (YAMAMOTO et al. 1987, 1988). *L. pneumophila* has also been shown to multiply in various cell lines, including differentiated HL-60 (MARRA et al. 1990) and U-937 (CIANCIOTTO et al. 1989a) human macrophage-like cell lines, MRC-5 human embryonic lung fibroblasts (WONG et al. 1980; DAISY et al. 1981; OLDHAM and RODGERS 1985), Hep2 human epithelial laryngeal carcinoma cells (DAISY et al. 1981; OLDHAM and RODGERS 1985), HeLa human cervical carcinoma cells (DAISY et al. 1981), Vero African green monkey cells (OLDHAM and RODGERS 1985), and McCoy mouse synovial cells (DAISY et al. 1981). Although it can be cultured on artificial media, *L. pneumophila* multiplies exclusively intracellularly under tissue culture conditions (HORWITZ and SILVERSTEIN 1980) and, presumably, this is also the case in vivo. In the environment, *L. pneumophila* evidently multiplies in protozoa. The bacterium has been demonstrated to multiply in vitro in association with *Acanthamoeba palestinensis* (ANAND et al. 1983), *Acanthamoeba royreba* (TYNDALL and DOMINGUE 1982), *Acanthamoeba castellanii* Neff (HOLDEN et al. 1984), *Naegleria lovaniensis* (TYNDALL and DOMINGUE 1982), *Naegleria fowleri* (NEWSOME et al. 1985), *Tetrahymena pyriformis* (FIELDS et al. 1984), *Cylidium sp.* (BARBAREE et al. 1986), and *Hartmannella sp.* (WADOWSKY et al. 1988).

Initial studies of *L. pneumophila*–mononuclear phagocyte interaction that followed recognition of the bacterium as an important pathogen in 1977 revealed its capacity to multiply intracellularly in human mononuclear phagocytes, described its unusual form of entry into phagocytes, defined its intracellular pathway in mononuclear phagocytes (including its capacity to inhibit phagosome–lysosome fusion and phagosome acidification), and delineated the roles of humoral and cell-mediated immunity in host defense against *L. pneumophila*. More recent studies have begun to elucidate the molecular basis for *L. pneumophila*–mononuclear phagocyte interaction. This review will summarize major advances in our understanding of these interactions.

2 Phagocytosis

2.1 Morphology of Entry

Legionella pneumophila enters phagocytes, including monocytes, alveolar macrophages, polymorphonuclear leukocytes, and differentiated HL-60 cells, by an unusual process termed "coiling phagocytosis," in which long phagocyte pseudopods coil around the organism as it is internalized (HORWITZ 1984; MARRA et al. 1990) (Fig. 1). Treatment of *L. pneumophila* with high-titer anti-*L. pneumophila* antiserum neutralizes the coiling phenomenon. Such antibody-coated organisms are ingested by "conventional phagocytosis" in which phagocyte pseudopods move circumferentially and more or less symmetrically about the organism until they meet and fuse at the distal side (HORWITZ 1984). Whether anti-*L. pneumophila* antibody masks surface determinants that mediate coiling phagocytosis or provides ligands that allow Fc receptor-mediated conventional phagocytosis to dominate internalization of *L. pneumophila* is unknown. In any case, the end result of both phagocytic processes is the same: the organism comes to reside in a membrane- bound phagosome in which it subsequently multiplies.

Viability is not a determinant of the phagocytic process. Both live and killed *L. pneumophila* enter by coiling phagocytosis (HORWITZ 1984).

Coiling phagocytosis may not be unique to *L. pneumophila*. *Leishmania donovani* (CHANG 1979) and *Chlamydia psittaci* (WYRICK and BROWNRIDGE 1978) may also enter by coiling phagocytosis. However, other intracellular pathogens including *Mycobacterium leprae* (SCHLESINGER and HORWITZ 1990), *Mycobacterium tuberculosis* (PAYNE et al. 1987; SCHLESINGER et al. 1990), *Trypanosoma cruzi* (NOGUEIRA and COHN 1976; TANOWITZ et al. 1975), and *Toxoplasma gondii* (JONES and HIRSCH 1972) enter by conventional phagocytosis. Thus, the significance of the coiling phenomenon to intracellular pathogenesis, if any, is not clear.

2.2 Receptors Mediating Entry

Complement receptors CR1 and CR3 on human monocytes, which primarily recognize complement fragments C3b and C3bi, respectively, mediate phagocytosis of *L. pneumophila* (PAYNE and HORWITZ 1987). Monoclonal antibodies against these receptors block uptake of *L. pneumophila* into monocytes, and consequently, intracellular multiplication of *L. pneumophila* in these phagocytes (PAYNE and HORWITZ 1987).

As monocytes differentiate to so-called monocyte-derived macrophages in culture, they become increasingly permissive to *L. pneumophila* infection (HORWITZ and SILVERSTEIN 1981a). This may in part reflect the enhanced CR1 and CR3 expression (ESPARZA et al. 1986; FIRESTEIN and ZVAIFLER 1987) and function (NEWMAN et al. 1980) that accompany in vitro maturation.

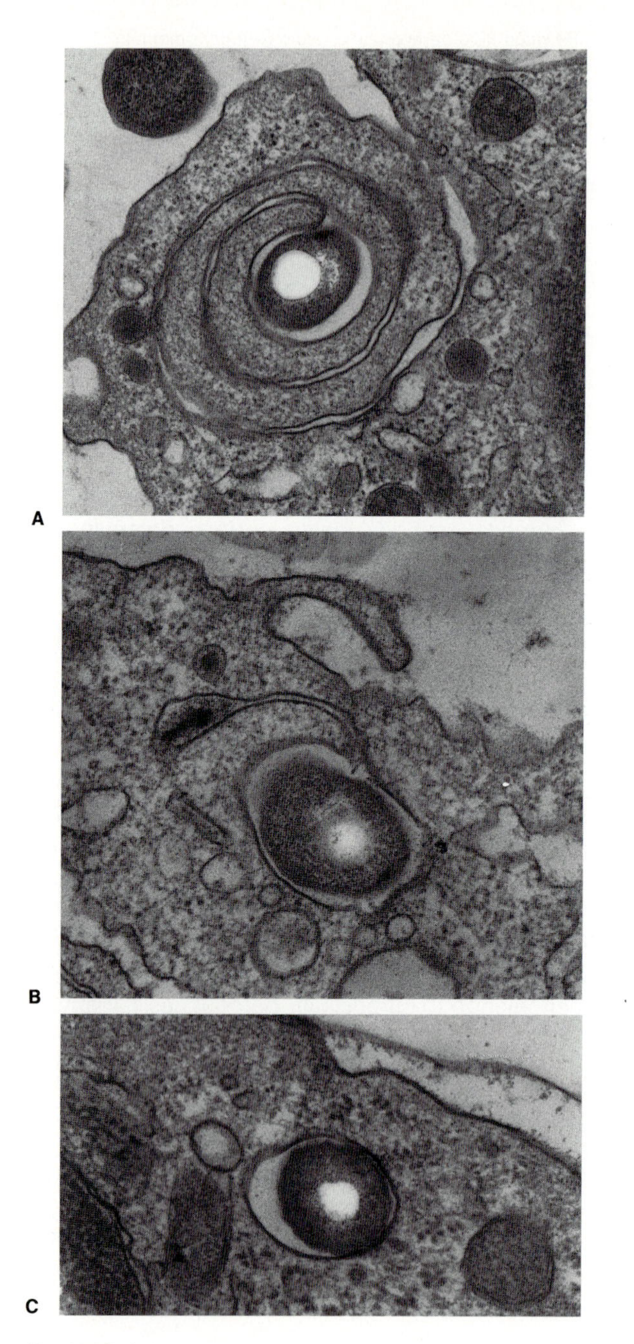

Fig. 1A–C Coiling phagocytosis. **A** Long monocyte pseudopods are coiled around the bacterium, which contains a large lucent fat vacuole. x 33 000. The monocyte pseudopod coils have broken down, presumably as a result of fusing together, so that the bacterium is enclosed in an intermediate-stage intracellular vacuole. Incomplete fusion of monocyte plasma membrane has resulted in the temporary formation of an intracellular sinus cavity. x 50 000. **C** The bacterium residues in a membrane-bound intracellular vacuole. x 52 000. (**A** and **B**: HORWITZ 1984; **C**: HORWITZ 1983a)

Complement receptors appear to provide a general pathway for entry of intracellular pathogens into mononuclear phagocytes (PAYNE and HORWITZ 1987). These receptors have been shown to play a role in the uptake of *Leishmania donovani* (BLACKWELL et al. 1985), *Leishmania major* (MOSSER and EDELSON 1985), *Mycobacterium tuberculosis* (PAYNE et al. 1987; SCHLESINGER et al. 1990), *Mycobacterium leprae* (SCHLESINGER and HORWITZ 1990), and *Histoplasma capsulatum* (BULLOCK and WRIGHT 1987). Complement receptors may provide such pathogens safe passage into mononuclear phagocytes because ligation of these receptors by particles coated with fragments of complement component C3 does not trigger an oxidative burst and the release of toxic oxygen metabolites (WRIGHT and SILVERSTEIN 1983; YAMAMOTO and JOHNSTON 1984). Consistent with this hypothesis, *L. pneumophila* (JACOBS et al. 1984) and several other intracellular pathogens, including *M. leprae* (HOLZER et al. 1986), *T. gondii* (WILSON et al. 1980), and *H. capsulatum* (EISSENBERG and GOLDMAN 1986), have been reported to elicit little or no oxidative burst upon entry into mononuclear phagocytes.

Legionella pneumophila's capacity to avoid toxic oxygen molecules may be important to its intracellular survival, because the bacterium is susceptible to relatively low concentrations of hydrogen peroxide (LOCKSLEY et al. 1982; HORWITZ 1985). This may reflect its low stores of scavengers that might protect it from the toxic effects of hydrogen peroxide, including catalase, glutathiose peroxidase, glutathione reductase, and glutathione (LOCKSLEY et al. 1982). As is typical of microorganisms, *L. pneumophila* is susceptible to lower concentrations of hydrogen peroxide in the presence of a peroxidase, including myeloperoxidase (LOCKSLEY et al. 1982; LOCHNER et al. 1983), eosinophil peroxidase (LOCKSLEY et al. 1982), and lactoperoxidase (HORWITZ 1985), and a halide. In addition to hydrogen peroxide, *L. pneumophila* is sensitive to hydroxyl radical (LOCKSLEY et al. 1982), a more distal product of the oxidative burst. However, the bacterium is relatively resistent to superoxide, perhaps reflecting its relatively abundant stores of superoxide dismutase (LOCKSLEY et al. 1982).

Aside from avoiding it, *L. pneumophila* may possess other mechanisms for protecting itself from the toxic consequences of the oxidative burst. *L. pneumophila* has been reported to elaborate a toxin that inhibits polymorphonuclear leukocyte oxygen consumption, hexose monophosphate shunt activity, bacterial iodination, and bacterial killing (FRIEDMAN et al. 1980, 1982). Along these lines, *L. pneumophila* has also been reported to contain a factor that blocks myeloperoxidase-mediated protein iodination (PERRY et al. 1987). The identifies of molecules possessing these inhibitory capacities have not been determined.

2.3 Ligands Mediating Entry

Fragments of complement component C3 covalently bound to the bacterial surface mediate ingestion of *L. pneumophila* by complement receptors (PAYNE and HORWITZ 1987; BELLINGER-KAWAHARA and HORWITZ 1987a). Consequently, *L. pneumophila* uptake is markedly reduced in heat- inactivated serum or in the absence of serum (PAYNE and HORWITZ 1987). C3 fixes to *L. pneumophila* by the alternative pathway of complement activation (BELLINGER-KAWAHARA and HORWITZ 1987a).

Consistent with a general role for C3 fragments in phagocytosis of intracellular pathogens, serum has been shown to promote uptake of other intracellular pathogens including *Leishmania donovani* (BLACKWELL et al. 1985; WILSON and PEARSON 1987); *Leishmania major* (MOSSER and EDELSON 1984), *Mycobacterium tuberculosis* (PAYNE et al. 1987; SCHLESINGER et al. 1990), and *Mycobacterium leprae* (SCHLESINGER and HORWITZ 1990), and C3 has been shown to fix to these pathogens, generally by the alternative pathway of complement activation (BLACKWELL et al. 1985; MOSSER and EDELSON 1984; PUENTES et al. 1988; PAYNE et al. 1987; SCHLESINGER et al. 1990; SCHLESINGER and HORWITZ 1990).

As macrophages secrete the complement components of the alternative and classic complement pathways, complement may be generally available in tissues to mediate phagocytosis by complement receptors. Consistent with this idea, macrophages have been shown to deposit C3 on *Leishmania donovani* in the absence of serum (WOZENCRAFT et al. 1986). Along these lines, monoclonal antibodies against complement receptors have been found to block monocyte phagocytosis of *L. pneumophila* (PAYNE and HORWITZ, unpublished data) and *M. tuberculosis* (SCHLESINGER et al. 1990) in the absence of serum. Whether this effect of antibody is due to inhibition of complement receptor binding to C3 fragments deposited on the organisms by the monocytes or to other ligands on the organisms is unknown.

2.4 Acceptor Molecules for C3 on the L. pneumophila Surface

C3 fixes selectively to the *L. pneumophila* surface. The major outer membrane protein (MOMP), a 29-kDa cation selective porin (GABAY et al. 1985), is an acceptor molecule for C3 (BELLINGER-KAWAHARA and HORWITZ 1987b). MOMP fixes C3 on Western blots of whole *L. pneumophila*, *L. pneumophila* membranes, or purified MOMP, and it is the only molecule of *L. pneumophila* that does so. The lipopolysaccharide (LPS) of *L. pneumophila* does not fix C3 on Western blots (BELLINGER-KAWAHARA and HORWITZ 1987b).

The major outer membrane protein also fixes C3 when incorporated into liposomes. As measured by an ELISA for C3, liposome–MOMP complexes fix approximately 20 times more C3 than liposomes alone, which fix just over background levels of C3 (BELLINGER-KAWAHARA and HORWITZ, unpublished data).

When opsonized in serum, liposomes with MOMP incorporated into their membranes are ingested by monocytes. Phagocytosis of such liposomes is dose dependent upon serum. The liposome–MOMP–C3 complexes are ingested by conventional phagocytosis and come to reside in membrane-bound phagosomes (BELLINGER-KAWAHARA and HORWITZ, unpublished data).

Complement receptors CR1 and CR3, C3 fragments, and MOMP thus appear to constitute a complete receptor–ligand–acceptor molecule system mediating monocyte recognition and ingestion of *L. pneumophila*.

3 Intracellular Pathway

After phagocytosis, the *L. pneumophila* phagosome interacts sequentially with monocyte smooth vesicles, mitochondria, and ribosomes until a novel ribosome-lined replicative vacuole is formed (HORWITZ 1983a) (Figs. 2, 3). *L. pneumophila* then multiples within this phagosome with a doubling time at mid-log phase of about 2 h until the host cell becomes packed full with bacteria and ruptures (HORWITZ and SILVERSTEIN 1980).

The *L. pneumophila* phagosome does not fuse with monocyte lysosomes, as revealed by electron microscopic studies employing electron-opaque lysosomal markers and acid phosphatase cytochemistry (HORWITZ 1983b). The *L. pneumophila* phagosome also does not become acidified to the low pH levels characteristic of phagolysosomes (HORWITZ and MAXFIELD 1984). Quantitative fluorescence microscopy of individual *L. pneumophila* phagosomes has revealed that phagosomes containing live *L. pneumophila* have a mean pH of 6.1, which is 0.8 pH units higher than phagosomes containing formalin-killed *L. pneumophila*. The "defect" in acidification is localized to the *L. pneumophila* phagosome, as phagolysosomes containing IgG-coated erythrocytes in the same monocytes have a pH of <5 (HORWITZ and MAXFIELD 1984).

In contrast to live *L. pneumophila*, formalin-killed *L. pneumophila* do not enter phagosomes that interact with monocyte smooth vesicles, mitochondria, and ribosomes (HORWITZ 1983a). Instead, they enter phagosomes that fuse with lysosomes, and they are rapidly digested in the phagolysosome (HORWITZ 1983b).

Intracellular pathogens follow at least three distinct pathways through the mononuclear phagocyte: intraphagosomal, intraphagolysosomal, and extraphagosomal pathways. As noted, *L. pneumophila* takes the intraphagosomal pathway. *L. pneumophila* shares this pathway with *T. gondii* and *Chlamydia psittaci*, and to some extent with *M. tuberculosis*. All of these pathogens inhibit phagosome–lysosome fusion (HORWITZ 1983b; JONES and HIRSCH 1972; FRISS 1972; TODD and STORZ 1975; WYRICK and BROWNRIDGE 1978; EISSENBERG et al. 1983; ARMSTRONG and D'ARCY HART 1971). Like *L. pneumophila*, *T. gondii* (JONES and HIRSCH 1972) and *C. psittaci* (FRISS 1972) reside in phagosomes that interact with mitochondria, and *T. gondii* inhibits phagosome acidification (SIBLEY et al. 1985). The factors that determine the common selection of the intraphagosomal pathway by these phylogenetically disparate pathogens are unknown.

Intracellular pathogens are exquisitely adapted to their intracellular milieu. Presumably for this reason, a mutant *L. pneumophila* that enters monocytes like

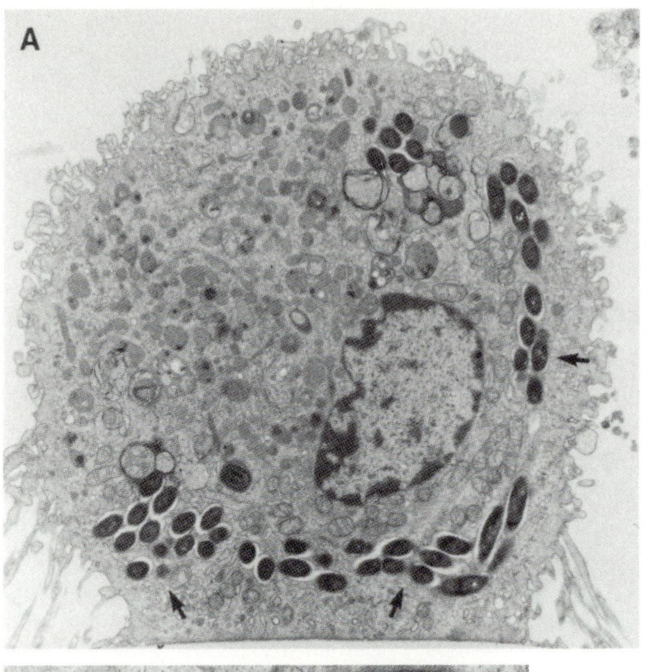

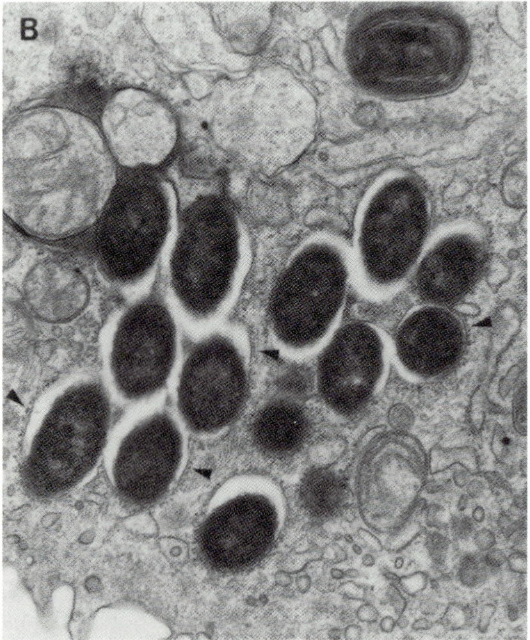

Fig. 2 A, B. Intracellular multiplication of *L. pneumophila* in human alveolar macrophages. Human alveolar macrophages were obtained by bronchoalveolar lavage, cultured as a monolayer, and infected with *L. peneumophila*. The macrophages were incubated for 24 h and processed for electron microscopy. **A** The alveolar macrophage is heavily infected with *L. pneumophila* (*arrows*). x 57 000. **B** At a higher magnification of a portion of the macrophage shown in **A**, *L. pneumophila* bacteria are observed in ribosome-lined phagosomes (*arrowheads*). The ribosomes are separated from the phagosome membrane by a gap of approximately 100Å. x 21 000. (NASH et al. 1984)

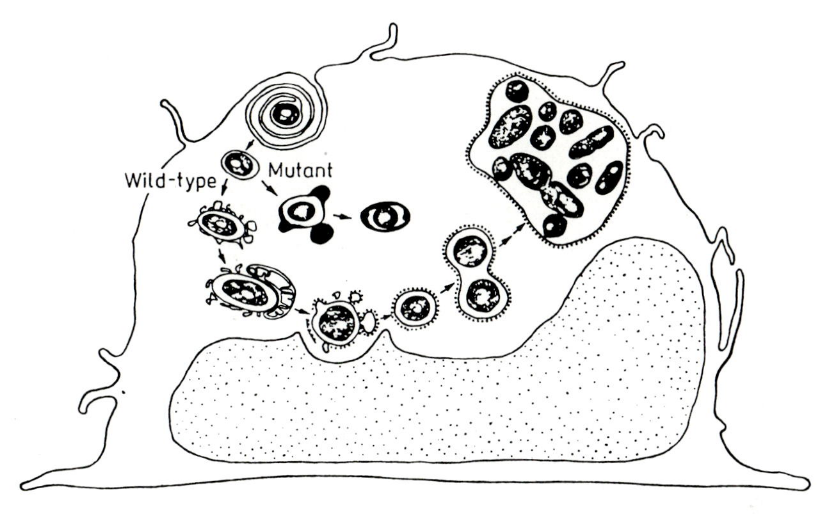

Fig. 3. Comparative intracellular biology of wild-type and mutant *L. pneumophila*. Wild-type and mutant *L. pneumophila* enter monocytes similarly—by coiling phagocytosis—but thereafter they follow different intracellular pathways. Wild-type *L. pneumophila* resides in a distinctive ribosome-lined phagosome formed after interaction of the phagosomal membrane with monocyte smooth vesicles, mitochondria, and ribosomes. The wild-type phagosome does not fuse with monocyte lysosomes. The wird-type bacterium multiplies within the ribosome-lined phagosome until the monocyte becomes packed full with bacteria and ruptures. In contrast, the mutant *L. pneumophila* does not interact with monocyte smooth vesicles, mitochondria, or ribosomes. The mutant *L. pneumophila* phagosome fuses with monocyte lysosomes. The mutant bacterium survives but does not multiply in a phagolysosome. (HORWITZ 1989)

the wild type, by coiling phagocytosis, but then resides in a phagosome that fails to inhibit phagosome–lysosome fusion, is avirulent (Fig. 3). The mutant survives but fails to multiply in the phagolysosome (HORWITZ 1987).

Although its precise role in pathogenesis has not been defined, the *mip* protein, the 24-kDa product of the *mip* gene, potentiates *L. pneumophila* infection of macrophages. An isogenic mutant lacking this protein exhibits diminished virulence for U937 cells and human alveolar macrophages in vitro (CIANCIOTTO et al. 1989a) and guinea pigs in vivo (CIANCIOTTO et al. 1989b).

4 Immunity

4.1 Humoral Immunity

4.1.1 Role in Host Defense

Patients respond to *L. pneumophila* infection with the production of antibody against the organism. In vitro studies indicate that humoral immunity does not play a primary role in host defense against *L. pneumophila* (HORWITZ and

SILVERSTEIN 1981a, b). First, anti-*L. pneumophila* antibody does not promote complement-mediated killing of *L. pneumophila* (HORWITZ and SILVERSTEIN 1981b); indeed, as discussed above *L. pneumophila* utilizes the complement system to gain entry into monocytes. Second, while anti-*L. pneumophila* antibody promotes uptake of *L. pneumophila* by monocytes, alveolar macrophages, and polymorphonuclear leukocytes, this antibody does not promote substantial killing of the bacteria (HORWITZ and SILVERSTEIN 1981 a–c; NASH et al. 1984). The net result is that more bacteria gain access to the intracellular milieu in which they multiply. Finally, anti-*L. pneumophila* antibody does not inhibit intracellular multiplication of *L. pneumophila*; antibody-coated bacteria multiply intracellularly at the same rate as bacteria phagocytized in the absence of antibody (HORWITZ and SILVERSTEIN 1981c).

4.1.2 Antigens Stimulating Humoral Immunity

The LPS of *L. pneumophila* is the dominant antigen recognized by anti-*L. pneumophila* antibody; greater than 98% of the antibody produced by patients with Legionnaires' disease recognizes this antigen (GABAY and HORWITZ 1985). It is serogroup specific (CIESIELSKI et al. 1986).

The major cytoplasmic membrane protein (MCMP), a 60- to 65-kDa protein, is the major protein antigen recognized by anti-*L. pneumophila* antibody (GABAY and HORWITZ 1985; SAMPSON et al. 1986). This protein is a genus common antigen and heat shock protein (LEMA et al. 1988) cross-reactive with *M. tuberculosis* 65-kDa and *E. coli* Gro EL heat shock proteins (SHINNICK et al. 1988). Interestingly, relatively little antibody is made to MOMP, the most abundant protein of the bacterium (GABAY and HORWITZ 1985).

4.2 Cell-Mediated Immunity

4.2.1 Role in Host Defense

Patients also respond to *L. pneumophila* infection by expanding the pool of lymphocytes that recognize the pathogen (HORWITZ 1983c). In vitro studies indicate that, in contrast to humoral immunity, cell-mediated immunity plays a primary role in host defense against *L. pneumophila.* Human monocytes or alveolar macrophages treated with *L. pneumophila* antigen- or mitogen-stimulated lymphocyte supernatant fluids become activated and inhibit the intracellular multiplication of *L. pneumophila* (HORWITZ and SILVERSTEIN 1981a; HORWITZ 1983c; NASH et al. 1984). Similarly, interferon-γ-activated human monocytes or alveolar macrophages inhibit the intracellular multiplication of *L. pneumophila* (BHARDWAJ et al. 1986; NASH et al. 1988; JENSEN et al. 1987).

Activated monocytes or alveolar macrophages inhibit *L. pneumophila* multiplication but do not kill the intracellular bacteria (HORWITZ and SILVERSTEIN 1981a; NASH et al. 1984), even at very high concentrations of interferon-γ (BHARDWAJ et al. 1986; NASH et al. 1988). Moreover, in the presence of anti-*L.*

pneumophila antibody, activated monocytes exhibit no more capacity to kill *L. pneumophila* than nonactivated monocytes (HORWITZ and SILVERSTEIN 1981a; BHARDWAJ et al. 1986). Indeed, interferon-γ-activated monocytes do not kill *L. pneumophila* even in the presence of high concentrations of antibiotic inhibitors of bacterial protein or RNA synthesis (BHARDWAJ and HORWITZ 1988), which by themselves are able only to inhibit but not kill intracellular bacteria (HORWITZ and SILVERSTEIN 1983).

Activated monocytes inhibit *L. pneumophila* intracellular multiplication in two general ways. First, activated monocytes phagocytize fewer organisms, thereby limiting access of the bacteria to the intracellular milieu in which they multiply (HORWITZ and SILVERSTEIN 1981a). Activated monocytes may accomplish this by down-regulating the function of complement receptors (WRIGHT et al. 1986; FIRESTEIN and ZVAIFLER 1987; ESPARZA et al. 1986), which mediate phagocytosis of *L. pneumophila*, as discussed above. Second, activated monocytes inhibit the multiplication of bacteria that are internalized (HORWITZ and SILVERSTEIN 1981a). They accomplish this by limiting the availability of iron to intracellular *L. pneumophila*, as will be discussed below.

4.2.2 Antigens Stimulating Cell-Mediated Immunity

The major secretory protein (MSP) of *L. pneumophila*, a 39-kDa metalloprotease (DREYFUS and IGLEWSKI 1986), is a potent stimulator of cell-mediated immunity to *L. pneumophila* in the guinea pig model of Legionnaires' disease (BREIMAN and HORWITZ 1987; BLANDER and HORWITZ 1989). Guinea pigs infected sublethally with *L. pneumophila* by aerosol exhibit strong splenic lymphocyte proliferation and cutaneous delayed-type hypersensitivity to MSP (BREIMAN and HORWITZ 1987; BLANDER and HORWITZ 1989). Guinea pigs immunized with MSP subcutaneously also develop a strong cell-mediated immune response to the molecule, as manifested by splenic lymphocyte proliferation and cutaneous delayed-type hypersensitivity (BLANDER and HORWITZ 1989). As will be discussed below, such animals also develop protective immunity.

Legionella pneumophila antigenic preparations lacking MSP, such as *L. pneumophila* membranes, also induce cell-mediated immune responses (BLANDER and HORWITZ 1990a). Therefore, other *L. pneumophila* molecules capable of inducing cell-mediated immune responses remain to be identified. Interestingly, the two major components of the outer membrane, LPS and MOMP, do not induce strong cell-mediated immune responses (BREIMAN and HORWITZ 1987).

4.3 Protective Immunity

4.3.1 The Guinea Pig Model

The guinea pig provides a superb animal model of Legionnaires' disease. When exposed to aerosols of *L. pneumophila*, guinea pigs develop a pneumonic illness that clinically and pathologically mimics Legionnaires' disease in humans.

Sublethally infected guinea pigs develop humoral and cell-mediated immune responses to *L. pneumophila* antigens and strong protective immunity to lethal aerosol challenge with *L. pneumophila.* Such animals limit the multiplication of *L. pneumophila* in their lungs (BREIMAN and HORWITZ 1986).

4.3.2 Antigens Stimulating Protective Immunity

Several antigenic preparations, in addition to live *L. pneumophila* in sublethal concentrations, induce protective immunity in guinea pigs to lethal aerosol challenge with *L. pneumophila*. First, the avirulent mutant *L. pneumophila* described above, which fails to inhibit phagosome–lysosome fusion and which survives but does not multiply in human monocytes, induces strong protective immunity in guinea pigs immunized with it by the aerosol route (BLANDER et al. 1989). Protection with the avirulent mutant is comparable to that provided by sublethal aerosol infection with the wild type (BLANDER et al. 1989).

Second, as noted above, the MSP of *L. pneumophila* induces strong protective immunity in guinea pigs immunized with it by subcutaneous inoculation (BLANDER and HORWITZ 1989). MSP can also induce cross-protective immunity between different serogroups of *L. pneumophila* (BLANDER and HORWITZ 1990b). Interestingly, this immunoprotective molecule is not a virulence determinant in the guinea pig model of Legionnaires' disease (BLANDER et al. 1990). Studies comparing a cloned isogenic mutant *L. pneumophila* that does not produce MSP with its progenitor MSP-producing strain have revealed that the two strains have comparable virulence in the guinea pig model. The two strains have equivalent LD_{50}s and LD_{100}s for guinea pigs, multiply in the lungs of challenged guinea pigs at the same rates, and produce indistinguishable pathologic lesions in guinea pig lungs (BLANDER et al. 1990).

Third, *L. pneumophila* membranes induce strong protective immunity in guinea pigs when administered either by aerosol or subcutaneously. Interestingly, in contrast, formalin-killed *L. pneumophila* does not induce strong cell-mediated immune responses or protective immunity (BLANDER and HORWITZ 1990a).

4.4 Processing and Presentation of L. pneumophila Antigens

Very little is known about the processing and presentation of parasite antigens by infected mononuclear phagocytes. It seems likely that immunodominant T cell antigens such as the MSP are processed intracellularly in mononuclear phagocytes to immunogenic epitopes that are displayed on the surface of these antigen-presenting cells. In support of this hypothesis, immunocytochemistry employing affinity-purified monospecific anti-MSP antibody has demonstrated that *L. pneumophila* produces MSP intracellularly and that this molecule colocalizes with *L. pneumophila* phagosomes (CLEMENS and HORWITZ 1990). Erythromycin, which inhibits *L. pneumophila* protein synthesis, abolishes MSP immunoreactivity. Furthermore, immunoelectron microscopic studies have revealed that MSP localizes to the *L. pneumophila* phagosome and cytoplasmic clusters of infected monocytes (CLEMENS and HORWITZ 1990).

5 Role of Iron in Monocyte Activation

Legionella pneumophila intracellular multiplication is iron-dependent. Multiplication is inhibited by deferoxamine, an iron chelator, and this inhibition is reversed by iron-saturated transferrin (BYRD and HORWITZ 1989).

Interferon-γ-activated monocytes inhibit *L. pneumophila* intracellular multiplication by limiting the availability of iron to the bacterium. The capacity of interferon-γ-activated monocytes to inhibit *L. pneumophila* intracellular multiplication is reversed by iron transferrin (BYRD and HORWITZ 1990a). The capacity of such activated monocytes to inhibit *L. pneumophila* multiplication is also reversed by iron lactoferrin and by nonphysiologic iron compounds (BYRD and HORWITZ 1990a). The reversing effect of iron lactoferrin raises the interesting possibility that release of lactoferrin by polymorphonuclear leukocytes at sites of inflammation in the *L. pneumophila*-infected lung may be counterproductive to host defense.

Interferon-γ-activated monocytes may limit intracellular iron availability to *L. pneumophila* in at least two ways. First, such monocytes markedly down-regulate transferrin receptors, which mediate internalization of iron transferrin by these cells (BYRD and HORWITZ 1989). Second, such monocytes markedly down-regulate the concentration of intracellular ferritin, the major iron storage protein in these cells (BYRD and HORWITZ 1990b).

The potential importance of transferrin receptors to *L. pneumophila* iron acquisition is underscored by the study of an individual whose monocytes have low numbers of transferrin receptors. This person's monocytes are uniquely nonpermissive to *L. pneumophila* intracellular multiplication. However, the addition of ferric ammonium citrate, which may enter monocytes independent of the transferrin receptor endocytic pathway, completely reverses the non-permissive state of these monocytes and allows *L. pneumophila* to multiply in them at the same rate as in normal monocytes (BYRD and HORWITZ 1990c).

6 Conclusion

Over the past decade, the *Legionella pneumophila* model of intracellular parasitism has come of age. There is now a basic understanding of the bacterium's interactions with mononuclear phagocytes and the host immune system. Key bacterial and host molecules governing these interactions are being identified, and the development of genetic systems for studying *L. pneumophila* should spur this process. With these elements in place, the next decade should see the study of this model providing important new insights into intracellular pathogenesis and the molecular basis of intracellular parasitism.

Acknowledgments. Dr. Horwitz is Gordon MacDonald scholar at UCLA and recipient of an American Cancer Society Faculty Research Award. This work is supported by grant AI 22421 from the National Institutes of Health.

References

Anand CM, Skinner AR, Malic A, Kurtz JB (1983) Interaction of *L. pneumophila* and a free living amoeba (*Acanthamoeba palestinensis*). J Hyg (Lond) 91: 167–178

Armstrong JA, D'Arcy Hart P (1971) Response of cultured macrophages to *Mycobacterium tuberculosis* with observations on fusion of lysosomes with phagosomes. J Exp Med 134: 713–740

Barbaree JM, Fields BC, Feeley JC, Gorman GW, Martin WT (1986) Isolation of protozoa from water associated with a legionellosis outbreak and demonstration of intracellular multiplication of *Legionella pneumophila*. Appl Environ Microbiol 51: 422–424

Bellinger-Kawahara CG, Horwitz MA (1987a) *Legionella pneumophila* fixes complement component C3 to its surface—demonstration by ELISA. *In*: Program of the 1987 Annual Meeting of the American Society of Microbiology, Atlanta, GA, March 1–6, p 86

Bellinger-Kawahara CG, Horwitz MA (1987b) The major outer membrane protein is a prominent acceptor molecule for complement component C3 on *Legionella pneumophila*. Clin Res 35: 468A

Bhardwaj N, Horwitz MA (1988) Gamma interferon and antibiotics fail to act synergistically to kill *Legionella pneumophila* in human monocytes. J Interferon Res 8: 283–293

Bhardwaj N, Nash T, Horwitz MA (1986) Gamma interferon-activated human monocytes inhibit the intracellular multiplication of *Legionella pneumophila*. J Immunol 137: 2662–2664

Blackwell JM, Ezekowitz RAB, Roberts MB, Channon JY, Sim RB, Gordon S (1985) Macrophage complement and lectin-like receptors bind *Leishmania* in the absence of serum. J Exp Med 162: 324–331

Blander SJ, Horwitz MA (1989) Vaccination with the major secretory protein of *Legionella pneumophila* induces cell-mediated and protective immunity in a guinea pig model of Legionnaires' disease. J Exp Med 169: 691–705

Blander SJ, Horwitz MA (1990a) Vaccination of guinea pigs with *Legionella pneumophila* membranes induces protective immunity against lethal aerosol challenge. Clin Res 38:269A

Blander SJ, Horwitz MA (1990b) Cross-protective immunity to Legionnaires disease induced by vaccination with the major secretory protein of *Legionella*. Clin Res 38:269A

Blander SJ, Breiman RF, Horwitz MA (1989) A live avirulent mutant *Legionella pneumophila* vaccine induces protective immunity against lethal aerosol challenge. J Clin Invest 83: 810–815

Blander SJ, Szeto L, Shuman HA, Horwitz MA (1990) The major seceretory protein of *Legionella pneumophila*, a protective immunogen, is not a virulence factor in a guinea pig model of Legionnaires' disease. Clin Res 38:589A

Breiman RF, Horwitz MA (1986) Guinea pigs sublethally infected with aerosolized *Legionella pneumophila* develop humoral and cell- mediated immune responses and are protected against lethal aerosol challange. A model for studying host defense against lung infections caused by intracellular pathogens. J Exp Med 164: 799–811

Breiman RF, Horwitz MA (1987) The major secretory protein of *Legionella pneumophila* stimulates proliferation of splenic lymphocytes from immunized guinea pigs. Clin Res 35: 469A

Bullock WE, Wright SD (1987) Role of the adherence-promoting receptors CR3, LFA-1, and p150, 95 in binding of *Histoplasma capsulatum* by human macrophages. J Exp Med 165: 195–210

Byrd TF, Horwitz MA (1989) Interferon gamma-activated human monocytes down-regulate transferrin receptors and inhibit the intracellular multiplication of *Legionella pneumophila* by limiting the availability of iron. J Clin Invest 83: 1457–1465

Byrd TF, Horwitz MA (1990a) Iron-lactoferrin and non-physiologic iron-chelates reverse the capacity of activated monocytes to inhibit *Legionella pneumophila* intracellular multiplication. Clin Res 38:368A

Byrd TF, Horwitz MA (1990b) Interferon gamma-activated human monocytes downregulate the intracellular concentration of ferritin: a potential new mechanism for limiting iron availability to *Legionella pneumophila* and subsequently inhibiting intracellular multiplication. Clin Res 38:481A

Byrd TF, Horwitz MA (1990c) An individual's monocytes which are uniquely non-permissive to *Legionella pneumophila* intracellular multiplication have low numbers of transferrin receptors, and iron reverses the non-permissive state. Clin Res 38:304A

Chang KP (1979) *Leishmania donovani* promastigote—macrophage surface interactions in vitro. Exp Parasitol 48: 175–189

Cianciotto NP, Eisenstein BI, Mody CH, Toews GB, Engleberg NC (1989a) A *Legionella pneumophila* gene encoding a species-specific surface protein potentiates initiation of intracellular infection. Infect Immun 57: 1255–1262

Cianciotto NP, Eisenstein BI, Engleberg NC (1989b) A site-directed mutation in *Legionella pneumophila* resulting in attenuation of both macrophage infectivity and virulence in guinea pigs. Clin Res 37: 561A

Ciesielski CA, Blaser MJ, Wong W-LL (1986) Serogroup specificity of *Legionella pneumophila* is related to lipopolysaccharide characteristics. Infect Immun 51: 397–404

Clemens DL, Horwitz MA (1990) Demonstration that *Legionella pneumophila* produces its major secretory protein in infected human monocytes and localization of the protein by immunocytochemistry and immunoelectron microscopy. Clin Res 38:480A

Daisy JA, Benson CE, McKitrick J, Friedman HM (1981) Intracellular replication of *Legionella pneumophila*. J Infect Dis 143: 460–464

Dreyfus LA, Iglewski BH (1986) Purification and characterization of an extracellular protease of *Legionella pneumophila*. Infect Immun 51: 736–743

Eissenberg LG, Goldman WE (1986) *Histoplasma capsulatum* fails to trigger release of superoxide from macrophages. Infect Immun 55: 29–34

Eissenberg LG, Wyrick PB, Davis CH, Rumpp JW (1983) *Chlamydia psittaci* elementary body envelopes: ingestion and inhibition of phagolysosome fusion. Infect Immun 40: 741–751

Elliott JA, Winn WC (1986) Treatment of alveolar macrophages with cytochalasin D inhibits uptake and subsequent growth of *Legionella pneumophila*. Infect Immun 51: 31–36

Esparza I, Fox RI, Schreiber RD (1986) Interferon-gamma-dependent modulation of C3b receptors (CRI) on human peripheral blood monocytes. J Immunol 136: 1360–1365

Fields BS, Shotts EB, Feeley JC, Gorman GW, Martin WT (1984) Proliferation of *Legionella pneumophila* as an intracellular parasite of the ciliated protozon *Tetrahymena pyriformis*. Appl Environ Microbiol 47: 467–471

Firestein GS, Zvaifler NJ (1987) Down regulation of human monocyte differentiation antigens by interferon gamma. Cell Immunol 104: 343–354

Fraser DW, Tsai TR, Orenstein W et al. (1977) Legionnaires' disease. Description of an epidemic of pneumonia. N Engl J Med 297: 1189–1197

Friedman RL, Iglewski BH, Miller RD (1980) Identification of a cytotoxin produced by *Legionella pneumophila*. Infect Immun 29: 271–274

Friedman RL, Lochner JE, Bigley RH, Iglewski BH (1982) The effects of *Legionella pneumophila* toxin on oxidative processes and bacterial killing of human polymorphonuehear leukocytes. J Infect Dis 146: 328–334

Friss RR (1972) Interaction of L-cells and *Chlamydia psittaci*—entry of the parasite and host response to its development. J Bacteriol 110: 706–721

Gabay JE, Horwitz MA (1985) Isolation and characterization of the cytoplasmic and outer membranes of the Legionnaires' disease bacterium (*Legionella pneumophila*). J Exp Med 161: 409–422

Gabay JE, Blake MS, Niles W, Horwitz MA (1985) Purification of the major outer membrane protein of *Legionella pneumophila* and demonstration that it is a porin. J Bacteriol 162: 85–91

Glick TH, Gregg MB, Berman B, Mallison GF, Rhodes WW Jr, Kassanoff I (1978) Pontiac fever: an epidemic of unknown etiology in a health department. I. Clinical and epidemiologic aspects. Am J Epidemiol 107: 149–160

Holden EP, Winkler HH, Wood DO, Leinbach ED (1984) Intracellular growth of *Legionella pneumophila* within *Acanthamoeba castellanii* Neff. Infect Immun 45: 18–24

Holzer TJ, Nelson KE, Schauf V, Crispen RG, Anderson BR (1986) *Mycobacterium leprae* fails to stimulate phagocytic cell superoxide anion generation. Infect Immun 51: 514–520

Horwitz MA (1983a) Formation of a novel phagosome by the Legionnaires' disease bacterium (*Legionella pneumophila*) in human monocytes. J Exp Med 158: 1319–1331
Horwitz MA (1983b) The Legionnaires' disease bacterium (*Legionella pneumophila*) inhibits phagosome-lysosome fusion in human monocytes. J Exp Med 158: 2108–2126
Horwitz MA (1983c) Cell-mediated immunity in Legionnaires' disease. J Clin Invest 71: 1686–1697
Horwitz MA (1984) Phagocytosis of the Legionnaires' disease bacterium (*Legionella pneumophila*) occurs by a novel mechanism: engulfment within a pseudopod coil. Cell 36:33–37
Horwitz MA (1985) Host resistance to *Legionella pneumophila*: interactions of *L. pneumophila* with leukocytes. In: Katz SM (ed) Legionellosis, vol 2. CRC Press, Boca Raton , FL, pp 143–157
Horwitz MA (1987) Characterization of avirulent mutant *Legionella pneumophila* that survive but do not multiply within human monocyte. J Exp Med 166: 1310–1328
Horwitz MA (1989) The immunobiology of *Legionella pneumophila* In: Moulder JW (ed) Intracellular Parasitism CRC Press, Boca Raton, FL, pp 141–156
Horwitz MA, Maxfield FR (1984) *Legionella pneumophila* inhibits acidification of its phagosome in human monocytes. J Cell Biol 99: 1936–1943
Horwitz MA, Silverstein SC (1980) The Legionnaires' disease bacterium (*Legionella pneumophila*) multiplies intracellularly in human monocytes, J Clin Invest 66: 441–450
Horwitz MA, Silverstein SC (1981a) Activated human monocytes inhibit the intracellular multiplication of Legionnaires' disease bacteria. J Exp Med 154: 1618–1635
Horwitz MA, Silverstein SC (1981b) Interaction of the Legionnaires' disease bacterium (*Legionella pneumophila*) with human phagocytes. I. *L. pneumophila* resists killing by polymorphonuclear leukocytes, antibody, and complement. J Exp Med 153: 386–397
Horwitz MA, Silverstein SC (1981c) Interaction of the Legionnaires' disease bacterium (*Legionella pneumophila*) with human phagocytes. II. Antibody promotes binding of *L. pneumophila* to monocytes but does not inhibit intracellular multiplication. J Exp Med 153: 398–406
Horwitz MA, Silverstein SC (1983) The intracellular multiplication of Legionnaires' disease bacteria (*Legionella pneumophila*) in human monocytes is reversibly inhibited by erythromycin and rifampin. J Clin Invest 71: 15–26
Jacobs RF, Locksley RM, Wilson CB, Haas JE, Klebanoff SJ (1984) Interaction of primate alveolar macrophages and *Legionella pneumophila*. J Clin Invest 73: 1515–1523
Jensen WA, Rose RM, Wasserman AS, Kalb TH, Anton K, Remond HG (1987) In vitro activation of the antibacterial activity of human pulmonary macrophages by recombinant gamma interferon. J Infect Dis 155:574–577
Jones TC, Hirsch JG (1972) The interaction between *Toxoplasma gondii* and mammalian cells. II. The absence of lysosomal fusion with phagocytic vacuoles containing living parasites. J Exp Med 136: 1173–1194
Jones TC, Yeh S, Hirsch JG (1972) The interaction between *Toxoplasma gondii* and mammalian cells. I. Mechanism of entry and intracellular fate of the parasite. J Exp Med 136: 1157–1172
Kishimoto RA, Kastello MO, White JD, Shirey FG, McGann VG, Larson EW, Hedlund KW (1979) In vitro interaction between normal cynomolgus monkey alveolar macrophages and Legionnaires' disease bacteria. Infect Immun 25: 761–763
Kishimoto RA, White JD, Shirey FG, McGann VG, Berendt RF, Larson EW, Hedlund KW (1981) In vitro response of guinea pig peritoneal macrophages to *Legionella pneumophila*. Infect Immun 31: 1209–1213
Lema MW, Brown A, Butler CA, Hoffman PS (1988) Heat shock response in *Legionella pneumophila*. Can J Microbiol 34: 1148–1153
Lochner JE, Friedman RL, Bigley RH, Iglewski BH (1983) Effect of oxygen-dependent antimicrobial systems on *Legionella pneumophila*. Infect Immun 38: 487–489
Locksley RM, Jacobs RF, Wilson CB, Weaver WM, Klebanoff SJ (1982) Susceptibility of *Legionella pneumophila* to oxygen-dependent microbicidal systems. J Immunol 129: 2192–2197
Marra A, Horwitz MA, Shuman HA (1990) The HL-60 model for the interaction of human macrophages with the Legionnaires' disease bacterium. J Immunol 144:2738–2744.
McDade JE, Shepard CC, Fraser DW, Tsai TR, Redus MA, Dowdle WR and the Laboratory Investigation Team (1977) Legionnaires' disease. Isolation of a bacterium and demonstration of its role in other respiratory disease. N Engl J Med 297: 1197–1203
Mosser DM, Edelson PJ (1984) Activation of the alternative complement pathway by Leishmania promastigotes: parasite lysis and attachment to macrophages. J Immunol 132: 1501–1505
Mosser DM, Edelson PJ (1985) The mouse macrophage receptor for C3bi (CR3) is a major mechanism in the phagocytosis of *Leishmania* promastigotes. J Immunol 135: 2785–2789
Muder RR, Yu VL, Woo AH (1986) Mode of transmission of *Legionella pneumophila*: a critical review. Arch Intern Med 146: 1607–1612

Myerowitz RL, Pasculle AW, Dowling JN et al. (1971) Opportunistic lung infection due to "Pittsburgh Pneumonia Agent". N Engl J Med 301: 953–958

Nash TW, Libby DM, Horwitz MA (1984) Interaction between the Legionnaires' disease bacterium (*Legionella pneumophila*) and human alveolar macrophages. Influence of antibody, lymphokines, and hydrocortisone. J Clin Invest 74: 771–782

Nash T, Libby DM, Horwitz MA (1988) Gamma interferon activated human alveolar macrophages inhibit the intracellular multiplication of *Legionella pneumophila*. J Immunol 140: 3978–3981

Newman SL, Musson RA, Henson PM (1980) Development of functional complement receptors during in vitro maturation of human monocytes into macrophages. J Immunol 125: 2236–2243

Newsome AL, Baker RL, Miller RD, Arnold RR (1985) Interactions between *Naegleria fowleri* and *Legionella pneumophila*. Infect Immun 50: 449–452

Nogueira N, Cohn ZA (1976) *Trypanosoma cruzi*: mechanism of entry and intracellular fate in mammalian cells. J Exp Med 143: 1402–1420

Oldham LJ, Rodgers FG (1985) Adhesion, penetration and intracellular replication of *Legionella pneumophila*: an in vitro model of pathogenisis. J Gen Microbiol 131: 697–706

Pau C-P, Plikaytis BB, Carlone GM, Warner IM (1988) Purification, partial characterization, and seroreactivity of a genuswide 60-kilodalton Legionella protein antigen. J Clin Microbiol 26: 67–71

Payne NR, Horwitz MA (1987) Phagocytosis of *Legionella pneumophila* is mediated by human monocyte complement receptors. J Exp Med 166: 1377–1389

Payne NR, Bellinger-Kawahara CG, Horwitz MA (1987) Phagocytosis of *Mycobacterium tuberculosis* by human monocytes is mediated by receptors for the third component of complement. Clin Res 35: 617A

Perry A, Engleberg NC, Eisenstein BI (1987) An inhibitor of myeloperoxidase-mediated protein iodination isolated from *Legionella pneumophila*. Clin Res 35: 617A

Puentes SM, Sacks DL, da Silva R, Joiner KA (1988) Complement binding by two developmental stages of *Leishmania major* promastigotes varying in expression of a surface glycolipid. J Exp Med 167: 887–902

Sampson JS, Plikaytis BB, Wilkinson HW (1986) Immunologic response to patients with legionellosis against major protein—containing antigens of *Legionella pneumophila* serogroup 1 as shown by immunoblot analysis. J Clin Microbiol 23: 92–99

Schlesinger LS, Horwitz MA (1990) Phagocytosis of leprosy bacilli is mediated by complement receptors CR1 and CR3 on human monocytes and complement component C3 in serum. J Clin Invest 85: 1304–1314

Schlesinger LS, Bellinger-Kawahara CG, Payne NR, Horwitz MA (1990) Phagocytosis of *Mycobacterium tuberculosis* is mediated by human monocyte complement receptors and complement component C3. J Immunol 144: 2771–2780

Shinnick TM, Vodkin MH, Williams JC (1988) The *Mycobacterium tuberculosis* 65-kilodalton antigen is a heat shock protein which corresponds to common antigen and to the *Escherichia coli* Gro EL protein. Infect Immun 56: 446–451

Sibley LD, Weidner E, Krahenbuhl JL (1985) Phagosome acidification blocked by intracellular *Toxoplasma gondii*. Nature 315: 416–419

Tanowitz H, Wittner M, Kress Y, Bloom B (1975) Studies of in vitro infection by *Trypanosoma cruzi*. I. Ultrastructural studies on the invasion of macrophages and L-cells. Am J Trop Med Hyg 25: 25–33

Todd WJ, Storz J (1975) Ultrastructural cytochemical evidence for the activation of lysosomes in the cytocidal effect of *Chlamydia psittaci* in macrophages. Infect Immun 12: 638–646

Tyndall RL, Domingue EL (1992) Cocultivation of *Legionella pneumophila* and free-living amoebae. Appl Environ Microbiol 44: 954–959

Wadowsky RM, Butler LJ, Cook MK et al. (1988) Growth-supporting activity for *Legionella pneumophila* in tap water cultures and implication of Hartmannellid amoebae as growth factors. Infect Immun 54: 2677–2682

Wilson CB, Tsai V, Remington JS (1980) Failure to trigger the oxidative burst by normal macrophages. Possible mechanism for survival of intracellular pathogens. J Exp Med 151: 328–346

Wilson ME, Pearson RC (1987) Roles of CR3 and mannose receptors in the attachment and ingestion of *Leishmania donovani* by human mononuclear phagocytes. Infect Immun 56: 363–369

Wong MC, Ewing EP Jr, Callaway CS, Peacock WL Jr (1990) Intracellular multiplication of *Legionella pneumophila* in cultured human embryonic lung fibroblasts. Infect Immun 28: 1014–1018

Wozencraft AO, Sayers G, Blackwell JM (1986) Macrophage type 3 complement receptors mediate serum-independent binding of *Leishmania donovani*. J Exp Med 164: 1332–1337

Wright SD, Silverstein SC (1983) Receptors for C3b and C3bi promote phagocytosis but not the release of toxic oxygen from human phagocytes. J Exp Med 158: 2016–2023

Wright SD, Detmers PA, Jong MTC, Meyer BC (1986) Interferon gamma depresses binding of ligand by C3b and C3bi receptors on cultured human monocytes, an effect reversed by fibronectin. J Exp Med 163: 1245–1259

Wyrick PB, Brownridge EA (1978) Growth of *Chlamydia psittaci* in macrophages. Infect Immun 19: 1054–1060

Yamamoto K, Johnston RB Jr (1984) Dissociation of phagocytosis from stimulation of the oxidative metabolic burst in macrophages. J Exp Med 159: 405–416

Yamamoto Y, Klein TW, Newton CA, Widen R, Friedman H (1987) Differential growth of *Legionella pneumophila* in guinea pig versus mouse macrophage cultures. Infect Immun 55: 1369–1374

Yamamoto Y, Klein TW, Newton CA, Widen R, Friedman H (1988) Growth of *Legionella pneumophila* in thioglycolate-elicited peritoneal macrophages from A/J mice. Infect Immun 56: 370–375

Yoshida S-I, Mizuguchi Y (1986) Multiplication of *Legionella pneumophila* Philadelphia-1 in cultured peritoneal macrophages and its correlation to susceptibility of animals. Can J Microbiol 32: 438–442

Chronic Granulomatous Disease: An Update and a Paradigm for the Use of Interferon-γ as Adjunct Immunotheraphy in Infectious Diseases

R. A. B. EZEKOWITZ

1 Introduction

Chronic granulomatous disease (CGD) is a rare inherited disorder in which the phagocyte NADPH oxidase is disabled. The failure of the enzyme to generate superoxide and related oxygen intermediates renders the patients with this disease susceptible to recurrent bacterial and fungal infections. The clinical syndrome (reviewed FORREST et al. 1988) usually presents within the first years of life with a history of recurrent infections, mostly pneumonias, abscesses of the liver or lungs, or subcutaneous bacterial infections. Despite the use of high dose antibiotics, the phagocytes' failure to kill ingested organisms serves as a stimulus for the chronic inflammatory state and granuloma formation. The granulomas are attempts to wall off infectious foci and if they occur in vital organs, like the gastrointestinal tract, kidney, liver, and brain, they contribute to the mortality and morbidity of this disorder.

Early attempts at the characterization of the phagocyte oxidase were hindered by its complexity (reviewed in SMITH and CURNUTTE 1991). Despite disparate findings on the molecular weights and relative importance of individual components, a consensus held that most, although not all, X-linked CGD kindred lacked a spectrally detectable cytochrome b_{559} (SEGA 1988). Although 60% of CGD patients follow the X-linked dominant pattern of

Division of Hematology/Oncology, The Children's Hospital and Dana-Farber Cancer Institute, Department of Pediatrics, Harvard Medical School, Boston, MA 02115, USA

Current Topics in Microbiology and Immunology, Vol. 181

inheritance, 40% have an autosomal recessive pattern of inheritance. Recently, genetic approaches, imporved biochemical purification procedures, and the development of cell-free systems which enabled segregation of subcellular fractions required to reconstitute the oxidase system in vitro have led to the characterization and elucidation of most of the components of the NADPH oxidase (reviewed by EZEKOWITZ and NEWBURGER 1988).

2 The NADPH Oxidase

It has long been recognized that the NADPH oxidase is a multicomponent system (CURNUTTE and BABIOR 1987). Recent evidence has elucidated at least four different components for which genetic defects have been described and result in the CGD phenotype. A phagocyte-specific cytochrome b heterodimer believed to be the terminal electron donor in the oxidase complex is absent in two different genetic forms of CGD (SEGAL 1988). Most commonly, X-linked recessive disease (xb-CGD, accounting for 60% of cases) is due to mutations in the gene encoding a 91-kDa membrane glycoprotein (referred to as gp91-*phox*) that is the larger subunit of cytochrome b (ORKIN 1989). The molecular cloning of gp91-*phox* gene was performed without knowledge of the gene product, based on the correct chromosomal assignment to Xp21.1 (ORKIN 1989; ROYER-POKORA et al. 1986). By reverse genetics, a cDNA with a translation product of 571 amino acids was identified and found to correspond to a phagocyte-specific 91-kDa protein that was absent in X-CGD. Formal proof that this was indeed the heavy chian of the b cytochrome rested on the N-terminal sequence from the purified protein, (SEGAL 1987; PARKOS et al. 1987). In addition to classic X-CGD, the phagocyte b cytochrome is also lacking in this subgroup of CGD due to mutations in the gene for the 22-kDa subunit (p22-*phox*) of cytochrome b (PARKOS et al. 1988).

The majority of patients with autosomal recessive CGD have normal levels of b cytochrome, but are lacking cytosolic factors required for activation of the NADPH oxidase (VOLPP et al. 1988; NUNOL et al. 1988; CURNUTTE et al. 1989; BOLSCHER et al. 1989). In a recent study of 25 autosomal recessive CGD patients, 22 lacked a 47-kDa phosphoprotein p47-*phox* and the remainder lacked a 67-kDa protein p67-*phox* (CLARK et al. 1989). cDNA for both proteins have been isolated and recombinant and purified proteins reconstitute the corresponding cytosol deficiencies in cell-free oxidase assays (VOLPP et al. 1989; LOMAX et al. 1989).

Figure 1 is a current model [as proposed by SMITH and CURNUTTE (1991) of the NADPH oxidase and its components. The model predicts assembly of cytosolic and membrane-bound components to constitute an electron transfer chain allowing molecular oxygen to be converted into superoxide and other oxygen intermediates like hydrogen peroxide. The precise interactions of each component are not known for certain, but studies in progress are aimed at mutating various sites of each specific component to address this problem in

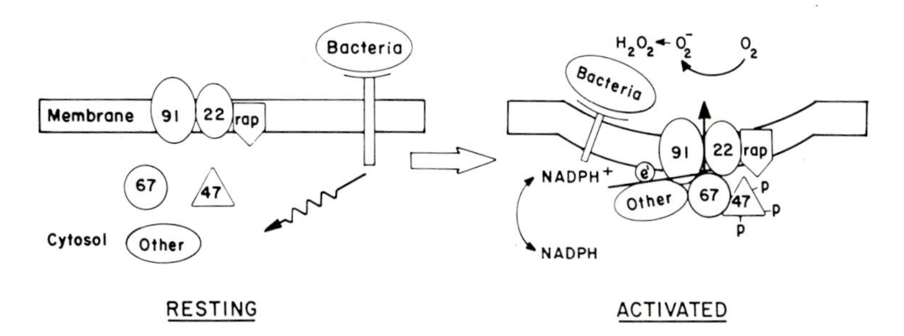

Fig. 1. A current model of the phagocyte NADPH oxidase based on a model proposed by SMITH and CURNUTTE (1991)

reconstituted in vitro systems. Such analyses have indicated the necessity for components other than those characterized, which are indicated in the diagram as other components.

3 Molecular Basis of Chronic Granulomatous Disease

Recent progress in elucidating the molecular genetic basis of all CGD has included the isolation of cDNAs for a group of essential oxidase proteins. It is now recognized that CGD phenotype can arise from genetic defects in at least four different components of the NADPH oxidase, as described in Table 1. Defects in the gene encoding for gp91-*phox* most commonly lead to the absence of the gene product without any obvious abnormality in the gene structure as determined by Southern blot and restriction fragment length polymorphism studies (DINAUER and ORKIN 1988). This suggests that point mutations rather than gene deletions result in a defective gp91-*phox* protein. In fact, a point mutation in the gp91-*phox* gene resulting in a Pro > His at position 415 (DINAUER et al. 1989) has been found in an Xb^+ CGD patient. (Xb^+ refers to the presence of spectrally normal cytochrome although functionally disabled with respect to its electron transfer capacity). In another study (BOLSCHER et al. 1991), point mutations in the gp91-*phox* gene were found in Xb^- and Xb^+ CGD patients. In one patient, the point mutation resulted in a premature termination signal at codon 73, providing an unequivocal explanation for the CGD phenotype. In the other five patients, distinct mutations resulted in a signal amino acid substitution. Although the reason why these changes should result in a dysfunctional gene product is not clear, in two patients with Xb^- CGD, who are RNA^+, a histidine was replaced by either a tyrosine or an arginine. It is possible that the histidine may be critical in heme binding and that loss of this residue could account for the CGD phenotype. The other substitutions are scattered throughout the protein and may alter its structure and hence its stability or interactions with other oxidase components, particularly p22-*phox*, the cytochrome light chain.

Table 1. CGD: genetic defects in phagocyte NADPH-oxidase proteins (DINAUER and EZEKOWITZ 1991)

Inheritance	Cyt b	Affected gene	Chromosomal location	Protein product	References
X-linked recessive	Neg. (rarely pos.)	CYBB	Xp21.1	gp 91-*phox* (91-kDa subunit cyt b)	ROYER-POKORA et al. 1986; DINAUER et al. 1989; TEAHAN et al. 1987; FRANCKE 1984; FRANCKE et al. 1985; BAEHNER et al. 1986
Autosomal recessive	Neg.	CYBA	16q24	p22-*phox* (22-kDa subunit cyt b)	DINAUER et al. 1990; NATHAN et al. 1986; WEENING et al. 1985
Autosomal recessive	Pos.	NCF1	7q11.23	p47-*phox*	VOLPP et al. 1989; LOMAXet al. 1989; FRANCKE et al. 1990
Autosomal	Pos.	NCF2	1q25	p67-*phox*	FRANCKE et al. 1990; LETO et al. 1990

Most autosomal recessive forms of CGD have normal cytochrome b levels and lack either the p47-*phox* or the p67-*phox*, two soluble proteins required for oxidase function. It appears that in some patients the identical mutation on both alleles of the p47-*phox* gene occurs at a splice junction, accounting for the CGD phenotype (CASMEYER et al. 1991). It is likely that mutations other than this might also occur, but have yet to be reported. A rare subgroup of autosomal recessive CGD resembles the X-linked form in that both chains of the cytochrome b are absent, so-called A-CGD. Specific genetic mutations in the gene of the 22-kDa light chain of the cytochrome, p22-*phox*, have been identified (DINAUER et al. 1990). In one patient there is a 10-kb homozygous deletion which accounts for the lack of the gene product. In yet another patient there is a frame shift mutation in one allele and a missense mutation in the other allele. In a third patient, a homozygous missence mutation appears to account for the genetic defect.

4 Interferon-γ: The Rationale for Its Use in Chronic Granulomatous Disease

The convergence of three independent but related events led to the idea that recombinant interferon-γ may be of use in the treatment of a rare group of inherited disorders of phagocytes that result in the phenotype of CGD. Firstly there was the presentation of a patient with an atypical presentation of the disease. Secondly there was the characterization of the molecular defect underlying the X-linked form of CGD which accounts for two-thirds of the patients (ROYER-POKORA et al.1986). The triad was completed by the realization that intradermal administration of interferon-γ was able to affect the oxidative capacity of circulating phagocytes as shown in patients with lepromatous leprosy (NATHAN et al. 1986). In the rest of this review, I will elaborate on each of these aspects and point out how they relate to one another. The end result is that interferon-γ is highly effective in reducing the time to serious infection in all genetic types of CGD and that its effects are due to more than its ability to up-regulate the respiratory burst in phagocytes and hence may have more general applications in adjunct thereapy for infectious disease.

4.1 The Patient

Chronic granulomatous disease is characterized by recurrent severe bacterial and fungal infections usually within the first few years of life (FORREST et al. 1988). Although phagocytes from patients with CGD ingest microorganisms normally, killing is deficient due to the failure of a membrane-associated NADPH oxidase to produce superoxide and related toxic derivatives (TAUBER et al. 1983). The index patient presented with a life-threatening pneumonia due to *Aspergillus fumigatus* as the first manifestatioin of CGD at age 14 (autumn of 1984). The patient had been otherwise well except for severe pustular acne. The patient was exposed to an unusually high inoculum of *Aspergillus fumigatus* from wood shavings and this infectious event first drew attention to his underlying immunodeficiency. The diagnosis of CGD was suspected given the unusual nature of the organisms and was confirmed by a negative nitroblue tetrazolium (NBT) test. The question therefore was why did this patient survive 14 years before his first serious infection? The hypothesis generated was that the patient's phagocytes were able to generate reduced amounts of reactive oxygen intermediates which under many circumstances were sufficient. However, on this occasion the inoculum of *Aspergillus* was so great that it overwhelmed his crippled oxidase. This raised questions as to the nature of the phagocyte oxidase and the molecules that regulate its function.

4.2 Interferon-γ

Interferon-γ, a glycoprotein secreted by activated T cellls and NK cells, has a powerful role as an immunomodulatory cytokine (PESTKA et al. 1987). The biologic action of this lymphokine is mediated via a specific 90-kDa receptor present on a wide variety of cell type (AUGET et al. 1988). Interferon-γ induces many pleiotypic effects on target cells but its predominant role is as a macrophage-activating factor (NATHAN and TSUNAWAKI 1986). The administration of this cytokine in vivo and in vitro enhances phagocyte killing of bacterial and protozoan pathogens (MURRAY 1988). The enhanced killing correlates with increased production of reactive oxygen intermediates by phagocytes, which accounts for most but not all of the cytocidal function. The recent advances in the characterization of the molecular components of the NADPH oxidase provided us with the opportunity to examine the molecular mechanisms underlying the augmentation of phagocyte superoxide production by interferon-γ. Initial studies revealed that in vitro treatment of human monocyte-derived macrophages and neutrophils with interferon-γ increased gp91-*phox* RNA transcripts although levels of p22-*phox* RNA were unaffected (NEWBURGER et al. 1988). Preliminary work indicates that p47-*phox* gene expression is also augmented by interferon-γ (ABRAHAMSON et al. 1990). It is noteworthy that interferon-γ up-regulation of transcripts of NADPH oxidase components alone does not account for the activation phenotype. However, the interferon responsiveness of these oxidase genes provides an experimental basis for attempts to correct the functional deficiency in phagocytes in CGD patients.

4.3 The Patient, the Oxidase, and Interferon-γ

How did the index patient survive for 14 years before his first serious infection? Our hypothesis was that under normal circumstances of low bacterial load, he was able to transiently up-regulate his NADPH oxidase to produce sufficient reactive oxygen intermediates. However, exposure to a large inoculum of *Aspergillus* overwhelmed his already crippled oxidase (NEWBURGER et al. 1986). The question therefore was, would interferon-γ, the physiologic up-regulator of the phagocyte NADPH oxidase, and specifically gp91-*phox*, in this instance the aberrant gene, affect this patient's phagocyte function? If this proved to be the case, would this apply to other CGD patients? Initial in vitro studies showed that addition of interferon-γ to granulocytes and monocytes from this patient and his affected brother resulted in a partial correction of both the defect in superoxide production and very low levels of gp91-*phox* transcripts (EZEKOWITZ et al. 1987), while two other X-CGD patients' cells failed to respond.

The availability of recombinant interferon-γ and its extensive use in cancer trials provided safety information and the expectation that in vivo administration of interferon-γ may be a rational approach. The immunomodulatory dose and route of administration became obvious after it was shown that intradermal injection into lesions in lepromatous leprosy resulted in a systemic effect on circulating monocytes (NATHAN et al. 1986). It was observed in this study that the depressed oxidative capacity of phagocytes was enhanced after treatment (NATHAN et al. 1986). This led us to administer 0.1 mg/m^2 of interferon-γ subcutaneously to our index patients. Treatment resulted in a near-normal level of superoxide production in granulocytes and monocytes. Granulocyte bactericidal capacity was indistinguishable from normal control treatment and there was a small but detectable increase in gp-91-*phox* protein, which was not detectable before treatment (EZEKOWITZ et al. 1988). Surprisingly, the improvement in phagocyte function peaked at 2 weeks and was sustained for 4 weeks after one subcutaneous administration of interferon-γ, despite an estimated 4- to

6-h circulating time of neutrophils. Of note is that at the dose of subcutaneous interferon-γ administered, serum levels were undetectable.

These observations suggested that the in vivo effect of interferon-γ cannot be explained by its action on mature phagocytes alone and implies an effect on immature cells. We set out to test the idea that the lymphokine affects progenitor cells which later give rise to mature progeny that express the corrected phenotype. These experiments are possible as in this CGD kindred the phagocyte responses to interferon-γ fall within a range that is easily detectable by in situ NBT dye reduction. Progenitor-derived colonies from peripheral blood examined before treatment were unable to generate superoxide as visualized by lack of NBT reduction compared with normal controls. By contrast, colonies derived 7 days after a single interferon-γ injection were able to generate superoxide, as shown by increased NBT reduction. Colonies harvested 21 days after treatment contained only rare cells capable of NBT reduction (EZEKOWITZ et al. 1990). When one relates the kinetics observed in these progenitor-derived colonies to the kinetics of the circulating phagocytes, it is clear that interferon-γ must act on progenitor cells and their mature progeny. The effect on circulating cells was not immediate and was first observed 36 h after treatment. Moreover, it was maximal at 14 days. Our favored explanation is that interferon-γ acts on progenitor cells and their differentiated products that begin to repopulate the circulating pool. This effect is first observed on mature cells in the bone marrow, which egress within 24–36 h and are found in the circulation. There appears to be a lag period of 5–7 days before a corrected progenitor can be assayed in the peripheral blood. It would appear, therefore, that these studies more broadly represent basic determinants of maturation from myeloid progenitors and provide a unique estimate of the kinetics of myeloid differentiation. Interferon-γ has been shown to influence myeloid differentiation and commitment and probably exerts its action in vivo in conjuction with other growth factors like GM-CSF and tumor necrosis factor (CASSATELLA et al. 1989). These factors have been shown to act synergistically with interferon-γ in augmenting cytochrome b heavey chain expression in vitro (CASSATELLA et al. 1989). The molecular mechanisms by which interferon-γ is able to reprogram the phenotype of early progenitor cells are unknown but must represent a fundamental step in determination of differentiated phenotype.

5 Other CGD Patients

In our original in vitro studies, we also included a few selected CGD patients who represented the spectrum of the clinical coruse, with the kindred described above at the upper end and classic X-CGD at the lower end. We were able to show that these patients had an intermediate to low in vitro and in vivo response (EZEKOWITZ et al. 1987, 1988). This work was extended by SECHLER et al. (1988), who examined the effects of the lymphokine on 30 patients, including some with autosomal recessive inheritance. These workers observed enhancement of bacterial killing in responders as well as nonresponders, suggesting that interferon-γ therapy may benefit CGD patients even in the absence of demonstrable extracellular production. These studies formed the basis for a phase III study to evaluate the efficacy and potential toxicity of interferon-γ in CGD. An international, multicenter, randomized, double-blind placebo-controlled study was undertaken in which 128 eligible patients were enrolled, including patients with different patterns of inheritance and requiring, in most cases, prophylactic antibiotic therapy. Patients received interferon-γ or placebo 3 times weekly for up to 1 year. Time to serious infection as defined as an event requiring hospitalization and intravenous antibiotics was the primary end point. Seventy-nine percent of interferon-γ-treated patients were free of serious infection at 1 year, as compared with 30% of those receiving placebo. The latter group required three times as many in-patient hospital days and had more multiple infections (International Chronic Granulomatous Disease Cooperative Study Group 1991). Of particular note was the fact that there appeared to be no statistically significant correlation between improved in vitro function and clinical course. This probably reflects the genetic heterogeneity underlying the CGD phenotype and raises questions as to the precise molecular mechanisms underlying the dramatic clinical benefit. Clearly, a subset of patients did show enhanced oxidative capacity; however, the majority appear to utilize other interferon-γ-triggered responses, suggesting that interferon-γ may have more general applications as an adjunct to conventional antimicrobicidal therapies in the treatment of infectious diseases. The effective use of this lymphokine in leprosy and leishmaniasis (BADARO et al. 1990) opens the way for investigating its efficacy in other infectious disease settings like septic neonates. In this way, the study of an unusual patient with an esoteric genetic disease may lead to a more general understanding of the clinical usefulness of interferon-γ in clinical practice.

Acknowledgments. I am greatful for the help and support of my colleagues Drs. Nathan, Orkin, Sieff, Dinauer, Newburger, Lux, and Platt. Special thanks to Marsha Kartzman for excellent assistance in the preparation of this manuscript.

References

Abrahamson SL, Lomax KJ, Malech HL, Gallin JI (1990) Recombinant interferon gamma and IL 4 regulate gene expression of several phagocyte oxidase components. Clin Res 38: 2367

Auget M, Dembic Z, Merlan G (1988) Molecular cloning and expression of the human interferon gamma receptor. Cell 55: 273–280

Badaro R et al. (1990) Treatment of visceral leishmaniasis with pentavalent antimony and interferon gamma. N Engl J Med 322: 16–20

Baehner RL, Kunkel LM, Monaco AP et al. (1986) DNA linkage analysis of X-chromosome-linked chronic granulomatous disease. Proc Natl Acad Sci USA 83: 3398–3401

Bolscher BGJM, Van Zweitjen R, Kramer IM, Weening RS, Verhoeven AJ, Roos D (1989) A phospoprotein of M,47 000 defective in autosomal chronic granulomatous disease, copurifies with one of two soluble components required for NADPH:O_2 oxidoreductase activity in human neutrophils. J Clin Invest 83: 757–763

Bolscher BGJM, deBoer M, deKlein A, Weening RS, Roos D (1991) Point mutations in the β subunit of cytochrome b_{558} leading to X-linked chronic granulomatous disease. Blood 77: 2482–2487

Casmeyer CM, Bu-Gahrin HN, Roadawy ARF et al. (1991) Autosomal recessive chronic granulomatous disease caused by a deletion of the dinucleotide repeat. Proc Natl Acad Sci USA 88: 2753–2757

Cassatella MA, Hartman L, Perussia B, Trinchieri G (1989) Tumor necrosis factor and immune interferon synergistically induce cytochrome b sub 245 heavy chain gene expression of nicotinamide-adenine dinucleotide phosphate hydrogenase oxidase in human leukemic myeloid cells. J Clin Invest 83: 1570–1579

Clark RA, Malech HL, Gallin JI et al. (1989) Genetic variants of chronic granulomatous disease: prevalence of deficiencies of two discrete cytosolic components of the NADPH oxidase system. N Engl J Med 321: 647–652

Curnutte JT, Babior BM (1987) Chronic granulomatous disease. In: Harris H, Hirschhorn K (eds) Advances in human genetics. Plenum, New York, pp 229–297

Curnutte JT, Scott PJ, Mayo LA (1989) Cytosolic components of the respiratory burst oxidase: resolution of four components, two of which are missing in complementing types of chronic granulomatous disease. Proc Natl Acad Sci USA 86: 825–829

Dinauer MC, Ezekowitz RAB (1991) Interferon gamma and chronic granulomatous disease. In: Current opinion in immunology. Current Biology, London, pp 61–64

Dinauer MC, Orkin SH (1988) Chronic granulomatous disease: molecular genetics. In: Curnutte JT (ed) Phagocytic defects II: abnormalities of the respiratory burst. WB. Saunders, Philadelphia (Hematology/Oncology Clinics of North America, vol 2), no. 2, pp 225–240

Dinauer MC, Curnutte JT, Rosen H, Orkin SH (1989) A missense mutation in the neutrophil cytochrome b heavy chain in cytochrome positive X-linked chronic granulomatous disease. J Clin Invest 84: 2012–2016

Dinauer MC, Pierce EA, Bruns GAP, Curnutte JT, Orkin SH (1990) Human neutrophil cytochrome b light chain (p22-*phox*). Gene structure, chromosomal location, and mutations in cytochrome-negative autosomal recessive chronic granulomatous disease. J Clin Invest 86: 1729–1737

Ezekowitz RAB, Newburger PE (1988) New perspectives in chronic granulomatous disease. J Clin Immunol 8: 419–425

Ezekowitz RAB, Orkin SH, Newburger PE (1987) Recombinant interferon gamma augments phagocyte superoxide production and X-chronic granulomatous disease gene expression in X-linked variant chronic granulomatous disease. J Clin Invest 80: 1009–1016

Ezekowitz RAB, Dinauer MC, Jaffe HS, Orkin SH, Newburger PE (1988) Partial correction of the phagocyte defect in patients with X-linked chronic granulomatous disease by subcutaneous interferon gamma. N Engl J Med 319: 146–151

Ezekowitz RAB, Sieff CA, Dinauer MC, Nathan DG, Orkin SH, Newburger PE (1990) Restoration of phagocyte function by interferon gamma in X-linked chronic granulomatous disease occurs at the level of a progenitor cell. Blood 76: 2443–2448

Forrest CB, Forehand JR, Axtell RA, Roberts RL, Johnston Jr RB (1988) Clinical features and current management of chronic granulomatous disease. In: Curnutte JT (ed) Phagocytic defects II: abnormalities of the respiratory burst. WB. Saunders, Philadelphia (Hematology/Oncology Clinics of North America, vol 2), no. 2, pp 253–266

Francke U (1984) Random X inactivation resulting in mosaic nullisomy of region Xp21.1–p21.3 associated with heterozygosity for ornithine transcarbamylase deficiency and for chronic granulomatous disease. Cytogenet Cell Genet 38: 298–307

Francke U, Ochs HD, DeMartinville B et al. (1985) Minor Xp21 chromosome deletion in a male associated with expression of Duchenne muscular dystrophy, chronic granulomatous disease, retinitis, pigmentosa, and McLeod syndrome. Am J Hum Genet 37: 250–267

Francke U, Hsieh CL, Foellmer BE, Lomax KJ, Malech HL, Leto TL (1990) Genes for two autosomal recessive forms of chronic granulomatous disease assigned to Iq25 (NCF2). Am J Hum Genet 47: 483–492

International Chronic Granulomatous Disease Cooperative Study Group (1991) A phase III study establishing efficacy of recombinant human interferon gamma for infectious prophylaxis in chronic granulomatous disease N Engl J Med 324: 509–516

Leto TL, Lomax KJ, Volpp BD et al. (1990) Cloning of a 67 K neutrophil oxidase factor with similarity to a noncatalytic region of p60c-src. Science 248: 727–730

Lomax KJ, Leto TL, Nunoi H, Gallin JI, Malech HL (1989) Recombinant 47 kilodalton cytosol factor restores NADPH oxidase in chronic granulomatous disease. Science 245: 409–412

Murray HW (1988) Interferon-γ, the activated macrophage, and host defense against microbial challenge. Ann Intern Med 108: 595–608

Nathan CF, Tsunawaki S (1986) Secretion of toxic oxygen production by macrophages. Regulatory cytokines and their effects on the oxidase. Ciba Found Symp 118: 211–230

Nathan CF, Kaplan G, Levis W et al. (1986) Local and systemic effects of intradermal recombinant interferon gamma in patients with lepromatous leprosy. N Engl J Med 315: 6–15

Newburger PE, Luscinskas FW, Ryan T et al. (1986) Variant chronic granulomatous disease: modulation of the neutrophil defect by severe infection. Blood 68: 914–919

Newburger PE, Ezekowitz RAB, Whitney C, Wright J, Orkin SH (1988) Induction of phagocyte cytochrome b heavy chain gene expression by interferon gamma. Proc Natl Acad Sci USA 85: 5215–5219

Nunoi H, Rotrosen D, Gallin JI, Malech HL (1988) Two forms of autosomal chronic granulomatous disease lack distinct neutrophil cytosol factors. Science 242: 1298–1301

Orkin SH (1989) Molecular genetics of chronic granulomatous disease. In: Paul WE (ed) Annu Rev Immunol 7: 277–307

Parkos CA, Allen RA, Cochrane CG, Jesaitis AJ (1987) Purified cytochrome b from human granulocyte plasma membrane is comprised of two polypeptides with relative molecular weights of 91,000 and 22,000. J Clin Invest 80: 732–742

Parkos CA, Dinauer MC, Walker LE, Allen RA, Jesaitis AJ, Orkin SH (1988) The primary structure and unique expression of the 22 kilodalton light chain of human neutrophil cytochrome b. Proc Natl Acad Sci USA 85: 3319–3323

Parkos CA, Dinauer MC, Jesaitis AJ, Orkin SH, Curnutte JT (1989) Absence of both the 91 kDa and 22 kDa subunits of human neutrophil cytochrome b in two genetic forms of chronic granulomatous disease. Blood 73: 1416–1420

Pestka S, Langer JA, Zoon KE, Samuel SE (1987) Interferons and their actions. Annu Rev Biochem 56: 727–777

Royer-Pokora B, Kunkel LM, Monaco AP et al. (1986) Cloning the gene for an inherited human disorder—chronic granulomatous disease—on the basis of its chromosomal location. Nature 322: 32–38

Sechler JMG, Malech HL, White CJ, Gallin JI (1988) Recombinant human interferon gamma reconstitutes defective phagocyte function in patients with chronic granulomatous disease of childhood. Proc Natl Acad Sci USA 85: 4874–4878

Segal AW (1987) Absence of both cytochrome b-245 subunits from neutrophils in X-linked chronic granulomatous disease. Nature 326: 88–91

Segal AW (1988) Cytochrome b-245 and its involvement in the molecular pathology of chronic granulomatous disease. In: Curnutte JT (ed) Phagocytic defects II: abnormalities of the respiratory burst. W.B. Saunders, Philadelphia (Hematology/Oncology Clinics of North America, vol 2, no. 2, pp 213–223)

Smith RM, Curnutte JT (1991) Molecular basis of chronic granulomatous disease. Blood 77: 673–686

Tauber AI, Borregaard N, Simons E, Wright J (1983) Chronic granulomatous disease: a syndrome of phagocyte oxidase deficiencies. Medicine 62: 286–309

Teahan C, Rowe P, Parker P, Torry N, Segal AW (1987) The X-linked chronic granulomatous disease gene codes for the β chain of cytochrome b_{245}. Nature 327: 720–723

Volpp BD, Nauseef WM, Clark RA (1988) Two cytosolic neutrophil oxidase components absent in autosomal chronic granulomatous disease. Science 242: 1295–1297

Volpp BD, Nauseef WM, Donelson JE, Moser DR, Clark RA (1989) Cloning of the cDNA and functional expression of the 47 kilodalton cytosolic component of human neutrophil respiratory burst oxidase. Proc Natl Acad Sci USA 86: 7195–7199

Weening RS, Corbell L, de Boer M et al. (1985) Cytochrome b deficiency in an autosomal form of chronic granulomatous disease recognized by monocyte hybridization. J Clin Invest 75: 915–920

Subject Index

Current Topics in Microbiology and Immunology

Volumes published since 1988 (and still available)

Vol. 137: **Mock, Beverly; Potter, Michael (Ed.):** Genetics of Immunological Diseases. 1988. 88 figs. XI, 335 pp. ISBN 3-540-19253-0

Vol. 138: **Goebel, Werner (Ed.):** Intracellular Bacteria. 1988. 18 figs. IX, 179 pp. ISBN 3-540-50001-4

Vol. 139: **Clarke, Adrienne E.; Wilson, Ian A. (Ed.):** Carbohydrate-Protein Interaction. 1988. 35 figs. IX, 152 pp. ISBN 3-540-19378-2

Vol. 140: **Podack, Eckhard R. (Ed.):** Cytotoxic Effector Mechanisms. 1989. 24 figs. VIII, 126 pp. ISBN 3-540-50057-X

Vol. 141: **Potter, Michael; Melchers, Fritz (Ed.):** Mechanisms in B-Cell Neoplasia 1988. Workshop at the National Cancer Institute, National Institutes of Health, Bethesda, MD, USA, March 23–25, 1988. 1988. 122 figs. XIV, 340 pp. ISBN 3-540-50212-2

Vol. 142: **Schüpach, Jörg:** Human Retrovirology. Facts and Concepts. 1989. 24 figs. 115 pp. ISBN 3-540-50455-9

Vol. 143: **Haase, Ashley T.; Oldstone Michael B. A. (Ed.):** In Situ Hybridization 1989. 22 figs. XII, 90 pp. ISBN 3-540-50761-2

Vol. 144: **Knippers, Rolf; Levine, A. J. (Ed.):** Transforming. Proteins of DNA Tumor Viruses. 1989. 85 figs. XIV, 300 pp. ISBN 3-540-50909-7

Vol. 145: **Oldstone, Michael B. A. (Ed.):** Molecular Mimicry. Cross-Reactivity between Microbes and Host Proteins as a Cause of Autoimmunity. 1989. 28 figs. VII, 141 pp. ISBN 3-540-50929-1

Vol. 146: **Mestecky, Jiri; McGhee, Jerry (Ed.):** New Strategies for Oral Immunization. International Symposium at the University of Alabama at Birmingham and Molecular Engineering Associates, Inc. Birmingham, AL, USA, March 21–22, 1988. 1989. 22 figs. IX, 237 pp. ISBN 3-540-50841-4

Vol. 147: **Vogt, Peter K. (Ed.):** Oncogenes. Selected Reviews. 1989. 8 figs. VII, 172 pp. ISBN 3-540-51050-8

Vol. 148: **Vogt, Peter K. (Ed.):** Oncogenes and Retroviruses. Selected Reviews. 1989. XII, 134 pp. ISBN 3-540-51051-6

Vol. 149: **Shen-Ong, Grace L. C.; Potter, Michael; Copeland, Neal G. (Ed.):** Mechanisms in Myeloid Tumorigenesis. Workshop at the National Cancer Institute, National Institutes of Health, Bethesda, MD, USA, March 22, 1988. 1989. 42 figs. X, 172 pp. ISBN 3-540-50968-2

Vol. 150: **Jann, Klaus; Jann, Barbara (Ed.):** Bacterial Capsules. 1989. 33 figs. XII, 176 pp. ISBN 3-540-51049-4

Vol. 151: **Jann, Klaus; Jann, Barbara (Ed.):** Bacterial Adhesins. 1990. 23 figs. XII, 192 pp. ISBN 3-540-51052-4

Vol. 152: **Bosma, Melvin J.; Phillips, Robert A.; Schuler, Walter (Ed.):** The Scid Mouse. Characterization and Potential Uses. EMBO Workshop held at the Basel Institute for Immunology, Basel, Switzerland, February 20–22, 1989. 1989. 72 figs. XII, 263 pp. ISBN 3-540-51512-7

Vol. 153: **Lambris, John D. (Ed.):** The Third Component of Complement. Chemistry and Biology. 1989. 38 figs. X, 251 pp. ISBN 3-540-51513-5

Vol. 154: **McDougall, James K. (Ed.):** Cytomegaloviruses. 1990. 58 figs. IX, 286 pp. ISBN 3-540-51514-3

Vol. 155: **Kaufmann, Stefan H. E. (Ed.):** T-Cell Paradigms in Parasitic and Bacterial Infections. 1990. 24 figs. IX, 162 pp. ISBN 3-540-51515-1

Vol. 156: **Dyrberg, Thomas (Ed.):** The Role of Viruses and the Immune System in Diabetes Mellitus. 1990. 15 figs. XI, 142 pp. ISBN 3-540-51918-1

Vol. 157: **Swanstrom, Ronald; Vogt, Peter K. (Ed.):** Retroviruses. Strategies of Replication. 1990. 40 figs. XII, 260 pp. ISBN 3-540-51895-9

Vol. 158: **Muzyczka, Nicholas (Ed.):** Viral Expression Vectors. 1992. 20 figs. IX, 176 pp. ISBN 3-540-52431-2

Vol. 159: **Gray, David; Sprent, Jonathan (Ed.):** Immunological Memory. 1990. 38 figs. XII, 156 pp. ISBN 3-540-51921-1

Vol. 160: **Oldstone, Michael B. A.; Koprowski, Hilary (Ed.):** Retrovirus Infections of the Nervous System. 1990. 16 figs. XII, 176 pp. ISBN 3-540-51939-4

Vol. 161: **Racaniello, Vincent R. (Ed.):** Picornaviruses. 1990. 12 figs. X, 194 pp. ISBN 3-540-52429-0

Vol. 162: **Roy, Polly; Gorman, Barry M. (Ed.):** Bluetongue Viruses. 1990. 37 figs. X, 200 pp. ISBN 3-540-51922-X

Vol. 163: **Turner, Peter C.; Moyer, Richard W. (Ed.):** Poxviruses. 1990. 23 figs. X, 210 pp. ISBN 3-540-52430-4

Vol. 164: **Bækkeskov, Steinnun; Hansen, Bruno (Ed.):** Human Diabetes. 1990. 9 figs. X, 198 pp. ISBN 3-540-52652-8

Vol. 165: **Bothwell, Mark (Ed.):** Neuronal Growth Factors. 1991. 14 figs. IX, 173 pp. ISBN 3-540-52654-4

Vol. 166: **Potter, Michael; Melchers, Fritz (Ed.):** Mechanisms in B-Cell Neoplasia 1990. 143 figs. XIX, 380 pp. ISBN 3-540-52886-5

Vol. 167: **Kaufmann, Stefan H. E. (Ed.):** Heat Shock Proteins and Immune Response. 1991. 18 figs. IX, 214 pp. ISBN 3-540-52857-1

Vol. 168: **Mason, William S.; Seeger, Christoph (Ed.):** Hepadnaviruses. Molecular Biology and Pathogenesis. 1991. 21 figs. X, 206 pp. ISBN 3-540-53060-6

Vol. 169: **Kolakofsky, Daniel (Ed.):** Bunyaviridae. 1991. 34 figs. X, 256 pp. ISBN 3-540-53061-4

Vol. 170: **Compans, Richard W. (Ed.):** Protein Traffic in Eukaryotic Cells. Selected Reviews. 1991. 14 figs. X, 186 pp. ISBN 3-540-53631-0

Vol. 171: **Kung, Hsing-Jien; Vogt, Peter K. (Eds.):** Retroviral Insertion and Oncogene Activation. 1991. 18 figs. X, 179 pp. ISBN 3-540-53857-7

Vol. 172: **Chesebro, Bruce W. (Ed.):** Transmissible Spongiform Encephalopathies. 1991. 48 figs. X, 288 pp. ISBN 3-540-53883-6

Vol. 173: **Pfeffer, Klaus; Heeg, Klaus; Wagner, Hermann; Riethmüller, Gert (Ed.):** Function and Specificity of γ/δ T Cells. 1991. 41 figs. XII, 296 pp. ISBN 3-540-53781-3

Vol. 174: **Fleischer, Bernhard; Sjögren, Hans Olov (Eds.):** Superantigens. 1991. 13 figs. IX, 137 pp. ISBN 3-540-54205-1

Vol. 175: **Aktories, Klaus (Ed.):** ADP-Ribosylating Toxins. 1992. 23 figs. IX, 148 pp. ISBN 3-540-54598-0

Vol. 176: **Holland, John J. (Ed.):** Genetic Diversity of RNA Viruses. 34 figs. IX, 226 pp. ISBN 3-540-54652-9

Vol. 177: **Müller-Sieburg, Christa; Torock-Storb, Beverly; Visser, Jan; Storb, Rainer (Eds.):** Hematopoietic Stem Cells. 18 figs. XIII, 143 pp. ISBN 3-540-54531-X

Vol. 178: **Parker, Charles J. (Ed.):** Membrane Defenses Against Attack by Complement and Perforins. 26 figs. VIII, 188 pp. ISBN 3-540-54653-7

Vol. 179: **Rouse, Barry T. (Ed.):** Herpes Simplex Virus. 9 figs. Approx. X, 180 pp. ISBN 3-540-55066-6

Vol. 180: **Sansonetti, P. J. (Ed.):** Pathogenesis of Shigellosis. 15 figs. X, 143 pp. ISBN 3-540-55058-5